T0325173

Oxide Semiconductors and Thin Films

MATERIALS RESEARCH SOCIETY
SYMPOSIUM PROCEEDINGS VOLUME 1494

Oxide Semiconductors and Thin Films

Symposia held November 25–30, 2012, Boston, Massachusetts, U.S.A.

EDITORS

André Schleife
Lawrence Livermore National Laboratory
Livermore, California, U.S.A.

Martin Allen
University of Canterbury
Christchurch, New Zealand

Craig B. Arnold
Princeton University
Princeton, New Jersey, U.S.A.

Steven M. Durbin
University at Buffalo, The State University of New York
Buffalo, New York, U.S.A.

Nini Pryds
Technical University of Denmark
Roskilde, Denmark

Christof W. Schneider
Paul Scherrer Institute
Villigen PSI, Switzerland

Tim Veal
University of Liverpool
Liverpool, U.K.

Materials Research Society
Warrendale, Pennsylvania

CAMBRIDGE
UNIVERSITY PRESS

CAMBRIDGE
UNIVERSITY PRESS

Shaftesbury Road, Cambridge CB2 8EA, United Kingdom

One Liberty Plaza, 20th Floor, New York, NY 10006, USA

477 Williamstown Road, Port Melbourne, VIC 3207, Australia

314–321, 3rd Floor, Plot 3, Splendor Forum, Jasola District Centre, New Delhi – 110025, India

103 Penang Road, #05–06/07, Visioncrest Commercial, Singapore 238467

Cambridge University Press is part of Cambridge University Press & Assessment, a department of the University of Cambridge.

We share the University's mission to contribute to society through the pursuit of education, learning and research at the highest international levels of excellence.

www.cambridge.org
Information on this title: www.cambridge.org/9781605114712

Materials Research Society
506 Keystone Drive, Warrendale, PA 15086
http://www.mrs.org

First published 2013

Army Research Office (ARO) support was provided under Grant W911NF-13-1-0021. The views, opinions, and/or findings contained in this report are those of the author(s) and should not be construed as an official Department of the Army position, policy, or decision, unless so designated by other documentation.

A catalogue record for this publication is available from the British Library

CODEN: MRSPDH

ISBN   978-1-605-11471-2   Hardback

CONTENTS

ZnO AND RELATED MATERIALS

*Invited Paper

NON-ZnO OXIDES

DEVICES AND APPLICATIONS

PREFACE

Symposium Z, "Oxide Semiconductors," and Symposium F, "Oxide Thin Films for Renewable Energy Applications," were held Nov. 25–Nov. 30 at the 2012 MRS Fall Meeting in Boston, Massachusetts.

Oxide materials are attracting considerable attention both as semiconductors for a wide range of potential device applications but also in energy research spanning from photo- and electro-catalysis, to electrolytes and electrodes used in batteries or fuel cells. This symposium proceedings volume collects recent reports from the meeting aimed at providing a fundamental understanding of bulk oxide materials as well as thin films and nano-structures. The topics covered in this volume are quite broad and include such areas as growth and doping, defects and characterization, and device applications. For convenience, the papers are divided into three sections: (1) ZnO and Related Materials, (2) Non-ZnO Oxides, and (3) Devices and Applications.

All the contributions to the symposia focused on solving pressing issues and providing scientific insight. We hope that the reader finds this collection of papers to convey the multidisciplinary approach of physics, chemistry, materials science, and engineering needed to advance this field.

André Schleife
Martin Allen
Craig B. Arnold
Steven M. Durbin
Nini Pryds
Christof W. Schneider
Tim Veal

June 2013

ACKNOWLEDGMENTS

The papers published in this volume result from two MRS Fall 2012 symposia—Z and F. We sincerely thank all of the oral and poster presenters of the symposia who contributed to this proceedings volume. In particular, we are grateful to the many invited speakers all of whom provided well-attended and valuable additions to the meeting. We also thank the reviewers of these manuscripts, who provided valuable feedback to the editors and to the authors. It is an understatement to say that the symposia and the proceedings would not have happened without the organizational help of the Materials Research Society and its staff, particularly the publications staff for guiding us smoothly through the submission/review process. The organizers of Symposium Z thank the Air Force Research Laboratory for its financial support under grant W911NF-13-1-0021.

MATERIALS RESEARCH SOCIETY SYMPOSIUM PROCEEDINGS

MATERIALS RESEARCH SOCIETY SYMPOSIUM PROCEEDINGS

MATERIALS RESEARCH SOCIETY SYMPOSIUM PROCEEDINGS

Volume 1534E — Low-Dimensional Semiconductor Structures, 2012, T. Torchyn, Y. Vorobie, Z. Horvath, ISBN 978-1-60511-511-5

Prior Materials Research Society Symposium Proceedings available by contacting Materials Research Society

ZnO and Related Materials

Mater. Res. Soc. Symp. Proc. Vol. 1494 © 2012 Materials Research Society
DOI: 10.1557/opl.2012.1574

Acceptor Dopants in Bulk and Nanoscale ZnO

Matthew D. McCluskey,[1] Marianne C. Tarun,[1] and Samuel T. Teklemichael[1]
[1]Department of Physics and Astronomy, Washington State University
Pullman, WA 99164-2814, U.S.A.

ABSTRACT

Zinc oxide (ZnO) is a semiconductor that emits bright UV light, with little wasted heat. This intrinsic feature makes it a promising material for energy-efficient white lighting, nano-lasers, and other optical applications. For devices to be competitive, however, it is necessary to develop reliable p-type doping. Although substitutional nitrogen has been considered as a potential p-type dopant for ZnO, recent theoretical and experimental work suggests that nitrogen is a deep acceptor and will not lead to p-type conductivity. In nitrogen-doped samples, a red photoluminescence (PL) band is correlated with the presence of deep nitrogen acceptors. PL excitation (PLE) measurements show an absorption threshold of 2.26 eV, in good agreement with theory. The results of these studies seem to rule out group-V elements as shallow acceptors in ZnO, contradicting numerous reports in the literature. Optical studies on ZnO nanocrystals show some intriguing leads. At liquid-helium temperatures, a series of sharp IR absorption peaks arise from an unknown acceptor impurity. The data are consistent with a hydrogenic acceptor 0.46 eV above the valence band edge. While this binding energy is still too deep for many practical applications, it represents a significant improvement over the 1.4-1.5 eV binding energy for nitrogen acceptors. Nanocrystals present another twist. Due to their high surface-to-volume ratio, surface states are especially important. In our model, the 0.46 eV level is shallow with respect to the surface valence band, raising the possibility of surface hole conduction.

INTRODUCTION

ZnO is an electronic material with desirable properties for a range of energy applications.[1] ZnO is a wide band gap (3.4 eV) semiconductor that emits light in the near-UV region of the spectrum. The high efficiency of the emission, thanks in part to stable excitons at room temperature,[2] makes ZnO a strong candidate for efficient solid-state white lighting. Reports of stimulated emission in ZnO nanowires[3,4] and multicrystallite thin films[5,6] suggest the feasibility of UV lasers made from this material. ZnO is already used as a transparent conductor[7] in solar cells, a UV-absorbing component in sunscreens,[8] and the active material in varistors.[9] Researchers have also fabricated transparent transistors, invisible devices that could find widespread use in products such as liquid-crystal displays.[10]

Besides its fundamental optical and electrical properties, ZnO has other benefits that could make it a dominant material for energy applications. In contrast to GaN, large single crystals can be grown routinely.[11] ZnO is relatively benign environmentally and is actually used as a dietary supplement in animal feed.[12] From an economic perspective, the low cost of zinc versus indium provides an advantage over indium tin oxide for use as a transparent conductor.[7]

Despite these advantages, the lack of fundamental knowledge about dopants and defects presents an obstacle to the development of practical devices. Reliable p-type doping, required for high-performance transistors, lasers, or light-emitting diodes (LEDs), has been elusive.[13] As reviewed by McCluskey and Jokela,[14] the scientific literature contains numerous reports that nitrogen doping can produce *p*-type ZnO. However, Look and Chaflin[15] and Bierwagen *et al.*[16] have pointed out the difficulty in determining carrier type from Hall-effect measurements on low-mobility samples.

More recently, Lyons, Janotti, and Van de Walle[17] calculated the properties of substitutional nitrogen in ZnO using density functional theory (DFT) with hybrid functionals. Their calculations show that nitrogen is a *deep* acceptor, with the acceptor level 1.3 eV above the valence-band maximum. Lany and Zunger,[18] using generalized Koopmans DFT, obtained an even deeper level. Such a deep level would yield an insignificant hole concentration at room temperature.

In our experimental work,[19] we provided evidence that nitrogen is indeed a deep acceptor and therefore cannot produce *p*-type ZnO. A broad PL emission band near 1.7 eV, with an excitation onset of ~2.2 eV, was observed, in agreement with the deep-acceptor model[17] of the nitrogen defect. The deep level can be explained by considering the low energy of the ZnO valence band relative to the vacuum level. In our "universal acceptor model," we hypothesize that acceptor levels are roughly constant, irrespective of the host semiconductor. As shown in Fig. 1, these levels tend to be much higher than the ZnO valence band and therefore are "deep." This intrinsic feature of ZnO and other oxide semiconductors, which stems from the large electronegativity of oxygen, will prove a formidable challenge to *p*-type doping.

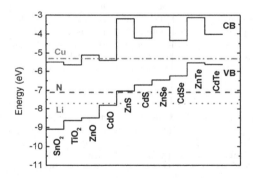

Figure 1. Proposed universal acceptor levels for copper, nitrogen, and lithium. Valence-band (VB) offsets are from theory[20,21] and experiment,[22] while conduction band (CB) minima are from the experimental band gaps. Energy values are relative to the vacuum level.[23] Levels for copper,[24] nitrogen,[17] and lithium[25] acceptors are their values in ZnO and are assumed to be constant relative to vacuum.

EXPERIMENT

Bulk single crystals were grown via seeded chemical vapor transport in an ammonia (NH₃) ambient, which provides nitrogen and hydrogen dopants.[26] An undoped reference sample was obtained by growing the crystal in an argon ambient. ZnO nanopowder was purchased from Sigma-Aldrich. Scanning electron microscopy (SEM) indicated that the particles have an average diameter of ~90 nm. The particles were pressed into pellets 7 mm in diameter with a thickness of 0.25 mm.

Infrared spectroscopy was performed using a Bomem DA8 vacuum Fourier transform infrared (FTIR) spectrometer with a globar light source, a KBr beamsplitter, and a liquid-nitrogen-cooled indium antimonide (InSb) detector. PL and PLE measurements were obtained using a JY-Horiba FluoroLog-3 spectrofluorometer equipped with double-grating excitation and emission monochomators (1200 grooves/mm; 2.1 nm/mm dispersion) and an R928P photomultiplier tube (PMT). The excitation source was a 450-W xenon CW lamp. An instrumental correction was performed to correct for the wavelength-dependent PMT response, grating efficiencies, and the variation in output intensity from the lamp.

DISCUSSION

Nitrogen in bulk ZnO

Figure 2 shows the red PL band (~1.7 eV) from N-doped and undoped samples. The spectra were obtained with excitation energy of 2.53 eV at room temperature. The inset shows the IR absorption peak of the N-H complex.   The samples were annealed in O₂ at 775°C to dissociate some of the N-H bonds and activate isolated nitrogen acceptors. The intensity of the red PL band is proportional to the concentration of activated N acceptors. The N-doped sample clearly shows a much higher intensity.

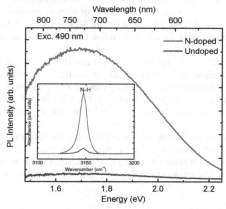

Figure 2. Red PL band from N-doped and undoped ZnO samples, at an excitation energy of 490 nm (2.53 eV). Inset: IR absorption peak of the N-H complex from the two respective samples.

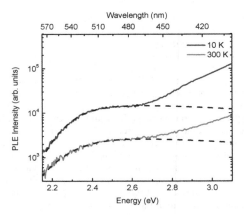

Figure 3. PLE spectrum of the red band (720 nm) of N-doped sample at 10 K and 300 K. Fits from the model[27], with parameters E_{opt} = 2.26 eV and ΔE = 0.14 eV for 10 K and E_{opt} = 2.24 eV and ΔE = 0.18 eV for 300 K, are shown by the broken lines.

In Fig. 3, the PLE spectrum for the red band from the N-doped sample was obtained by monitoring the intensity of the red emission at 720 nm as a function of excitation energy. The PLE signal is proportional to the absorption cross section of the N_O^- defect. An explicit formula for the absorption of deep levels was obtained by Jaros.[27] In that model, the key parameters are E_{opt}, the vertical transition energy from the deep level to the conduction band minimum, and ΔE, the broadening due to vibrational overlap between the initial and final states. Other parameters, which do not affect the absorption onset significantly, were obtained from Ref. 28. As shown in Fig. 3, the model produces good fits to the PLE spectrum for photon energies below 2.6 eV. The model parameters are E_{opt} = 2.26 eV and ΔE = 0.14 eV for 10 K, and E_{opt} = 2.24 eV and ΔE = 0.18 eV for 300 K. The increase above 2.6 eV may be due to other defect levels or band-tail states.

These results are consistent with the configuration coordinate diagram for optical transitions based on the nitrogen deep-acceptor model.[17] Specifically, we find a Stokes shift of 2.24 − 1.68 = 0.56 eV at room temperature. Assuming that the Frank-Condon shift is roughly half this value, we estimate that the nitrogen acceptor level is 1.9-2.0 eV below the conduction-band minimum. This places the acceptor level 1.4-1.5 eV above the valence-band maximum, in good agreement with theory.

The peak energy position of the red band as a function of temperature is shown in Fig. 4. At low temperatures, a band in the near-IR (1.5 eV) grows and dominates, leading to an apparent shift in the PL peak. The origin of this near-IR band is not known. One possibility is that N-H has a donor level in the gap.[29] The transition of an electron to this donor level may yield a PL peak.

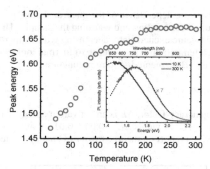

Figure 4. Peak energy position of the red band as a function of temperature. Inset: PL spectra at 10 K and 300 K, respectively. Gaussian fits are shown by the broken lines.

Acceptors in ZnO nanocrystals

Previous work[30,31] showed a low temperature (10 K) series of IR absorption peaks, in the energy range of 0.425–0.457 eV, for "as-grown" ZnO nanoparticles of average diameter ~ 20 nm. The result was characteristic of a hydrogenic acceptor spectrum with a hole binding energy of 0.4–0.5 eV. Although the identity of the acceptor was not determined, electron paramagnetic resonance measurements suggested that it may be a vacancy complex. In this work, as received ZnO nanoparticles from Sigma-Aldrich with larger diameter (~90 nm) showed similar results as the as-grown sample.

Figure 5 shows strong IR absorption spectrum for the as-received (Sigma-Aldrich) ZnO nanoparticles, consistent with previous work.[30,31] The IR absorption spectrum was calculated using absorbance = $\log_{10}(I_R/I)$, where I_R and I are the transmission spectra for no sample (blank) and an as-received sample, respectively. A quadratic baseline was then subtracted from the absorbance spectrum. The IR absorption peaks disappear (not shown here) after exposure to formic acid (HCOOH) vapor, consistent with electrical compensation of the acceptor by the formate ion. This observation also agrees with our previous results on as-grown ZnO nanocrystals.[32]

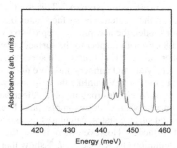

Figure 5. Low temperature (10 K) IR spectrum of as-received ZnO nanocrystals, showing electronic transitions.

Figure 6 shows the temperature dependence of the integrated area, for one of the IR peaks, indicating consistent results upon cooling and warming the sample. The inset shows the IR peak, corresponding to a hydrogenic excited state, decreasing with temperature. The disappearance of the peak at high temperatures is evidence of thermal ionization of the acceptors. The solid line is a fit according to a Boltzmann distribution function,

$$\alpha(T) = \alpha_0 / [1 + g \exp(-\Delta E/k_B T)] , \tag{1}$$

where α_0 is a constant, g is a degeneracy factor, ΔE is an activation energy, k_B is the Boltzmann constant and T is temperature (K). The fit yields $\Delta E = 0.09 \pm 0.01$ eV and $g = 132 \pm 94$. This result is consistent with the previous report on the as-grown ZnO nanoparticles, $\Delta E = 0.08$ eV.[31]

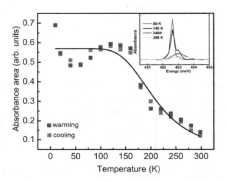

Figure 6. IR absorbance area of an acceptor excited state peak upon cooling and warming of the sample. The solid line is a fit to the Boltzmann distribution function [Eq. (1)]. Inset: IR absorption spectra of an acceptor peak, showing a decrease with temperature.

The activation energy ΔE is much lower than the hole binding energy of 0.46 eV. We proposed that the holes are thermally excited from the acceptor ground state to a band of surface states that lay 0.38 eV above the valence-band maximum.[31] This model is qualitatively consistent with *ab initio* calculations that predict the existence of surface states 0.5 eV above the valence-band maximum.[33] According to our model, the acceptor is deep (0.46 eV) with respect to the bulk valence band but shallow (0.08 eV) with respect to the surface states.

Figure 7 shows a low-temperature (10 K) PL emission spectrum at an excitation wavelength of 325 nm, for the as-received ZnO nanocrystals. We observed a broad green emission band centered around 524 nm (2.37 eV), unlike the as-grown sample which showed a broad red luminescence. In general, surface states may involve defects and impurities. Hydroxides on the surface of ZnO nanocrystals,[34] for example, have been correlated with green emission.[35] The previously mentioned surface states, calculated to be 0.5 eV above the valence band, arise from oxygen-deficient surfaces.[33] However, there is also theoretical evidence that the surface states may be intrinsic. Calculations by Kresse *et al.*[36] show that valence-band and conduction-band surface states originate from the O-terminated and Zn-terminated *c*-face ZnO surfaces, respectively.

8

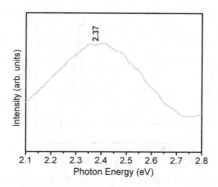

Figure 7. Low temperature (10 K) PL emission spectrum of ZnO nanocrystals at an excitation wavelength of 325 nm, showing green emission at 2.37 eV.

Figure 8 shows a low temperature (10 K) PL spectrum of the as-received ZnO nanocrystals under an excitation wavelength of 325 nm. Peaks in the near-band-gap range of 3.18 eV – 3.35 eV have previously been attributed to bound excitons and their phonon replicas. The PL peak at 3.36 eV is due to the neutral donor bound exciton (D^0, X).[37] We noticed this emission peak is blue shifted compared to the previous PL emission (3.35 eV) observed in the as-grown sample.[31] Low temperature PL peaks at 3.31 and 3.22 eV have been attributed to the TO and TO + LO phonon replica of donor bound excitons.[38, 39] Fonoberov *et al*[40] attributed the PL peak around 3.31 eV observed at low temperature (T < 150 K) in ZnO nanocrystals to acceptor bound excitons (A^0, X). They suggested that zinc vacancies or surface defects could act as acceptor impurities. Fallert *et al*[41] assigned the same peak (around 3.31 eV), observed in ZnO powders at 5 K, to excitons bound to defect states at the particle surface. The PL peak at 3.22 eV, recorded at 4.2 K, has been assigned to donor-acceptor pair transitions involving a shallow donor and shallow acceptor.[42] The emission peak at 3.17 eV could be compared with the previous emission peak observed at 3.18 eV in the as-grown ZnO nanocrystals. This PL peak could be due to LO phonon replica of the donor bound exciton.[43] The PL emission at 3.11 eV, not observed in the as-grown sample, could perhaps be due to a phonon replica of the donor bound exciton.

In our previous work on the as-grown ZnO nanocrystals,[31] the 2.97 eV emission peak was attributed to the transition of a free electron to the neutral acceptor. In the present work, the observation of this emission peak on the as-received ZnO nanocrystals indicates consistent result. In the PL experiment, above-gap light excites electrons into the conduction band. Electrons then fall from the conduction-band minimum to the acceptor level, emitting a photon of energy 3.43 − 0.46 = 2.97 eV. This PL peak supports the argument that ZnO nanocrystals contain acceptors with a hole binding energy of 0.46 eV.

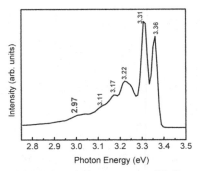

Figure 8. Low temperature (10 K) PL emission spectrum of ZnO nanocrystals at an excitation wavelength of 325 nm.

CONCLUSIONS

Our experimental evidence suggests that nitrogen is a deep acceptor in ZnO, with a hole ionization energy of 1.4-1.5 eV. The center shows a Stokes shift of 0.56 eV, consistent with large lattice relaxation. Clearly, such a deep acceptor is impractical for p-type doping. ZnO nanocrystals appear to contain an acceptor with a hole ionization energy of 0.46 eV. While the identity of this acceptor is unknown, we speculate that it could involve a zinc vacancy.

ACKNOWLEDGMENTS

The authors acknowledge helpful discussions with Leah Bergman, Evan Glaser, Eugene Haller, Stephan Lany, Chris Van de Walle, and Wladek Walukiewicz. Funding for this work was provided by the National Science Foundation Grant No. DMR-1004804 (MDM, MCT) and Department of Energy Grant No. DE-FG02-07ER46386 (STT).

REFERENCES

1. S.J. Pearton, D.P. Norton, K. Ip, Y.W. Heo, and T. Steiner, Journ. Vacuum Sci. Tech B **22**, 932 (2004).
2. D.C. Look, "Recent advances in ZnO materials and devices," Mater. Sci. Engin. B **80**, 383 (2001).
3. M.H. Huang, S. Mao, H. Feick, H.Q. Yan, Y.Y. Wu, H. Kind, E. Weber, R. Russo, and P.D. Yang, Science **292**, 1897 (2001).
4. S. Chu, G. Wang, W. Zhou, Y. Lin, L. Chernyak, J. Zhao, J. Kong, L. Li, J. Ren and J. Liu, Nature Nanotechnology **6**, 506 (2011).
5. Z.K. Tang, G.K.L. Wong, P. Yu, M. Kawasaki, A. Ohtomo, H. Koinuma, and Y. Segawa, Appl. Phys. Lett. **72**, 3270 (1998).

6. P. Zu, Z.K. Tang, G.K.L. Wong, M. Kawasaki, A. Ohtomo, H. Koinuma, and Y. Segawa, Solid State Commun. **103**, 459 (1997).
7. T. Minami, MRS Bulletin **25** (8), 38 (2000).
8. G.P. Dransfield, Radiat. Prot. Dosimetry **91**, 271 (2000).
9. D.R. Clarke, Journal of the American Ceramic Society **82**, 485 (1999).
10. J.F. Wager, Science **300**, 1245 (2003).
11. J.M. Ntep, S.S. Hassani, A. Lusson, A. Tromson-Carli, D. Ballutaud, G. Didier, and R. Triboulet, Journ. Crystal Growth **207**, 30 (1999).
12. J.W. Smith, M.D. Tokach, R.D. Goodband, J.L. Nelssen, and B.T. Richert, Journal of Animal Science **75**, 1861 (1997).
13. S.J. Jokela and M.D. McCluskey, Phys. Rev. B **76**, 193201 (2007).
14. M.D. McCluskey and S.J. Jokela, J. Appl. Phys. **106**, 071101 (2009).
15. D.C. Look and B. Claflin, Phys. Stat. Sol. (b) **241**, 624 (2004).
16. O. Bierwagen, T. Ive, C.G. Van de Walle, and J.S. Speck, Appl. Phys. Lett. **93**, 242108 (2008).
17. J.L. Lyons, A. Janotti, and C.G. Van de Walle, Appl. Phys. Lett. **95**, 252105 (2009).
18. S. Lany and A. Zunger, Phys. Rev. B **81**, 205209 (2010).
19. M.C. Tarun, M. Zafar Iqbal, and M.D. McCluskey, AIP Advances **1**, 022105 (2011).
20. C.G. Van de Walle and J. Neugebauer, Nature **423**, 626 (2003).
21. Ç Kiliç and A. Zunger, Appl. Phys. Lett. **81**, 73 (2002).
22. J. Wang, X.-L. Liu, A.-L. Yang, G.-L. Zheng, S.-Y. Yang, H.-Y. Wei, Q.-S. Zhu, and Z.-G. Wang, Appl. Phys. A **103**, 1099 (2011).
23. R.L. Lichti, K.H. Chow, and S.F.J. Cox, Phys. Rev. Lett. **101**, 136403 (2008).
24. F.A. Selim, M.C. Tarun, D.E. Wall, L.A. Boatner, and M.D. McCluskey, Appl. Phys. Lett. **99**, 202109 (2011).
25. O.F. Schirmer and D. Zwingel, Solid State Commun. **8**, 1559 (1970).
26. S.J. Jokela, M.C. Tarun, and M.D. McCluskey, Physica B **404**, 4801 (2009).
27. M. Jaros, Phys. Rev. B **16**, 3694 (1977).
28. McCluskey, M.D., N.M. Johnson, C.G. Van de Walle, D.P. Bour, M. Kneissl, and W. Walukiewicz, Phys. Rev. Lett. **80**, 4008 (1998).
29. J.L. Lyons, A. Janotti, and C.G. Van de Walle, Phys. Rev. Lett. **108**, 156403 (2012).
30. S. T. Teklemichael, W. M. Hlaing Oo, M. D. McCluskey, E. D. Walter, and D. W. Hoyt, Appl. Phys. Lett. **98,** 232112 (2011).
31. S. T. Teklemichael and M. D. McCluskey, Nanotechnology **22,** 475703 (2011).
32. S. T. Teklemichael and M. D. McCluskey, J. Phys. Chem. C **116,** 17248 (2012).
33. J. D. Prades, A. Cirera, J. R. Morante, and A. Cornet, Thin Solid Films **515,** 8670 (2007).
34. H. Zhou, H. Alves, D.M. Hofmann, W. Kriegseis, B.K. Meyer, G. Kaczmarczyk, and A. Hoffmann, Appl. Phys. Lett. **80**, 210 (2002).
35. N.S. Norberg and D.R. Gamelin, J. Phys. Chem. B **109**, 20810 (2005).
36. G. Kresse, O. Dulub, and U. Diebold, Phys. Rev. B **68,** 245409 (2003).
37. D. C. Reynolds, D. C. Look, B. Jogai, C. W. Litton, T. C. Collins, W. Harsch, and G. Cantwell, Phys. Rev. B **57,** 12151 (1998).
38. T. Matsumoto, H. Kato, K. Miyamoto, M. Sano, E. A. Zhukov, and T. Yao, Appl. Phys. Lett. **81,** 1231 (2002).
39. Y. Zhang, B. Lin, X. Sun, and Z. Fu, Appl. Phys. Lett. **86,** 131910 (2005).

40. V. A. Fonoberov, K. A. Alim, A. A. Balandin, F. Xiu, and J. Liu, Phys. Rev. B **73**, 165317 (2006).
41. J. Fallert, R. Hauschild, F. Stelzl, A. Urban, M. Wissinger, H. Zhou, C. Klingshirn, and H. Kalt, J. Appl. Phys. **101**, 073506 (2007).
42. K. Thonke, Th. Gruber, N. Teofilov, R. Schönfelder, A. Waag, and R. Sauer. Physica B **308-310**, 945 (2001).
43. X. Wang, S. Yang, X. Yang, D. Liu, Y. Zhang, J. Wang, J. Yin, D. Liu, H. C. Ong, and G. Du, J. Cryst. Growth **243**, 13 (2002).

Mater. Res. Soc. Symp. Proc. Vol. 1494 © 2012 Materials Research Society
DOI: 10.1557/opl.2012.1648

Microstructural and Optical Properties of Nitrogen Doped ZnO Nanowires

Ahmed Souissi[1,2], Nadia Hanèche[1], Corinne Sartel[1], Abdel Meftah[2], Alain Lusson[1], Meherzi Oueslati[1], Jean-Marie Bluet[3], Bruno Masenelli[3], Vincent Sallet[1] and Pierre Galtier[1]

[1]Groupe d'Etude de la Matière Condensée (GEMaC), Université de Versailles Saint-Quentin-en Yvelines / CNRS, 45 avenue des Etats-Unis, 78035 Versailles, France
[2]Département de Physique, Université El-Manar, Tunis, Tunisie
[3]Institut des Nanotechnologies de Lyon, INSA, 69621 Villeubanne, France

ABSTRACT

Nanowires with different nitrogen concentrations were grown by Metal-Organic Chemical Vapor Deposition (MOCVD) using DEZn, N_2O and NH_3 as zinc, oxygen and nitrogen doping sources respectively. Low temperature photoluminescence, Raman spectroscopy and Transmission Electron Microscopy are combined to study the incorporation of nitrogen in the wires. The observation of donor-acceptor pair band confirms that the incorporation nitrogen in ZnO nanowires is responsible for the creation of acceptor centers. The additional peaks observed in Raman are correlated to nano-sized inter-atomic distance fluctuations observed in TEM. These domains combined with a resonance effect are probably the explanation of the huge Raman cross section observed for the impurity related peaks.

INTRODUCTION

The achievement of efficient and reliable p-type ZnO in bulk or thin films is still currently a serious drawback for the development of optoelectronic devices. This has been related to the presence of residual n-type impurities, the creation of structural defects during growth, the difficulty to incorporate the dopant in the adequate atomic site and/or in a shallow acceptor state or the creation of complexes favored by thermodynamics. Among all the possible candidates for p-type doping, nitrogen is a potentially interesting candidate when it is incorporated in substitution to oxygen because its size is comparable and it should not distort to much the ZnO matrix. Although a large amount of studies have reported on the incorporation of nitrogen in bulk or thin film, very few convincing results demonstrating a p-type activity have been published. One of the reasons commonly invoked to explain this is that most of the studies have been performed on Sapphire substrates which generate a lot of extended defects and/or a noticeable diffusion of Al (a very efficient n-type dopant) from the substrate to the ZnO layer. On the other hand theoretical studies predict that nitrogen leads to a deep acceptor center in substitution to oxygen [1] or is preferentially incorporated in a complicated form like, for example, molecule or interstitial complexes [2]. In this context, the realization of doped nanowires has been proposed (i) to reduce the number of extend or localized defects and (ii) to exploit surface effects in order to improve the efficiency of the dopant incorporation in electrically active sites.

EXPERIMENTAL

Nitrogen-doped ZnO nanowires have been grown on (0001) sapphire substrates in a horizontal MOCVD reactor working at low pressure with DEZn, N_2O and NH_3 as zinc, oxygen and nitrogen sources. Growth temperature was ranging from 775 to 900°C. The carrier gas was helium. Low temperature photoluminescence (PL) was performed with a 351 nm line of an argon laser. Raman measurements were performed at room temperature with a T64000 system equipped with a microscope in back-scattering configuration. The 488 nm line from an Ar+ laser was used for micro-Raman spectroscopy. The morphology of the wires and their internal structure were assessed with Scanning Electron Microscopy (SEM) and Transmission Electron Microscopy (TEM).

RESULTS AND DISCUSSION

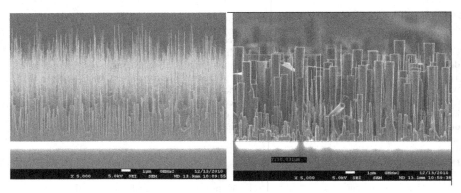

Figure 1: SEM image in cross section of ZnO wires. left) undoped sample, right) grown with the highest NH_3 flow used in this study (13.4 μmole/min).

The influence of the dopant precursor (NH_3) on the morphology of the wires is demonstrated in figure 1. The growth was performed at 800°C with a O/Zn molar ratio ($R_{VI/II}$) of 540. Most of the wires grow along their C-axis and parallel to the C-axis of the sapphire substrate. Convergent Beam Electron Diffraction (CBED) performed in TEM informs on polarity indicates that the wires effectively grow along the +C axis (Zn face). This orientation is usually observed on ZnO wires grown on sapphire substrates [3] and is not modified by the NH_3 flux. However, the introduction of nitrogen dramatically modify the aspect ratio of the wires which appear larger (and with a larger top) at high nitrogen flux and clearly faceted.

The low temperature PL spectra (T = 4K) of ZnO nanowires with different nitrogen concentrations are shown in figure 2. At the high energy side, the PL spectra are dominated by excitons bound to donors at 3.361 and 3.367 eV [4]. This suggests that the wires are still probably n-type. At the low energy side, the PL spectra are dominated by the donor-acceptor pair band (DAP) at 3.23-3.24 eV and its LO phonon replicas. The DAP assignation is justified by the blue shift observed when nitrogen concentration or excitation power are increased [5]. Thus, PL results confirm the creation of acceptor states in relation with the incorporation of nitrogen.

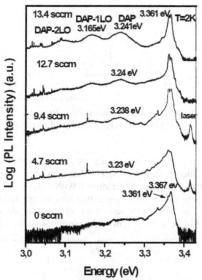

Figure 2: Low temperature photoluminescence spectra of nitrogen doped ZnO nanowires with NH$_3$ flux.

Raman spectroscopy has been used to detect the additional modes which are usually observed on doped ZnO thin layers. The results obtained for increasing NH$_3$ flows in the z(--)z configuration are shown in figure 3. On undoped samples, the Raman spectrum is dominated by the E$_2$ mode at 439 cm^{-1} in agreement with the selection rules. At least two additional peaks, I$_1$ and I$_2$ at 275 and 580 cm^{-1}, appear upon doping in agreement with previous results [6]. Whereas the intensity of the E$_2$ peak remains mostly constant, a clear dependence of the intensity of these additional peaks with NH$_3$ flow is observed. Following the work cited previously where SIMS was used for calibration on 2D thin layers, the nitrogen concentration of our samples is ranging from $2 \cdot 10^{17}$cm^{-3} to $1.2 \cdot 10^{19}$cm^{-3}. Interestingly, the intensity behavior of both I$_1$ and I$_2$ are not identical (Fig. 3). This confirms that the intensity of the I$_1$ and I$_2$ modes is related not only to the incorporation of nitrogen, but also to their excitation mechanism, through a resonance effect [7], which plays an important role due to their surprising high intensity (with respect to the estimated doping level). Furthermore, it has been shown that similar peaks are also observed in samples doped with Al, Fe, Ga and Sb indicating that they cannot simply be assigned to local impurity modes [8]. This was confirmed by L. Artùs at al [9] who showed that the incorporation of ^{14}N and ^{15}N isotopes lead to additional modes at the same energy which indicates that the observed

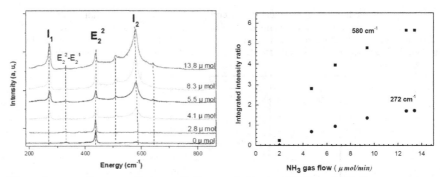

Figure 3: macro-Raman scattering on ZnO nanowires versus NH_3 flow. Left): Raman spectra. Right) Intensity ratio of I_1 (272cm^{-1}) and I_2 (580cm^{-1}) additional peaks with respect to bulk E_2 peak.

modes are not directly related to the vibration of N atoms but more probably to distortions/defects associated to the incorporation of N. This has also been confirmed recently by N. Nickel et al. [10] who performed similar experiments with natural ^{nat}Zn and ^{68}Zn and observed an isotopic shift associated to Zn for the I_1 mode that was interpreted in term of Zn_I-N_O and Zn_I-O_I complexes. This suggests that both resonance effect, the existence of specific complexes and local distortions should be at the origin of the observed additional modes.

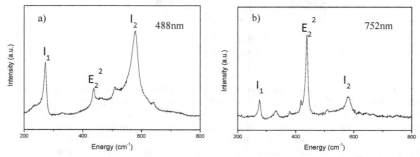

Figure 4: Room temperature Raman spectra of ZnO:N doped sample (N~10^{19}cm^{-3}) obtained using two different excitation lines a): λ =488nm, b): λ=758nm.

In order to check the origin of any possible resonance effect, we performed a comparison of the Raman spectra obtained using two different excitation lines (Figure 4). Clearly the observation of both I_1 and I_2 peaks is associated to a resonance effect. Thus, the electronic transition involved in the resonance observed for I_1 and I_2 is probably the intense and broad (0.5 eV large) green luminescence previously observed at 2.2 eV on nitrogen doped samples [11]. When the 488 nm line ($\hbar\omega_i$ = 2.541 eV) of an Ar$^+$ laser is used, it is likely that the intensity of the Raman peaks is enhanced by both incoming resonance and outgoing resonance due to the overlap between the excitation energy of the laser and a defect absorption band associated to nitrogen doping. More precisely, for a phonon mode $\hbar\omega_0$ and an optically allowed electronic transition E_a, the ingoing resonance should occurs when the incident photon energy $\hbar\omega_i$ is equal to E_a and the outgoing resonance should occurs when the energy of the scattered photon, $\hbar\omega_s = \hbar\omega_i - \hbar\omega_0$, is equal to E_a [12]. Ingoing resonance is thus a possible explanation of the global intensity increase of I_1 and I_2 with NH_3 flux through the rise of the 2.2 eV band related to nitrogen doping. The difference in behavior between I_1 and I_2 must be found in the outgoing resonance which should occur for a value of Ea of 2.507 eV and 2.47 eV for respectively I_1 and I_2, thus closer to the maximum of the defect band for I_2. Thus the resonance should be more efficient for I_2 than for I_1 and this is what we observe. The resonance could also be enhanced for the I_2 peak thanks to the Fröhlich coupling due to the LO character of this mode. On the other hand, the intensity of I_1 and I_2 is dramatically lowered with respect to E_2 bulk mode with the 752nm excitation line because in that case there is not overlapping between the laser line and the defect band. Interestingly, we do not observe any resonance effect on the E_2 bulk mode of ZnO. This suggests that there is a strong connection between the additional modes I_1 and I_2 and the 2.2 eV defect band and that the E_2 mode is arising from area in the samples which are not perturbed by the incorporated nitrogen and its possible related complexes.

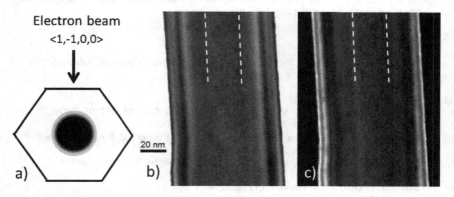

Figure 5: TEM observation of nitrogen doped ($\sim 10^{19}$cm^{-3}) ZnO nanowire. a): observation geometry, b): Bright field image, c): Dark field image formed with the (002) reflection.

TEM has been performed in order to check the internal structure of nanowires (Figure 5). We have not been able to observe any plane defects like dislocations or stacking faults on both undoped and doped nanowires. Due to the hexagonal shape of the wire (Figure 5.a), the intensity is oscillating in both bright field (BF) and dark field (DF) images when moving from the wire edge to the center (i.e. upon increasing the thickness passed by the electron beam) (Figure 5.b and 5.c). This is associated to dynamical effects (intensity transfer between transmitted and diffracted beams). The central part of the wire should exhibit an even contrast because its thickness is constant. However, a close examination of this area in both BF and DF reveal weak contrast fluctuations, at the nanometer size, on doped nanowires not observed on undoped

(Figure 5.b and 5.c). The fact that no fluctuation is observed on the edge of the doped wire demonstrates that the effect is not related to surface roughness (also confirmed by high resolution imaging). Thus the contrast fluctuation is localized in the core of the wire. Because we are here in strong Bragg condition these fluctuations reflect some local perturbation of the interatomic distances or local disorientation of the atoms in the lattice. They break the symmetry of the crystal and contribute probably to the activation of the additional mode observed in Raman scattering. These domains are not observed in the external part of the doped wires suggesting that elastic relaxation occurs close to the surface or that dopants are concentrated in the core of the wires.

CONCLUSIONS

Nitrogen doped ZnO nanowires were fabricated by MOCVD. The incorporation of nitrogen gives rise to an acceptor center as demonstrated by low temperature photoluminescence through the observation of donor-acceptor pairs. Raman scattering reveal the well known intense additional modes whose observation is explained in term of both resonance effect on the "green band" observed at 2.2 eV in PL and local lattice parameter fluctuations observed in TEM. The correlation between the Raman peak intensity of the additional modes and the defect band observed in PL strongly suggest that they have the same origin.

ACKNOWLEDGMENTS

This work was supported by the Agence Nationale de la Recherche (ANR) through the project MAD-FIZ.

REFERENCES

1. J. L. Lyons, A. Janotti, and C. G. Van de Walle, , *Appl. Phys. Lett.*, **95**, 252105 (2009).
2. N. Nickel and M.A. Gluba, Phys. Rev. Lett., 103, 145501 (2009).
3. G. Perillat-Merceroz, PH Jouneau, G Feuillet, R Thierry, M Rosina and P Ferret, *Journal of Physics: Conference Series*, **209**, 012034 (2010).
4. D. C. Reynolds, D. C. Look, B. Jogai, C. W. Litton, T. C. Collins, W. Harsch, and G. Cantwell, *Phys. Rev. B* **57**, 12151 (1998).
5. S. Yamauchi, Y. Goto and T. Hariu, J. Crystal Growth, **260**, 1 (2004).
6. Kaschner, U. Haboeck, M. Strassburg, M. Strassburg, G. Kaczmarczyk, A. Hoffmann, C. Thomsen, A. Zeuner, H. R. Alves, D. M. Hofmann, and B. K. Meyer, *Appl. Phys. Lett.*,**80**, 1909 (2002).
7. F. Friedrich and H. Nickel, *Appl. Phys. Lett.*, **91**, 111903 (2007).
8. C. Bundesmann, N. Ashkenov, M. Schubert, D. Spemann, T. Butz, E.M. Kaidashev, M. Lorenz and M. Grundmann, *Appl. Phys. Lett.*, **83**, 1974 (2003).
9. L. Artùs, R. Cusco, E. Alarcon-Llado, G. Gonzalez-Diaz, I. Martil, J. Jimenez, B. Wang,M. Callahan, *Appl. Phys. Lett.*, **90**, 181911 (2007).
10. F. Friederich, M.A. Gluba and N. H. Nickel, *Appl. Phys. Lett.*, **95**, 141903 (2009).
11. K. Iwata, P. Fons, A. yamada, K. Matsubara and S. Niki, J. Crystal Growth, **209**, 526 (2000).
12. M. Cardona ang G. Güntherodt, Light Scattering in Solids II (Springer, Berlin/Heidelberg/New-York, 1982).

Mater. Res. Soc. Symp. Proc. Vol. 1494 © 2013 Materials Research Society
DOI: 10.1557/opl.2013.157

Growth and Characterization of ZnO Thin Film by RF Magnetron Sputtering for Photoacoustic Tomography Sensor

Takuya Matsuo[1], Shuhei Okuda[1] and Katsuyoshi Washio[1]
[1]Graduate School of Engineering, Tohoku University, Sendai, Miyagi 980-8579, Japan

ABSTRACT

To apply thin ZnO film to photoacoustic tomography sensors, we investigated methods to improve its piezoelectricity with high optical transmittance. ZnO film was deposited by RF magnetron sputtering on a quartz substrate with various changes of the following conditions: RF sputtering power, Ar gas pressure, and substrate temperature (T_{SUB}). The preliminary optimization of sputtering conditions is to form the ZnO film with good c-axis crystalline alignment. The results of X-ray diffraction measurement and cross-sectional observations indicated that the high-T_{SUB} condition was preferable. This was because the desorption of Zn due to high-T_{SUB} during the deposition process induced the formation of excellent columnar grains normal to the substrate. To enhance the piezoresponse, the substitution of Zn with different crystal-radius atoms was investigated, the aim being to increase the electrically neutral dipole moment by the partial displacement of the Zn-O bond. The transition metal V, with the potential to have the various configurations and coordination numbers, was selected as the dopant. As a result, it was confirmed that the diffraction peak from the (002) plane shifted to low angles with small degradation of the diffraction intensities.

INTRODUCTION

Photoacoustic tomography with strong contrast and high resolution is one of the most attractive technologies in medical diagnoses. This imaging principle is based on the detection of ultrasonic waves accompanied by thermal expansion at the tissue irradiated by a short-pulsed laser beam. Thus, to enable the laser beam to pass through the photoacoustic tomography sensors (PTS) and transduce the ultrasonic wave into an electric signal, the materials composing the PTS should have the features of both good optical transmittance and piezoelectricity. From this point of view, ZnO film is one of the most promising candidates.

Here, to achieve good piezoelectricity, the ZnO film should have excellent c-axis crystalline orientation [1]. This is because the orientation of the crystalline axis is directly related to the electro-optic and acousto-optic effects of the film. Furthermore, to achieve strong piezoelectricity by enhancing the dipole moments of the wurtzite structure, experiments on the replacement of Zn atoms with other transition metals of different crystal radius were performed. Because the transition metals vary in charge and coordination numbers, doping them into ZnO films seems to be suitable. Atoms of different crystal or ionic radius placed at Zn sites enables loss of balance of the wurtzite structure, so it helps to polarize the dipole moments due to the ease of the bond rotation [2].

EXPERIMENT

The thin ZnO films about 500 nm in thickness were deposited on quartz substrates by RF magnetron spattering under various conditions. The sputtering target was a ceramic ZnO disc and the gas used during processing was only Ar. Sputtering conditions such as RF sputtering power (P_{RF}), Ar gas pressure (P_{Ar}), and substrate temperature (T_{SUB}) were set as processing parameters. V was doped into ZnO films by co-sputtering ZnO target with V chips.

In this study, the effects of the sputtering conditions on the ZnO films regarding crystalline quality and optical transmittance were investigated, and crystal orientations and qualities were characterized. The crystalline properties of ZnO films were investigated by out-of-plane X-ray diffraction (XRD) and the structural properties were observed by scanning electron microscopy (SEM) and transmission electron microscopy (TEM). Optical transmittance was also measured to characterize the optical properties of the ZnO films.

ZnO Growth rate

Figure 1 shows the growth rate of some samples under various deposition conditions. The sputtering parameter of P_{Ar} was optimized so as not to influence the growth rate. As seen in Fig. 1 (a), the growth rate increases proportionally with P_{RF}. This indicates that thick ZnO films which transduce slow ultrasonic waves can be easily deposited at high P_{RF}. However, the growth rate becomes slower above around $T_{SUB}=500°C$, as shown in Fig. 1 (b). This phenomenon is considered as follows. Zn atoms tend to desorb from the surface at high temperature and O atoms cause difficulty in adsorption on the surface. Therefore, transport and diffusion of Zn atoms to the proper atomic site restricts the growth rate.

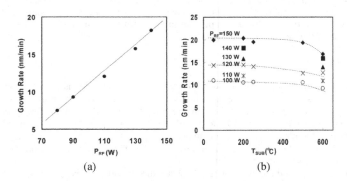

(a)   (b)

Figure 1. Dependence of ZnO growth rate on P_{RF} ($T_{SUB}=200°C$) (a), and on T_{SUB} ($P_{RF}=100$-150 W) (b).

XRD measurements

Out-of-plane XRD measurements rotating around the $2\theta/\omega$ axis of the sample stage from $30°$ to $80°$ were conducted to analyze the crystal face orientation of ZnO films. To confirm good c-axis alignment, the angle of the diffraction peak which should be focused on is around $34.4°$ of the ZnO (002) plane. In addition, rocking curve measurements were carried out to characterize the crystal orientation quality of ZnO films. Fig. 2 (a) shows one of the measured results of $2\theta/\omega$

scans and ω scans of the rocking curve around 2θ/ω=34.4° in the inset (P_{RF}=150 W, T_{SUB}=200, 400, 500, 600°C) and Fig. 2 (b) shows the dependence of the diffraction peak intensity from the (002) plane and the full width at half-maximum (FWHM) of the rocking curve on T_{SUB}. The intensity of the (002) diffraction peak becomes strong at a high T_{SUB} condition. On the other hand, in the case of low T_{SUB}, some peaks diffracted from the (101) and (103) planes were observed. These indicate that the crystal of wurtzite structure tends to align along the c-axis in high-T_{SUB} samples. Furthermore, the FWHM of the rocking curve measurements became narrower in higher T_{SUB} samples. This also shows good c-axis alignment in high-T_{SUB} samples.

From these results, a sputtering condition of high T_{SUB} is seen to be preferable to obtain ZnO polycrystalline film consisting of grains with the c-axis of wurtzite structure cells strictly oriented along the normal direction to the surface of the substrate. There was no dependence of the intensity on P_{RF}.

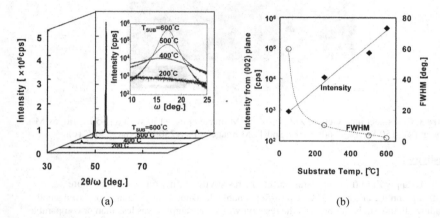

(a)   (b)

Figure 2. Out-of-plane XRD measured results of 2θ/ω scans from 30° to 80° and ω scans from 10° to 25° of rocking curve measurements (2θ/ω=34.4°) (a), and dependence of the diffraction peak intensity from the (002) plane and FWHM of rocking curve on T_{SUB} (b).

Cross-sectional structure

Cross-sectional views of ZnO films were observed by SEM and TEM as shown in Fig. 3: ZnO in conditions of P_{RF}=150 W, T_{SUB}=50°C (a), and P_{RF}=150 W, T_{SUB}=600°C (b). As predicted in XRD measurements, columnar grains aligned along c-axis could be seen. From Fig. 3 (a), it is seen that there were grains with various sizes and orientations, while it seems that columnar grains were oriented in the same direction in Fig. 3 (b). The TEM image also shows columnar grains of ZnO in Fig. 3 (c). Near the surface of films, however, some clusters of grains are seen. This result indicates that the degradation of crystalline quality occurred in the case of thick ZnO films. This is a problem involved in attempting to thicken the films for application to devices. We need to consider the balance of ZnO deposition and Zn volatilization by controlling P_{RF} and T_{SUB} to obtain thicker ZnO films with perfect c-axis oriented grains.

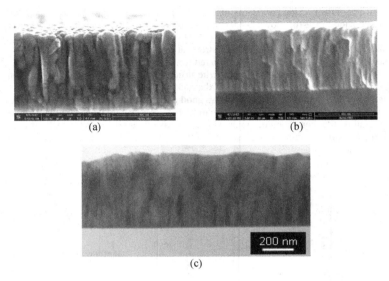

Figure 3. Cross-sectional view of ZnO films. SEM image at P_{RF}=150 W, T_{SUB}=50°C (a), SEM image at P_{RF}=150 W, T_{SUB}=600°C (b), and TEM image at P_{RF}=250 W, T_{SUB}=250°C (c).

Resistivity

To apply ZnO film to the transducer, the resistivity of ZnO film should be more specifically, on the order of 10^6 $\Omega \cdot$ cm [3-5] to enable the detection of signals from even small vibrations. In this study, it observed the resistivity of some samples was less than or comparable to on the order of 10^3 $\Omega \cdot$ cm. This seems to have been caused by the deficiency of O atom and it indicates that introduction of O_2 gas during the sputtering or the post annealing in O_2 atmosphere should be considered.

Optical transmittance

Samples deposited at high T_{SUB} were transparent while those of low T_{SUB} were dark with a metallic luster. Optical transmittance of ZnO films at a thickness of about 500 nm with a good c-axis orientation was 93%, including quartz substrate 625 nm thick. There were not any dependencies on P_{RF}, i.e., on the growth rate of ZnO. Samples with weak intensity of the (002) plane diffraction peak that appeared to be dark and metallic also had a transmittance of about 80%.

Effect of transition metal doping

To improve piezoelectricity, some V-doped ZnO samples were deposited. V, a transition metal, is thought to be sputtered neutrally, and thus it can possibly to change the charge and coordination numbers from 2 to 5 and from 4 to 8, respectively, in the ZnO crystal. Ideally, the charge number of V should be +2 and the coordination number should be 4, the same as that of the Zn atom in the ZnO crystal. X-ray fluorescence (XRF) was used to quantify the weight percentage (wt %) of V in the ZnO films. Fig. 4 shows a comparison of (002) diffraction peaks of undoped ZnO with that of V-doped (~4 wt %) ZnO (a) and the dependence of the intensity of the (002) diffraction peak of V doped samples and undoped samples on the diffraction angle (b). The diffraction peak intensity from the (002) plane in the V-doped ZnO film becomes weak because of the lattice deformation by replacement of Zn with V of different crystal radius. Although the V-doped samples showed a lower intensity at the (002) diffraction peak than that of the undoped ZnO, they did not become so much weaker and their intensities were stronger than those of low T_{SUB} samples of undoped ZnO. A characteristic change in V-doped ZnO film is a shift of (002) diffraction peaks to lower angles on the $2\theta/\omega$ axis, as clearly shown in Fig. 4 (b). These results show that the c-axis lattice constant in V-doped ZnO films becomes wider than that in the undoped ZnO lattice.

The result of resistivity measurement shows a low value, on the order of 10^{-3} $\Omega\cdot$cm for V-doping of 2-4 wt %, compared with undoped ZnO films. Thus, V doping leads to a reduction of resistivity. This means that the replacement of Zn by V seemed to have charge of more than 2 and the coordination numbers such as 6.

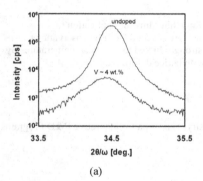
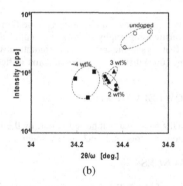

(a)   (b)

Figure 4. Comparison of (002) diffraction peaks of undoped ZnO with V-doped (~4 wt %) ZnO (a) and dependence of the intensity of (002) diffraction peak of V-doped samples and undoped samples on the diffraction angle.

DISCUSSION

While the growth rate above T_{SUB}=500°C decreased, the XRD peak of diffraction from the (002) plane in the sample of T_{SUB}=600°C had the highest intensity, and cross sections of high XRD-intensity samples clearly showed excellent columnar grains. From these results, it could be considered that Zn atoms were desorbed from the surface and then adsorbed again at the other energetically stable sites at high T_{SUB}. As a consequence, columnar grains were constructed normal to the substrate, i.e., c-axis oriented crystalline growth occurred. However, the result that some samples showed low resistivity indicates two problems.

First is an oxygen defect in ZnO films. This might be caused by the adsorption of Zn atoms at the oxygen vacant sites. The second is an instability and imperfection of the grain boundary. Therefore, to solve these problems, active free electrons both in grains and inter-grains should be diminished. From this point of view, O atoms should be added by some methods such as mixture of O_2 gas in the sputtering and annealing in O_2 atmosphere.

Concerning V doping aimed at improving piezoelectricity, the diffraction peak position of the (002) plane shifts toward a small diffraction angle that indicates the expansion of the lattice along c-axis, and its intensity of the (002) plane decreases that means the deterioration of the crystal quality. From the point of view of the peak shift toward the smaller angle, it can be thought that the crystal or ionic radius of V atoms doped in ZnO films is larger than that of the Zn atom (0.60Å). However, the ionic radius of V^{5+} with a coordination number of 4 is 0.355 Å. Therefore, considering the deterioration of the crystal quality, the V ions in ZnO film must be thought to be other states, such as V^{2+} and V^{3+}. Here, the ionic radii of V^{2+} and V^{3+} with a coordination number of 6 are 0.79 Å and 0.64 Å, respectively [6]. As a result, the shifting of the peak position and the decrease of peak intensity are thought to have a chance to induce imbalance of the dipole moment and enhance piezoresponse by the doping of transition metal.

CONCLUSIONS

We investigated the sputtering conditions of ZnO thin films for application to photoacoustic tomography sensors. XRD intensities and cross-sectional SEM observations showed that crystalline quality and orientation were strongly linked to substrate temperature. By V doping, (002) diffraction peaks of films shifted due to lattice deformations.

ACKNOWLEDGMENTS

This study was supported in part by the Industry-Academia Collaborative R&D Programs of the Japan Science and Technology Agency.

REFERENCES

1. F. S. Hickernell, IEEE Trans. Sonics Ultrason. SU-32, 621 (1985).
2. P. Feng, L. J. Ting, Y. Y. Chao, W. X. Bo, Z. Fei, Sci. China, Tech. Sci. 55, 2, 421-436 (2012).
3. W. J. Jeong, G. C. Park, Solar Energy Materials & Solar Cells 65, 37-45 (2001).
4. F. S. Hickernell, Proc. IEEE Ultrason. Symp. 785-794 (1980).
5. Y. C. Yang, C. Song, X. H. Wang, F. Zeng, F. Pan, Appl. Phys. Lett. 92, 012907 (2008).
6. R. D. Shannon, Acta Cryst. A32, 751 (1976).

Mater. Res. Soc. Symp. Proc. Vol. 1494 © 2013 Materials Research Society
DOI: 10.1557/opl.2013.124

Magnetic Interactions Study in ZnO Doped with Fe Ions Produced by Thermal Diffusion Processes

R. Baca[1], M. Galván[3], J. V. Méndez[2], J. A. Andraca[2] and R. Peña[3]
[1]Department of Electronics, National Polytechnic Institute, 07738, México City, México.
[2]Department of Nanoscience and Microtechnology, National Polytechnic Institute, 07738, México City, México.
[3]Department of Electrical Engineering, CINVESTAV, 07360, México City, México.

ABSTRACT

Recently, the oxides have received attention and great interest due to their magnetic ordering above of the room temperature by doping a very low amount of transition metal ions, which are very promising for applications such as biosensing, hyperthermia, doped magnetic semiconductors with lower energy losses and rapid response at alternating-magnetic fields. In this work the magnetic interactions on Fe doped ZnO thin-films was studied. Raman spectroscopy allowed the monitoring of iron ions diffusion and demonstrated that symmetry modes are crucial for understanding of the magnetic ordering. X-ray diffraction (XRD) was used to determine the oxidation state of the iron ions and stress into ZnO lattice. MFM confirmed that magnetic moments and magnetic forces on scanned surface depend on magnetic-domain structure formation.

INTRODUCTION

Spin-based multifunctional devices using electron spin in addition to charge may lead to semiconductor with greatly increased functionality being optically transparent, and having electrical and magnetic control by defect concentration [1, 2, 3], which would lead to devices such as biosensors with Fe doped ZnO thin-films and sensitive to glucose concentration, because Fe^{3+} ions introduces redox centre in ZnO [4]. Magnetic nanoparticles (MNP) can be made to generate local heat in vivo, which lead to their use as hyperthermia agents to destroy the tumor cells [5]. In this work we study the magnetic interactions in ZnO thin-films doped with Fe ions motivated on previous challenges.

THERMAL DIFFUSION PROCESSES

It is known that under air atmosphere ZnO can be formed, and Zn ions going into a sequence of jumps between occupied interstitial sites [6, 7], and $\alpha\text{-}Fe_2O_3$ (hematite phase) can be obtained, when iron foils from $200\,^\circ C$ are heated in humidity atmosphere [8]. Fe doped ZnO thin-films were deposited on quartz substrates by thermal diffusion processes in air atmosphere into two stages. In first stage, alternative layers of grain-oriented silicon-iron foils and zinc granules (4N purity) were deposited sequentially by vacuum evaporation. As a second stage, the deposited layers were heated in air atmosphere at $420\,^\circ C$, $520\,^\circ C$, and $620\,^\circ C$ with duration of 10min into resistively heated quartz tube furnace.

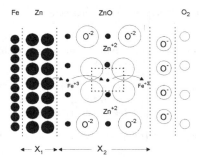

Figure 1. Schematic model of interstitial Fe^{3+} movement into ZnO layer.

The thickness of both films was 300nm and was measured by a Tencor profilometer. At high temperature ($>420°C$), small amounts of iron ions into ZnO occurs randomly, because the diffusion coefficient D_{Fe} of interstitial Fe^{3+} ions increases as a function of its activation energy U_{Fe} with the temperature [9]. Iron ions (Fe^{3+}) diffusion into ZnO occurs when Fe ions from its interface position (*Fe - Zn*) can be moved after that X_1 decrease at critical thickness of 10nm [6] as shown in Fig. 1.

DISCUSSION

The characterization method was assisted by Raman studies with a Horiba Jobin-Yvon micro-Raman system (HR800), with excitation line $\lambda = 632.8$ *nm* at 20 *mW* of He-Ne laser. Structure formation as a function of stress was investigated with a PANalytical X-ray diffractometer with *CuK* $_\alpha$ radiation ($\lambda = 0.15418$ *nm*). Magnetic images produced by magnetic domain structures were analyzed with a Digital Instrument (Veeco) Nanoscope.

Spinel structure $ZnFe_2O_4$ has been synthesized by high temperature methods with ZnO and Fe_2O_3 mixed together and heated at $1000°C$ [10]. For lower temperatures at melting point of ZnO ($1975°C$), the Fe^{2+} ions diffusion is negligible, because FeO phase has been obtained at high temperatures ($>1500°C$) [8,11]. Therefore, thermal diffusion processes in air atmosphere satisfies the structural formula written as $Zn_{1-x}Fe_xO$, where x represents the Fe atomic concentration, which can define the fraction of Fe^{3+} ions into ZnO lattice as interstitial ions or the substitution of Fe^{3+} for Zn^{2+} [7].

Monitoring of the Raman bands

Raman spectra of Fe-doped ZnO thin-films deposited on quartz substrates and heated at various temperatures are shown in Fig. 2(a). The position of E_2 band corresponding to ZnO films (x = 0) was shifted lightly to high frequency with increasing the temperature, which might be attributed to effects of films stress [7]. A_1 (LO) band due to macroscopic electric field associated with anisotropy in short-range interatomic forces became dispersed gradually with the temperature [12].

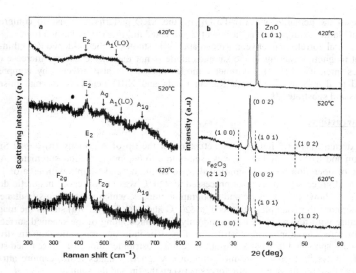

Figure 2. (**a**) Raman spectra of thermally diffused thin-films produced from $420\,^{\circ}C$ to $620\,^{\circ}C$ (**b**) XRD patterns of the same samples.

The A_{1g} band corresponding to hematite phase (α-Fe_2O_3) with compact-hexagonal ordering of Fe^{3+} ions occupied octahedral sites was identified at $520\,^{\circ}C$. With increasing the temperature A_{1g} band changes to F_{2g} band assigned to spinel phase with disordering of the Zn^{2+} ($0.74\,\text{Å}$) and Fe^{3+} ($0.67\,\text{Å}$) cations on tetrahedral and octahedral sites into two separated Raman modes of first order and located at $620\,^{\circ}C$ [13]. The increase of magnetic ordering with the increase of A_{1g} band is related to domain formation into ZnO as $Zn_{0.9}Fe_{0.1}O$ films (x = 0.1) for the increasing of temperature from $520\,^{\circ}C$ to $620\,^{\circ}C$ with a certain degree of crystalline disorder [10].

X-ray diffraction studies

XRD patterns of Fe-doped ZnO thin-films deposited on quartz substrates and heated at various temperatures are shown in Fig. 2(b). Only the diffraction peak of ZnO was observed in samples heated at $420\,^{\circ}C$. Samples processed at $520\,^{\circ}C$ possesses peaks corresponding to the hexagonal structure of ZnO [14]. Samples processed at $620\,^{\circ}C$ with the peak corresponding to the hexagonal structure of ZnO according to PANalytical card no. 00-005-0669, and peak to the rhombohedral phase of Fe_2O_3 according to PANalytical card 00-010-0288.

It could be seen that ZnO films processed at $520\,^{\circ}C$ and $620\,^{\circ}C$ only exhibited (0 0 2) diffraction peak, which indicated that Fe doped ZnO did not change the hexagonal wurtzite structure of ZnO films [7]. Full width at half maximum (FWHM) of (0 0 2) diffraction peak increased with the temperature was not observed. Since all films were prepared with different substrate temperature, the intrinsic stress originating from thermal mismatch between films and substrate should be included as well as stress from the film structure, which can induce interstitial Fe^{3+} ions clustering (lattice disordering) as shown in Fig. 1.

The presence of the hematite phase (α-Fe_2O_3) on the XRD patterns give rise to compression stress into ZnO lattice with peaks displacement as indicated in Fig. 2(b) by dashed lines.

Magneto-structural correlation between Raman and XRD studies of the Fe-doped ZnO thin-films indicates that magnetic ordering in the samples alone is not the only cause for increase of the Raman-modes intensity [12]. The symmetry modes intensity is also affected by compressive stress related to displacement of both Raman bands and XRD peaks by thermal mismatch between films and substrate [7].

Magnetic interactions

An external magnetic field to magnetization of the tip of Antimony (n) doped Si with 4.5μm x 40μm and 0.01−0.025 Ohm-cm was used before to beginning the measurements. A non-contact mode of operation has been employed between the tip and sample with area of 1μm × 1μm for scanned processes. In samples processed at 420°C was not observed magnetic domain structures. Fig. 3 shows the magnetic and topographic images, which were taken simultaneously at room temperature from samples produced at 520°C and 620°C. Fig. 3(a) shows the magnetic domain structure with few white regions randomly distributed along of the scanned surface and Fig. 3(c) shows its topographic microstructure. Fig. 3(b) indicates magnetic domain structure with several dark spots and few white regions randomly distributed along of the scanned surface and Fig. 3(d) shows its topographic microstructure. Magnetic images indicate more attractive magnetic forces than repulsive magnetic forces between the tip and the sample.

Using a width algorithm from Veeco Nanoscope software was measured the cantilever deflection as a function of height difference between two dominant features that can occur at distinct heights (h_i) on surface of the samples.

a. **Average magnetic force** as a function of a set of height difference on the scanned areas of the topographic image to obtain the mean sum of squares was estimated by the followed mean distribution

$$h\overline{F}_z = -k \frac{1}{n} \cdot \sum_{i=1}^{n} (h_{max} - h_i)^2 \tag{1}$$

b. **Average mutual-magnetic moment** between the surface and magnetized tip as a measure of the coupling effectiveness and related with the mutual inductance [15] between two coils was also studied by

$$\mu_m \approx -\left(\frac{k}{B_S}\right) \frac{1}{n} \cdot \sum_{i=1}^{n} (h_{max} - h_i)^2 \tag{2}$$

Evaluation of both average magnetic force and average mutual-magnetic moment as a function of the k constant of cantilever between 20 and 80 N/m and process temperatures was performed and their estimated values are given in Table 1.

The transport mechanisms involved into applications as glucose biosensors, and magnetic bipolar transistors can be related with MFM analysis.

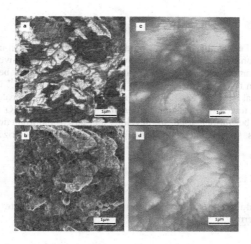

Figure 3. (**a**) Magnetic domain structure and (**c**) topographic microstructure of sample annealed at $520\,^\circ$C. (**b**) Magnetic domain structure and (**d**) topographic microstructure of sample annealed at $620\,^\circ$C.

Magnetic force (F_z) proportional to spin-polarized current and related to response of the films to controlling of spin-polarized current injection was evaluated, because semiconductor devices enhanced in functionality by the use of a magnetic semiconductor, which have the potential to generate spin-polarized current injection into nonmagnetic semiconductors. The novel result is a large current of spin-polarized electrons and ultrafast switching behavior [16].

Mutual-magnetic moment (μ_m) given in Table I was also studied and determine the possible magnetic moment that could be influenced in MFH studies. Magnetic fluid hyperthermia (MFH) involves dispersing MNP throughout the target tissue to cause heat by Néel relaxation [13]. Heat dissipation from MNP is caused by the delay in the relaxation of magnetic moment. Magnetic moment is a critical parameter of MFH from 42 to 45 °C [5].

Table 1. Average values of both magnetic force and mutual-magnetic moment.

Temperature	$\overline{F}_Z$ (pN)	μ_m (Am^2) x 10^{-13}
$520\,^\circ$C	*Min.* 0.546 *Max.* 2.187	*Min.* 2.787 *Max.* 11.13
$620\,^\circ$C	*Min.* 0.445 *Max.* 1.782	*Min.* 2.268 *Max.* 9.078

CONCLUSIONS

ZnO doped with Fe ions thin-films was produced by thermal diffusion in air atmosphere. Structural characterization has demonstrated that the films exhibit compressive stress related to displacement of both Raman bands and XRD peaks by thermal mismatch between films and substrate as well as from the film structure, which is connected with the Fe^{3+} ions clustering as tetrahedral and octahedral sites into ZnO lattice. Magnetic images have indicated both attractive and repulsive magnetic forces with relation to domains formation into ZnO lattice. Fe-doped ZnO thin-films could be further used to develop spin-based multifunctional devices for biomedicine as well as semiconductors with large current of spin-polarized electrons and ultrafast switching.

ACKNOWLEDGMENTS

The authors would like to acknowledge to the M. S. Adolfo Tavira for its help with the measuring of X-ray patterns.

REFERENCES

1. S. Pearton, Nature Materials, 3, 203 (2004).
2. J. Philip, et al., Nature Materials, Advanced Online Publication (2006).
3. S. D. Yoon, et al., J. Appl. Phys, 99 (2006).
4. S. Saha, M. Tomar and V. Gupta, J. Appl. Phys, 111 (2012).
5. S. Laurent, S. Dutz, U. O. Häfeli and M. Mahmoudi, Advanced in Colloid and Interface Science, 166, 8 (2011).
6. R. Baca, et al., in *Kinetics of the oxidation of Zn foils in air atmosphere*, edited by Paul Muralt, (IOP Conf. Series: Materials Science and Engineering, 8, Strasbourg, France, 2010).
7. C. Wang, Z. Chen , Y. He, L. Li and D. Zhang, A. Surface Science, 255, 6881 (2009).
8. A. Cotton, and G. Wilkinson, *Advanced inorganic chemistry*, 4th ed. (John Wiley & Sons, Inc, 2008) p. 715; p. 905.
9. A. S. Grove, *Physics and Technology of Semiconductor Devices*, 1st ed. (John Wiley & Sons, New York, 1967) p.35.
10. Z. Wang, D. Schiferl, Y. Zhao and H.St. C. O'Neill, J. Phys. and Chem. of Sol, 64, 2517 (2003).
11. M. A. Garcia-Lobato, A. Martinez, M.Castro-Roman, C.Falcony and L.Escobar-Alarcon, Physica B, 406, 1496 (2011).
12. J. P. Singh, R. C. Srivastava, H. M. Agrawal and R. Kumar, J. Raman Spectroscopy, 42, 1510 (2011).
13. S. Blundell, *Magnetism in Condensed Matter*, 2nd ed. (Oxford University in Press, Great Britain, 2008) p. 166.
14. A. B-Korczyc, et al., J Sol-Gel Sci Technol, (2011).
15. J.A. Edminister, *Theory and Problems of Electromagnetics*, 2nd ed. (Shaum´s Outline Series, McGraw-Hill, United States of America, 1995) p.169.
16. M. E. Flatté, Z. G. Yu, E. Johnston-Halperin and D. D. Awschalom, A. Phys. Letts, 82, 4740 (2003).

Mater. Res. Soc. Symp. Proc. Vol. 1494 © 2013 Materials Research Society
DOI: 10.1557/opl.2013.372

Ab initio study of the effect of oxygen vacancy on magnetism in Co doped ZnO

S. Lardjane[1,2], G. Merad[2], N. Fenineche[1], H.I. Faraoun[2] and A. Billard[1]

[1]IRTES-LERMPS, UTBM, Site de Montbéliard, 90010-Belfort cedex, France
[2] LEPM-URMER, University of Tlemcen, BP 119 13000, Tlemcen, Algeria.

ABSTRACT

The effect of oxygen vacancy (V_O) on the electronic and magnetic properties of ZnCoO was studied with first principle methods based on density functional theory (DFT). Calculations were performed, on a periodic $3\times3\times3$ wurtzite supercell of ZnO which consists of 108 atoms with two Co ions substituted for two Zn atoms, using the generalized gradient approximation with Hubbard U correction method (GGA+U). We have studied the interatomic exchange interaction with and without V_O for different configurations with different magnetic atom lattice arrangements. The total energies, electronic structures and magnetic moments were calculated for each configuration.

INTRODUCTION

The study of dilute magnetic semiconductors (DMSs) has been a topic of great interest in the past decade since both charge and spin of electrons can be used for novel spintronic devices. DMSs are formed by replacing a very small percentage of cations in a nonmagnetic semiconductor by magnetic transition metal (TM) ions and, therefore, they possess charge and spin degrees of freedom in a single material.
Due to the interesting optical and piezoelectric properties of ZnO, the realization of ZnO-based diluted magnetic semiconductors (DMSs) will lead to new technological innovations for multi-functional devices [1,2]. With a high Curie temperature (T_C), the ZnO:Co system is promising for applications requiring ferromagnetism above room temperature (RT) [3,4].
Despite a great deal of studies focusing on the Co-doped ZnO, there have been many contradictory reports on its magnetic states. Theoretically, Sato et al. [5] predicted that Co-doped ZnO might be ferromagnetic (FM) without any carrier doping and some other computational results also predicted the existence of high Tc for it [6-8]. Experimentally, Ueda et al. [9] reported ferromagnetic behavior of Co-doped ZnO synthesized by pulsed-laser deposition and with a Tc above 300 K and Janisch et al. [10] showed that magnetic moments of $Zn_{1-x}Co_xO$ films grown by various techniques spread over a relatively wide range from 0.56 to 2.6 μ_B/Co.
However, other reports claimed no trace of ferromagnetism in these systems [11,12]. Defects (O vacancy (V_O), Zn vacancy (V_{Zn}), Zn interstitial (Zn_i), n-type doping, p-type doping) were suggested to induce a high temperature ferromagnetism. Also, Seghier et al. [13] have suggested that the oxygen vacancy (Vo) is likely to be a main mediator for the ferromagnetism in the Co-doped ZnO. Therefore, it is necessary to investigate the effect of V_O on the magnetic properties of Co-doped ZnO.
In this paper, using a first principle method, we report the effect of V_O on the electronic structure and the magnetic coupling of Co-doped ZnO.

COMPUTATIONAL DETAILS

First principle calculations have been performed using the projector augmented wave (PAW) formalism [14] of the density functional theory (DFT) as implemented in the Vienna Ab-initio Simulation Package (VASP) [15]. The electrons exchange-correlation energy is described in the generalized gradient approximation with Hubbard U correction method (GGA+U). Typical values of U = 9 and 5 eV [16,17] are adopted for d electrons of Zn and Co atoms, respectively. The Brillouin zone is sampled using a $2\times2\times2$ gamma centred k-point mesh in the Monkhorst-Pack scheme. The convergence is considered complete if the changes of eigenvalues (total energy and band structure energy) between two steps are smaller than 10^{-5} eV. The error bar on the converged total energy is about 10^{-2} meV. The plane-wave cutoff energy in all calculations was set to 500 eV. A periodic wurtzite $(3\times3\times3)$ supercell of ZnO containing 108 atoms (54 Zn atoms and 54 O atoms) was employed. Two Zn atoms were replaced by two Co atoms on different sites in order to obtain (Zn,Co)O with a doping concentration of 3.7 at.%. We specified the ferromagnetic state by fixing the magnetic moment of the two atoms of cobalt in the unit cell at $+3\mu_B$ for atom number 1 and at $+3\mu_B$ for atom number 2. Whereas the anti-ferromagnetic state we fixed the magnetic moment of the cobalt atom number at $+3\mu_B$ and the magnetic moment of the Co atom number 2 at $-3\mu_B$.

RESULTS AND DISCUSSION

Using GGA+U scheme, calculations on Co-doped ZnO at a concentration of 3.7 at.%, reveal that the antiferromagnetic (AFM) state is the more stable for configurations, where the two Co impurities are first neighbours and lie on two adjacent Zn planes perpendicular to the hexagonal c axis (A1) and where they lie on two adjacent planes (A2) [18].
In order to study the effect of the oxygen vacancy (V_O) on the magnetic properties of Co-doped ZnO, different configurations of oxygen vacancy position have been adopted: O11, O12, O13, O21, O22 and O23 (fig. 1). The removed oxygen atom is directly connected to one Co atom in O11 and O21 configurations and to two Co atoms in O12 and O22 respectively, while it is connected by Zn-O bond to the two atoms of cobalt in O13 and O23. These configurations correspond to a concentration of 1.85 at.% of V_O. The magnetic coupling strength J for different configurations is calculated from the energy difference between the FM and AFM configurations given by: $\Delta E=(E_{FM} - E_{AFM})/2 = -(J/2)S_T(S_T + 1)$, where S_T is the total spin for two magnetic impurities of spin S ($S_T = 2S = 3$) [18].
Table 1 presents the total energies of the FM and AFM states and the exchange coupling between Co atoms for the supercell of $Zn_{0.963}Co_{0.037}O$ without defects [18], and in the presence of defects in different configurations of oxygen vacancies.
The antiferromagnetic coupling between the Co atoms in A2 becomes FM interaction with energy difference of 5.33 meV when the oxygen vacancy is nearest Co atoms (O21 configuration), but these ferromagnetic interactions are not enough strong to dominate the antiferromagnetic interaction in the system and to establish ferromagnetic state at room temperature. However, in O22 and O23, the AFM arrangement is favored by 6.63·meV and 6.65 meV respectively. In A1 configuration, the AFM state is more stable by 6.1 meV, 20.88 meV, 29.52 meV in O11, O12 and O13 respectively.

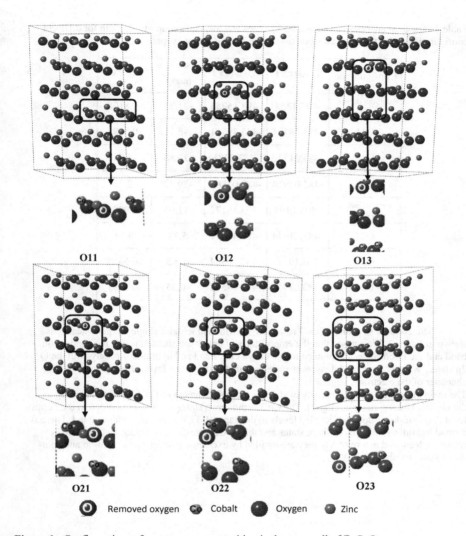

O11   O12   O13

O21   O22   O23

Removed oxygen   Cobalt   Oxygen   Zinc

Figure 1: Configurations of oxygen vacancy position in the supercell of ZnCoO.

33

Table 1: The total energies of FM and AFM states and magnetic coupling strength for different configurations of oxygen vacancy position in the $Zn_{0.963}Co_{0.037}O$ supercell.

Configuration	E FM (eV)	E$_{AFM}$ (eV)	E$_{AFM}$-E$_{FM}$ (meV)	J (meV)
without defect A1	-493.1358	-493.15753	-21.73	-1.8108
O11	-483,38701	-483,39311	-6,1	-0,5083
O12	-483,21722	-483,2381	-20,88	-1,74
O13	-483,04914	-483,07866	-29,52	-2,46
without defect (A2)	-493.1498	-493.16171	-11.91	-0.9925
O21	-483,41644	-483,41111	5,33	0,44
O22	-483,19721	-483,20384	-6,63	-0,5525
O23	-483,03960	-483,04595	-6,35	-0,5291

The electronic structure of the $Zn_{0.963}Co_{0.37}O$ with and without single oxygen vacancy is shown in figure 2. We can see that the empty states of cobalt are situated inside the conduction band and the filled states are situated inside the valence band and hybridize strongly with the O 2p states. Whereas no Co 3d density of states is found at the Fermi level defining an insulating character of the electronic structure.

The main effect of Vo on the density of states is the displacement of Co 3d states of spin-up to the highest binding energies (fig. 2). The Co 3d spin down states are also remounted to the Fermi level. V_O introduces doubly occupied levels in the gap at 1.3 eV above the valence band. There is a small hybridization between these states and the 3d Co states. The hybridized states Co-V_O retain the localized nature of V_O, they are divided by exchange interactions in spin-up and spin-down states by about 0.3 eV.

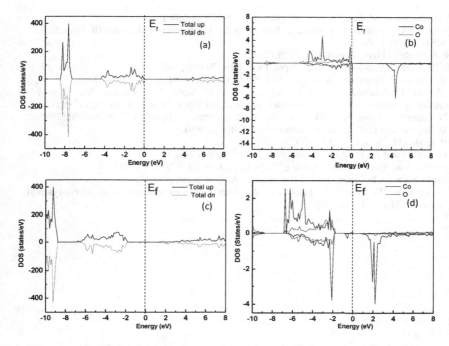

Figure 2: (a) Total DOS and (b) partial DOS of Co 3d and O 2p of $Zn_{0.963}Co_{0.037}O$ (c) Total DOS and (d) partial DOS of Co 3d and O 2p of $Zn_{0.963}Co_{0.037}O$ with V_O.

CONCLUSION

In this work, we have studied the electronic structure and magnetic properties of $Zn_{1-x}Co_xO$ (x=0.037) DMS in the presence of oxygen vacancy using the first principle spin polarized calculations. The results indicate that the the oxygen vacancies cannot be the origin of ferromagnetism in ZnCoO and the interaction between these defects and the cobalt atoms is very low.

REFERENCES

1. F. Pan, C. Song, X.J. Liu, Y.C Yang and F. Zeng Mater. Sci. Eng. R 62 1 (2008).
2. C. Liu, F. Yun and H. Morkoc J. Mater. Sci.-Mater. El. 16 555 (2005).
3. W. Prellier, A. Fouchet and B. Mercey 2003 J. Phys.: Condens. Matter 15 (1583).
4. S.A. Chambers, T.C. Droubay, C.M. Wang, K.M. Rosso, S.M. Heald, D.A. Schwartz, K.R. Kittilstved and D.R. Gamelin Mater. Today 92 (2006).
5. K. Sato and H. Katayama-Yoshida Jpn. J. Appl. Phys. Part 2 39 555 (2000).

6. P. Bruno and L. M. Sandratskii Phys. Rev. B 73, 045203 (2006).

7. L. Petit, T.C. Schulthess, A. Svane, Z. Szotek, W.M. Temmerman and A. Janotti Phys. Rev. B 73, 045107 (2006).

8. N. A. Spaldin Phys. Rev. B 69 125201 (2004).

9. K. Ueda, H. Tabata and T. Kawai Appl. Phys. Lett. 79 988 (2001).

10. R. Janisch, P. Gopal and N.A. Spaldin J. Phys.: Con- dens. Matter 17 R657 (2005).

11. Z. Jin, T. Fukumura, M. Kawasaki, K. Ando, H. Saito, T. Sekiguchi, Y.Z. Yoo, M. Murakami, Y. Matsumoto, T. Hasegawa and H. Koinuma Appl. Phys. Lett. 78 3824 (2001).

12. N. Jedrecy, H.J. von Bardeleben, Y. Zheng and J. Cantin L Phys. Rev. B 69 041308 (2004).

13. D. Seghier and H. P. Gislason Physica B 404 4800 (2009).

14. P. E. Blöchl, Phys. Rev. B 50 17953 (1994).

15. G. Kresse and J. Furthmüller, Phys. Rev. B 54 11169 (1996).

16. En-Zuo Liu, Yan He, and J. Z. Jiang. Appl. Phys. Lett. 93, 132506 (2008).

17. SuHo Na and Chul-Hong Park. J Korean Phys Soc 54, (2009) 86-872.

18. S. Lardjane, G. Merad, N. Fenineche, A. Billard, H.I. Faraoun. Journal of Alloys and Compounds 551 (2013) 306–311.

Mater. Res. Soc. Symp. Proc. Vol. 1494 © 2012 Materials Research Society
DOI: 10.1557/opl.2012.1696

Synthesis and Characterization of Cu-doped ZnO Film in Nanowire like Morphology Using Low Temperature Self-Catalytic Vapor-Liquid-Solid (VLS) Method

Ratheesh R. Thankalekshmi[1,2], Samwad Dixit[1,2], In-Tae Bae[3], Daniel VanHart[2] and A.C.Rastogi[1,2]

[1]Electrical and Computer Engineering Department,
[2]Center for Autonomous Solar Power (CASP),
[3]Small Scale Systems Integration and Packaging Center,
Binghamton University, State University of New York, Binghamton, NY-13902, U.S.A.

ABSTRACT

Cu-doped ZnO film in nanowire structure is synthesized by a closed space flux sublimation and periodic oxidation method at ~300°C over Si substrate. Oxidative process controlled self-catalytic VLS mechanism is proposed for the film growth. X-ray diffraction pattern establishes that Cu-doped ZnO nanowires retain the crystallite structure of the wurtzite ZnO. TEM studies indicate single crystal character of the Cu-doped ZnO nanowires. Optical absorption analysis of Cu-doped ZnO nanowires defines two direct energy band gaps. The low energy band gap at 3.2eV is intrinsic to the Cu-doped ZnO material. The higher energy band gap at 3.5eV is attributed to the nanosize, mediated by strong forward scattering of light from the nanowires. Sharp photoluminescence in Cu-doped ZnO corresponding to near bandgap free exciton emission is observed and a redshift of ~0.07 eV is consistent with the effect of Cu-doping. The visible emission band in both ZnO and Cu-doped ZnO shows a broad green emission band with Cu-substitution shifting the maximum visible luminescence towards the higher energy side.

INTRODUCTION

In the recent years, thin films of zinc oxide (ZnO) based materials system have attracted considerable interest due to their potential applications in various fields, such as ultraviolet optoelectronics, field effect devices, transparent conductor in display devices and solar cells. Doping ZnO with both isovalent and trivalent elements significantly alters its electronic and optical properties which can be tailored for multifunctional device applications. In this context, Cu-doped ZnO has received considerable attention, since Cu-doping has been shown to be a good candidate for producing both p-type and band gap reduced ZnO [1] as well as for realizing green emission in ZnO [2]. Cu-doped ZnO greatly facilitates incorporation of magnetic impurities in ZnO, which is important for spin FET devices using ZnO dilute magnetic semiconductor [3]. ZnO films in the nanostructured forms such as nanowires, nanoneedles and nanorods are important as these can significantly improve the device relevant electro-optical properties through quantum size effects. The most common method for synthesis of Cu-doped ZnO in 1-D structures is by vapor-liquid-solid (VLS) process using Au or Cu as catalyst at high temperatures [4-5]. For example, Cu-doped ZnO nanowires growth has been reported at 600°C [6]. We report on the synthesis of Cu-doped ZnO films with a single crystal nanowire structure at much lower ~300°C temperature by oxidative process controlled self-catalytic vapor-liquid-solid (VLS) method without secondary phases. This paper describes the growth, crystalline structure and optical properties of Cu-doped ZnO nanowire films.

EXPERIMENT

The Cu-doped ZnO nanowire thin films were synthesized by closed space flux sublimation of precursors in an inert low pressure environment followed by short periodic oxidation by exposure to oxygen in a tubular reactor [7]. The film growth involves sublimation of volatile BiI_3 mixed with CuI and fine (100 mesh) Zn powder in 1:1:3 weight ratio over substrates placed on top of the powdered precursor spaced at ~5 mm. The process is carried out at 0.3-0.5 Torr pressures and temperature ~300°C under 100 sccm Ar flow. Film formation follows a two-step approach, where flux sublimation is done for 3-5 min which produces a condensate layer of precursor solid-solution over substrate followed by a 30s exposure to oxygen at atmospheric pressure which by rapid oxidation, converts the condensate layer into a Cu-doped ZnO nanowire film. This process is repeated several times to build-up the Cu-doped ZnO nanowire film thickness on the downward polished faces of the substrates.

The crystal structure of Cu-doped ZnO samples was determined using PANalytical's X'Pert PRO Materials Research Diffractometer with Cu Kα radiation using a Ni filter. Morphological characterization was examined with Zeiss Supra 55 VP field emission scanning electron microscope. TEM analysis was performed using 200 kV JEOL JEM-2100F field emission TEM equipped with energy dispersive x-ray spectroscopy (EDS, Tokyo, Japan). The optical transmission measurements were carried out using Beckman DU 7400 spectrophotometer in the 280-800 nm range. The photoluminescence spectra were recorded at room temperature in the 360 to 650 nm range using the HORIBA Jobin Yvon iHR320 imaging spectrometer with excitation wavelength 350 nm to determine optical transition energies.

DISCUSSION
Nanostructure morphology and growth mechanism

SEM micrograph in figure 1(a) shows a typical nanowire like morphology of the Cu-doped ZnO film deposited over Si substrate. These nanowires are densely packed and are scattered randomly covering the entire Si substrate surface having an average 20-30 nm diameter with 3-5 μm lengths. Inset of figure 1(a) shows measured ~22 nm diameter of a single nanowire. Morphology of the nanowire cluster is revealed in a magnified SEM image in figure 1(b). The cluster comprises of several perfect hexagonal crystals from which the nanowire appear to have grown outwardly as more clearly seen from inset of figure 1(b).

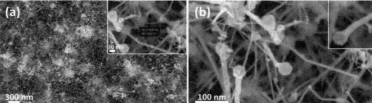

Figure 1. SEM images of Cu-doped ZnO film on Si substrate (a) nanowires emanating from clusters, nanowire of typical 22 nm shown by inset image (b) magnified image of a typical cluster with nanowires originating from a single crystal tip as shown clearly by inset image.

The conventional VLS process is initiated at the Zn-metal catalyst liquid droplets, which upon saturation form a solid precipitate that acts as a preferred site for further oxidative growth leading to nanowire formation. In the present technique, the nanowires grow from Zn-Bi solid solution which as phase diagram shows easily forms at 260°C [8]. The ZnO nanowire growth is initiated from Zn-Bi condensate over substrate surface on short exposure to oxygen as ZnO is immiscible in Zn-Bi solid solution phase. Therefore the base of ZnO-nanowires is inevitably a well formed hexagonal ZnO nanocrystal converted from Zn-Bi droplet as evidenced from figure 1(b). The formation of ZnO nanowires is therefore by a self-catalytic VLS process wherein Zn-Bi solid solution condensate functions as catalyst during the growth process. Bismuth content is far less than Zn and this keeps the liquid alloy saturated with Zn yielding the growth of ZnO. Incorporation of Cu into ZnO is through γ-Cu_5Zn_8 and β'-CuZn alloy which also forms at~250°C [9] and which gets incorporated in the Zn-Bi precursor condensate during flux sublimation stage. Based on the EDAX data, the relative elemental composition of nanowire is typically determined as Zn (52.5 at%), Cu (3.2 at%) and O (44.3 at%). Bi from precursor flux is also incorporated in the Cu-doped ZnO film but gives no secondary phases in the film.

Crystalline structure

Figure 2(a) shows the typical XRD pattern of a Cu-doped ZnO nanowire film on Si substrate in the $2\theta=30°$-$70°$ range. The XRD diffraction peaks of the as-deposited film are consistent with the reported standard values for a ZnO film (JCPDS No. 00-001-1136). Based on this interpretation it can be stated that the Cu-doped ZnO nanowire films form in the hexagonal wurtzite structure.

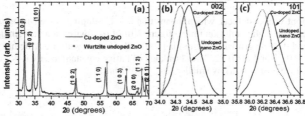

Figure 2. (a) XRD pattern of Cu-doped ZnO film on Si substrate. Standard diffraction peak locations from wurtzite undoped ZnO are indicated by vertical lines. (b) comparison of the (002) and (c) (101) lines of Cu-doped ZnO film and undoped nanostructured ZnO film on Si substrate.

However, the XRD pattern of the Cu-doped ZnO nanowire films generally show a shift towards the lower angle side compared with that of a standard wurtzite undoped ZnO. Both the nanostructure of the film as well as addition of Cu-dopant could be assigned for this shift. To clarify this in the figure 2(b) and (c), high resolution (002) and (101) diffractions from undoped nanostructured ZnO with Cu-doped ZnO nanowire film are compared. The Cu-doped ZnO nanowire films show ~0.12° shift towards higher scattering angle. As the ionic radius of Cu^{2+} (0.057 nm) is smaller than that of Zn^{2+} (0.06 nm), substitution of Cu^{2+} at Zn^{2+} sites will result in the lattice shrinkage and hence reduction in lattice constant. The observed higher angle shift is consistent with the interpretation that Cu^{2+} incorporation into the ZnO host lattice is by substitution at Zn^{2+} sites. The XRD pattern neither shows structural disorder by substitution of Cu for Zn in the nanowires and nor CuO phases in the film which also show homogeneous doping of Cu in ZnO. Furthermore, the Cu-doped ZnO nanowire film is highly crystalline as shown by sharp diffraction lines. The average grain size of the Cu-doped ZnO nanowires has been determined by Scherer's formula D= K*λ / β*cosθ, where D is the average grain size, K taken as 0.9 is the shape factor, λ is the x-ray wavelength, β is the line width at half maximum intensity and θ is the Bragg diffraction angle. The average size is estimated to be ~19.9 nm. This is comparable to the size determined from the SEM study.

Transmission electron microscopy studies

High resolution transmission electron microscopic imaging of lattice planes in Cu-doped ZnO nanowires was carried out. Cu-doped ZnO nanowires were carefully exfoliated from the Si substrate and transferred on a Ni grid coated with lacey carbon film. TEM image of nanowires in

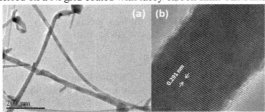

Figure 3. (a) Low-resolution TEM image of Cu-doped ZnO nanowires and (b) high-resolution image of the Cu-doped ZnO nanowire.

figure 3(a) are without visible defects, which means that the Cu-doping is well-integrated into the ZnO lattice sites. The high resolution image figure 3(b) of nanowire indicates single crystalline nature and that the growth direction is [0001]. ZnO has a wurtzite structure with lattice parameters a= 0.33 and c = 0.52. The shrinkage of lattice constant after Cu-doping is due to the substitution of Zn^{2+} (0.06 nm) by smaller Cu^{2+} (0.057 nm).

Optical Bandgap

Transmission spectra of the Cu-doped ZnO nanowire film and undoped nanostructured ZnO film on quartz substrates are shown in figure 4(a).

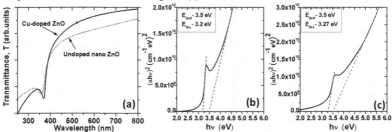

Figure 4. (a) Optical transmission spectrum of Cu-doped ZnO and undoped nanostructured ZnO film on quartz substrate. Tauc's plot showing (b) direct band gaps at 3.2 eV and 3.5 eV of the Cu-doped ZnO film (c) direct band gaps at 3.27 eV and 3.5 eV of the undoped nanostructured ZnO film.

Both undoped nanostructured ZnO and Cu-doped ZnO nanowire films show a steep absorption near the bandgap energy showing the crystalline nature of the films. Increased transmission in the lower wavelength beyond the band-gap energy for Cu-doped ZnO nanowire film and undoped nanostructured ZnO is caused by the forward scattering of light from the nanostructures analyzed using the Mie scattering theory [7]. Direct optical band gap for nanostructured undoped ZnO and Cu-doped ZnO films were determined by Tauc's relation $(\alpha h\nu)^2 = A(h\nu - E_G)$, where, absorption coefficient α is determined from the measured absorbance (Abs) and film thickness t, using the relation $(1-Abs)=exp(\alpha.t)$. This plot in figure 4(b) shows excellent straight line fits in the two energy regions each within a narrow photon energy range indicating possibility of two well defined direct energy band gaps. The lower energy direct band gap at $E_{G-L} \sim 3.2$ eV originates from the macroscopic structural feature in the Cu-doped ZnO film. Compared to undoped nanostructured ZnO film (figure 4(c)) which has a band gap ~3.27 eV, a lower optical band gap of 3.2 eV is observed in Cu-doped ZnO film. The bandgap lowering by 0.07 eV is attributed to Cu-doping in ZnO. It is well known that Cu-doping in ZnO tends to reduce the band gap due to the sp-d exchange interaction between the Cu- d electrons and the conduction band electrons of ZnO [10]. The higher energy band gap shown to occur at $E_{G-H} \sim 3.5$ eV for nanostructured undoped ZnO and Cu-doped ZnO nanowire films are attributed to the quantum size effects due to confinement of carriers in the nanostructures.

Photoluminescence Spectra

The room temperature PL spectra of undoped nanostructured ZnO and Cu-doped ZnO nanowire films on quartz substrate are shown in figure 5.

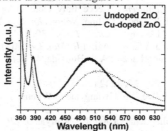

Figure 5. PL spectra of undoped nanostructured ZnO and Cu-doped ZnO nanowires on quartz substrate recorded at room temperature with an excitation wavelength at 350 nm.

Both PL spectra consist of broad green emission band and a strong UV emission. The UV emission peak in both films originates from excitonic recombination corresponding to the near-band-edge emission. The UV peak emission in undoped ZnO film at 375 nm (3.30 eV) is due to D^oX transition attributed to near band edge emission of bound neutral donors. The UV emission in Cu-doped ZnO nanowire films is red-shifted to 383 nm (3.23 eV). This 8 nm (0.07 eV) red shift is consistent with the reduction of the ZnO optical band gap by Cu-doping. The other possibility is the electronic transitions between the donor-acceptor pairs [11] since Cu-doping produces acceptor states in ZnO [12]. In undoped nanostructured ZnO films the green emission is centered ~510 nm. In ZnO this emission band is assigned to radiative recombination of electrons trapped in oxygen vacancy (V_O) defects with a photo-excited hole in the valence band.

In Cu-doped ZnO, this band is shifted to higher energy side to ~495 nm with increased intensity. This is consistent with several past studies in Cu-ZnO system and could be assigned to transition between V_O with ionized Cu impurity [13].

CONCLUSIONS

Cu-doped ZnO film in nanowire structure is synthesized by oxidative process controlled self-catalytic vapor-liquid-solid (VLS) mechanism. Based on XRD investigation Cu-doping at Zn site has been shown. The nanowires are single crystals and reduced lattice spacing compared to undoped ZnO is consistent with the Cu-doping effects. PL spectrum and optical bandgap studies also support these inferences.

ACKNOWLEDGMENT

This work was supported by the Office of Naval Research (ONR) under contract N00014-11-1-0658 which is gratefully acknowledged.

REFERENCES
1. K.S.Ahn, T.Deutsch, Y.Yan, Ch.Sh. Jiang, C.L.Perkins, J.Turner and M.Al-Jassim, *J. Appl. Phys.* **102,** 023517 (2007).
2. M.D.McCluskey and S.J.Jokela, *J. Appl. Phys.* **106,** 071101 (2009).
3. R. R. Thankalekshmi and A. C. Rastogi, *J. Appl. Phys.* **111,** 07D104 (2012).
4. E.Comini, M.Ferroni, N.Poli, G.Sberveglieri,S.Kaciulis, A.Mezzi and L.Pandolfi, *Proceedings of IEEE Sensors Conference,* Lecce, Italy, October pp. 835–838 (2008).
5. C.Xu, X.W.Sun, X.H.Zhang, L.Ke and S.J.Chua, *Nanotechnology* **15,** 856 (2004).
6. C.Xu , K.Yang , L.Huang and H.Wang, *J. Chem. Phys.* **130,** 124711 (2009).
7. R.R.Thankalekshmi, S.Dixit and A.C. Rastogi, *Adv. Mat. Lett.* **4(1),** 9 (2013).
8. Y.Djaballah, L. Bennour, F.Bouharkat and A. B.Bouzida, *Modelling Simul. Mater. Sci. Eng.* **13,** 361 (2005).
9. C.H.Yu and K.L.Lin, *J. Mater. Res.* **20,** 1242 (2005).
10. C. Xu,T.W. Koo,B.S. Kim,J.H. Lee,S.W.Hwang and D.Whang, *J.Nanosci. Nanotechnol.* **11,** 1 (2011).
11. A.C. Rastogi, S. B. Desu, P. Bhattacharya and R.S. Katiyar, *J. Electroceramics* **13,** 345 (2004).
12. J. B. Kim, D. Byun, S. Y. Ie, D. H. Park, W. K. Choi, J. W. Choi and B. Angadi, *Semicond. Sci. Technol.* **23,** 095004 (2008).
13. N.Y.Garces, L.Wang, L. Bai, N. C. Giles, L. E. Halliburton and G. Cantwell, *Appl. Phys. Lett.* **81,** 622 (2002).

Mater. Res. Soc. Symp. Proc. Vol. 1494 © 2013 Materials Research Society
DOI: 10.1557/opl.2012.1742

Copper Doped ZnO Thin Film for Ultraviolet Photodetector with Enhanced Photosensitivity

Akshta Rajan[1], Kashima Arora[1], Harish Kumar Yadav[2], Vinay Gupta[1] and Monika Tomar[3]
[1]Department of Physics and Astrophysics, University of Delhi, Delhi, INDIA.
[2] Department of Physics, St.Stephens College, University of Delhi, Delhi, INDIA.
[3] Physics Department, Miranda house, University of Delhi, Delhi, INDIA.

ABSTRACT

Ultraviolet photoconductivity in Copper doped ZnO (Cu:ZnO) thin films synthesized by sol-gel technique is investigated. Response characteristics of Pure ZnO thin film and Cu:ZnO thin film UV photodetector with 1.3 at. wt % Cu doping biased at 5 V for UV radiation of λ = 365 nm and intensity = 24 μwatt/cm^2 has been studied. Cu:ZnO UV photodetector is found to exhibit a high photocoductive gain (K = 1.5×10^4) with fast recovery ($T_{90\%}$ = 23s) in comparison to pure ZnO thin film based photodetector (K = 4.9×10^1 and $T_{90\%}$ = 41s). Cu^{2+} ions have been substituted in ZnO lattice which has been confirmed by X-ray diffraction (XRD) and Raman spectroscopy leading to lowering of dark current ($I_{off} \sim$ 1.44 nA). Upon UV illumination, more electron hole pairs are generated in the photodetector due to the high porosity and roughness of the surface of the film which favours adsorption of more oxygen on the surface of the photodetector. The photogenerated holes recombined with the trapped electrons, increasing the concentration of photogenerated electrons in the conduction band enhancing the photocurrent ($I_{on} \sim$ 0.02 mA) of the Cu:ZnO photodetector.

INTRODUCTION

Detection of ultra violet (UV) radiations is becoming very important in various commercial, military and scientific areas, which has made ultraviolet photodetectors reasonably an interesting field. In recent years, besides GaN which is well established commercial material for detection of UV rays, ZnO in the form of nanostructures and thin films have gained interest and proved to be advantageous material for selective UV photodetection, as well as solar blind UV detectors due to wide band gap (~3.3 eV), short carrier lifetime, large exciton binding energy (~60 meV), ease of fabrication of thin films by various deposition techniques, radiation hardness, low cost synthesis etc. Photoconductivity in ZnO involves adsorption and desorption of oxygen molecules leading to the process of trapping and releasing holes at the ZnO surface [1,2]. Photoconducting property in ZnO lattice can be improved by introducing defects within the band gap by intentionally doping foreign metals such as Al, N, Te, Li and Cu at Zn lattice site [3]. Copper doping in ZnO has been reported to form acceptor states within the bandgap of ZnO resulting in higher resistance of ZnO. Moreover, doping of copper in ZnO nanowires is known for strong multiplication of photocarriers which leads to dramatically enhanced sensitivity to optical radiation over multiple spectral ranges of UV and visible regions of the electromagnetic spectrum [3,4]. In the present work, we show the feasibility of realizing a highly sensitive UV photodetectors by exploring the photoconducting properties of Cu:ZnO using sol gel technique with 1.31 at. wt % of Copper. Changes in sensing parameters like dark current, photocurrent and photoconductive gain have been studied.

EXPERIMENT

Prior to the fabrication of UV photodetector, platinum interdigital electrodes (IDEs) were patterned on corning glass substrates using the conventional photolithographic technique. In order to prepare the Pt interdigital electrodes, Pt thin film of 90 nm thickness was deposited by RF sputtering. Finger width and gap width of the patterned electrodes were kept to be fixed at 500 μm each. In order to prepare the UV photodetector, ZnO and Cu:ZnO thin films were deposited on the top of patterned electrodes as the sensing layer and the measurements were carried out through the Pt contact pads of the patterned IDEs. The size of active device area is 1 cm^2. The schematic of the prepared photodetector is shown in Schematic 1.

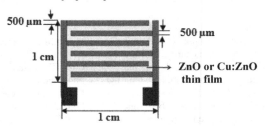

Schematic I: Schematic of the prepared photodetector

For the ZnO and Cu:ZnO thin film deposition, 0.227 M ZnO and Copper doped ZnO sols have been prepared. As a starting material, Zinc acetate dihydrate [Zn(CH$_3$COO)$_2$.2H$_2$O] purchased from Sigma-Aldrich (99.99% pure) was used. Methanol and Monoethanolamine (MEA) were used as a solvent and stabilizer respectively. Zinc acetate dihydrate [Zn(CH$_3$COO)$_2$.2H$_2$O] was first dissolved in a mixture of Methanol and copper sulphate (CuSO$_4$.5H$_2$O) (Fisher scientific, 99.99% pure) and then MEA was added. The prepared solution was spin coated over the IDE/Glass substrates. After each coating, the films were dried at 300°C for 3-5 min to evaporate the solvent and remove organic residuals. The total thickness of the films was maintained to be 113 nm. The films were annealed at 575°C for 1 hr in tube furnace to yield the desired crystallinity in deposited films.

Physical and optical properties of the films were analyzed by X-ray diffraction (XRD), Atomic Force microscopy (AFM) and UV-visible spectrophotometer. The Raman spectra was obtained at room temperature using Argon ion laser at 488 nm line to study the defect profile of the doped and pure ZnO thin films deposited in the present case. Steady state photoresponse of the deposited pure ZnO and Cu:ZnO thin films was measured at 5 V bias voltage by illuminating the samples using an UV lamp (λ = 365 nm, intensity = 24 μwatt/cm^2) as a radiation source. The photoresponse transients were recorded from semiconductor characterization system (Keithley 4200 SCS). The photoconductive Gain of a UV photodetector is defined as K = I$_{on}$/I$_{off}$, where I$_{on}$ is the photocurrent measured by illuminating the photodetector with UV radiations and I$_{off}$ is the dark current.

DISCUSSION

X-ray diffraction studies

Figure 1 shows the XRD patterns of ZnO and Cu:ZnO thin films. It may be seen that all the films exhibit polycrystalline nature having peaks corresponding to hexagonal wrutzite structure of ZnO. Both the films show dominant (002) oriented peak along with diffraction peaks corresponding to (100) and (101) planes of ZnO although it can be observed that intensity of (100) and (101) peaks has increased showing reduction in C-axis orientation after incorporating Cu^{2+} ion in ZnO lattice. There was no extra peak corresponding to CuO/Cu_2O in Cu:ZnO thin film due to the small concentration of Cu (1.3 at. wt%) in ZnO. However, an appreciable shift in all the three peak positions towards higher angle has been observed which indicates that the Cu^{2+} ions were substituted into the Zn^{2+} sites [5]. The estimated value of the crystallite size for 113 nm thin pure ZnO film was found to be around 25 nm. Interestingly it is found that on adding Cu as a dopant in ZnO thin film crystallite size decreased to 17 nm.

UV-Visible Spectroscopy

Figure 2 shows the UV-Visible spectra of the ZnO and Cu:ZnO thin films. Both the films (ZnO and Cu:ZnO) were found to be highly transparent having about 80% transparency in the visible region for the pure ZnO films which reduces to ~ 70 % for the Cu:ZnO thin film. Optical band gap (E_g) of films was evaluated by extrapolating the linear portion of Tauc plot between $(\alpha h\nu)^2$ versus $h\nu$ to $\alpha = 0$, where α is the absorption coefficient and $h\nu$ is the photon energy. Estimated value of the band gap was found to be 3.276 eV for the pure ZnO thin film and is in good agreement with the values reported by other workers for ZnO thin films grown by various techniques [6]. The band gap was found to be decreased from 3.276 eV to 3.242 eV with the incorporation of Cu in the ZnO thin film (inset of Figure 1). A lower value of band gap obtained for the Cu:ZnO film compared to ZnO is attributed to the fact that CuO has relatively low band gap (~ 1.2 eV) in comparison to that of ZnO (~ 3.3 eV). Hence, incorporation of Cu into ZnO reduces the band gap of Cu:ZnO composite thin film.

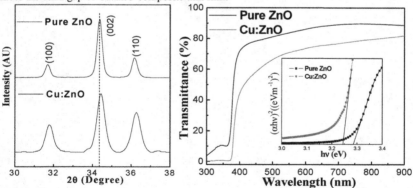

Figure 1. XRD spectra of ZnO and Cu:ZnO thin films.

Figure 2. UV-Visible spectra of ZnO and Cu:ZnO thin films. Inset showing Tauc plot.

Atomic force microscopy (AFM)

Figure 3 shows the AFM images of ZnO and Cu:ZnO thin films. It can be seen that 113 nm Cu:ZnO thin film possesses more surface roughness and porosity in comparison to pure ZnO thin film, which is advantageous for UV photodetection.

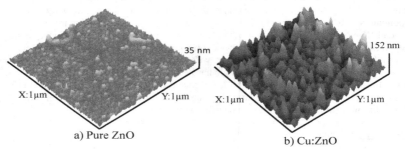

a) Pure ZnO

b) Cu:ZnO

Figure 3. AFM images of (a) ZnO and (b) Cu:ZnO thin films.

Raman scattering spectroscopy

Figure 4 shows the room temperature Raman spectra of pure ZnO and Cu:ZnO thin films. The characteristic optical modes of wurtzite ZnO at 99 and 437 cm^{-1} corresponding to the E_2^{low} and E_2^{high} modes respectively were observed. The peak at 303 cm^{-1} is due to scattering from the silicon substrate [7]. The oxygen sublattice vibrational optical mode (E_2^{high}) was found to shift (2 cm^{-1}) toward the lower frequency side in Cu:ZnO with respect to the ZnO thin film. This shift is due to the confinement of optical phonon in a finite region. The red shift of the E_2^{high} mode in Cu:ZnO thin films represents the substitution of Cu^{2+} ion at the Zn^{2+} site in the ZnO lattice [8].

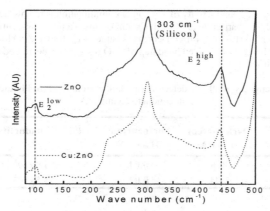

Figure 4. Raman Spectra of ZnO and Cu:ZnO thin films.

Sensing response Characteristics

UV-response characteristics of the pure and Cu:ZnO thin film photodetectors were investigated towards UV radiation of $\lambda = 365$ nm and intensity = 24 μwatt/cm^2 at a bias voltage of 5 V. Figure 5 represents the time-dependent on/off photoconduction measurements carried out for ZnO and Cu:ZnO thin film samples. It may be seen that when the UV is off, all the prepared UV-photodetectors show a low current (I_{off}) which increases and saturates at a particular level in the presence of UV radiation giving I_{on}. When the UV radiation is turned off, the current decreases slowly to attain its initial value of current showing persistence. Values of I_{off}, I_{on} and K obtained for both the samples is summarized in Table I. It may be seen from Table I that the I_{off} for pure ZnO UV photodetector at an applied bias of 5 V is found to be 816 nA and upon UV illumination (24 μwatt/cm^2 intensity) I_{on} increases to 0.04 mA. When 1.3 at. wt % of Cu is incorporated into ZnO lattice, a decrease in the I_{off} is obtained (1.44 nA) and upon UV illumination I_{on} increases to 0.02 mA thus giving a very high photoconductive gain of 1.5×10^4. Figure 5(b) gives the log(I)-Time plot from which the rise time and recovery time for the prepared ZnO and Cu:ZnO thin film based photodetector has been estimated. Rise time has been calculated as the time taken for the dark current to rise to 90 % of saturation value under UV illumination, and recovery time has been taken as time taken for the photo current to fall to 10 % of the saturation value when UV light is switched off. Rise time for pure ZnO thin film based photodetector has been calculated to be 21.5 seconds. However, the photoconductivity is persistent and does not return back to its original value. On the contrary, Cu:ZnO thin film based photodetector exhibits a faster response and recovery time of 16.4 seconds and 22.5 seconds respectively.

Photoconductivity in ZnO can be explained on the basis of adsorption and desorption of oxygen molecules on ZnO thin film surface. Adsorbed oxygen molecules on ZnO surface captures free electrons leading to a surface depletion and band bending towards the ZnO surface. Electron hole pairs are generated upon UV illumination, photogenerated holes move along bending and these migrated holes get trapped by negatively charged oxygen molecules reducing the barrier height. Increased conductivity of ZnO on UV illumination is mainly due to enhanced carrier density and reduced barrier height [9]. In Cu:ZnO, a low value of dark current (Figure 5, Table I) has been observed (I_{off} ~ 1.44 nA). Decreased value of I_{off} for Cu:ZnO thin film is due to formation of acceptor levels in forbidden band gap of ZnO resulting in decreased number of charge carrier concentration.

Table I: Sensing parameters of all UV detectors (a) Dark current (I_{off}) (b) Photocurrent (I_{on}) and (c) Photoconductive Gain (K)

Sample Prepared	Dark current (I_{off}) nA	Photocurrent (I_{on}) mA	Photo Conductive gain (K)
Pure ZnO	816	0.04	4.9×10^1
Cu:ZnO	1.44	0.02	1.5×10^4

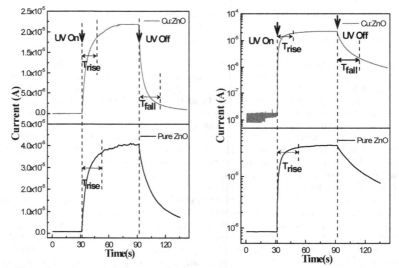

Figure 5: Time-dependent on/off photoconduction measurements on the pure and Cu:ZnO thin film based UV photodetector in the presence of UV radiation of 24 µwatt/cm^2 intensity (a) I-T plot (b) log (I)-T plot

Lowering of dark current might also be due to the formation of CuO-ZnO interfaces inside the bulk of ZnO. CuO is p-type, whereas ZnO is n type semiconductor. Therefore, formation of p-n junction inside the bulk of the sample is possible resulting in depletion region at the interface, thereby leading to narrow path for the conduction of charge carriers. As a result, resistance of the Cu:ZnO photodetector decreases and hence reduces the dark current. Also, the possibility of some reduction in dark current in ZnO/Pt and Cu/ZnO photodetector due to possible formation of junction at annealing temperature of 575 °C may not be ruled out. The surface states are expected to play crucial role in reducing the dark current due to trapping of free electrons. Probably, incorporation of Cu in ZnO thin film results in more surface states, thereby, trapping more free electrons and giving much lower value of dark current in comparison to that obtained in pure ZnO thin film based photodetector. When UV radiations interact with highly porous and rough Cu:ZnO thin film surface, more number of electrons and holes are generated due to more adsorption and desorption of O^{2-} species on the film surface (Figure 5, Table I). These photogenerated holes recombined with the trapped electrons, increasing the concentration of photogenerated electrons in the conduction band and participate in photocurrent giving a high value ($I_{on} \sim 0.02$ mA). Thus, the reduction in dark current is mainly responsible for the observed enhanced photoresponse.

CONCLUSIONS

The Cu:ZnO (0.13 At wt%) UV photodetector prepared by Sol-gel technique exhibits enhanced photoresponse and less persistence behaviour towards UV radiation of $\lambda = 365$ nm and intensity = 24 μwatt/cm^2 when biased at 5 volts. The value of photoconductive gain for Cu:ZnO thin film is found to be about three order higher (1.5×10^3) compared to its corresponding value obtained for pure ZnO thin film based photodetector (4.9×10^1). Substitution of Cu^{2+} ions at the Zn^{2+} sites in ZnO is found to be responsible for lowering of dark current which results in enhanced photo response. Doping of copper in ZnO has also improved rise time and fall time of the detector which is making this matrix suitable and attractive for fast UV detection. The results are encouraging for the fabrication of a UV photodetector using a simple fabrication technique.

ACKNOWLEDGMENTS

The authors acknowledge the financial support provided by the Department of Science and Technology (DST) and Department of Information Technology (DIT), Government of India. One of the authors (A R) is thankful to UGC for research fellowship.

REFERENCES

1. T. H. Moon, M. C. Jeong, W. Lee and J. M. Myoung, *Appl. Surf. Sci.*, **240**, 280, (2005).
2. H. K. Yadav, K. Sreenivas and V. Gupta, *J. Appl. Phys.*, **107**, 044507, (2010).
3. T. Ghosh and D. Basak, *J. Phys. D: Appl. Phys.*, **42**, 145304, (2009).
4. N. Kouklin, *Adv. Mater.* **20**, 2190, (2008).
5. N. E. Sung, I. J. Lee, A. Thakur, K. H. Chae, H. J. Shin and H. K. Lee, *Mater. Res. Bull.*, **47**, 2891, (2012).
6. V. Gupta and A. Mansingh, *J. Appl. Phys.*, **80**, 1063 (1996).
7. X. B. Wang, C. Song, K. W. Geng, F. Zeng and F. Pan, *Appl. Surf. Sci.*, **253**, 6905, (2007).
8. K. Samanta, P. Bhattacharya and R. S. Katiyar, *J. Appl. Phys.*, **105**, 113929 (2009).
9. Q. H. Li, T. Gao, Y. G. Wang and T. H. Wang, *Appl. Phys. Lett.*, **86**, 123117, (2005).

Mater. Res. Soc. Symp. Proc. Vol. 1494 © 2012 Materials Research Society
DOI: 10.1557/opl.2012.1697

Effects of Ga doping and nitridation on ZnO films prepared by RF Sputtering

Takumi Araki[1], Jun-ichi Iwata[1] and Hiroshi Katsumata[1]

[1]Department of Electronics and Bioinformatics, Meiji University, Kawasaki 214-8571, Japan

ABSTRACT

GaZnO and GaZnON thin films were deposited on both Si (100) and c-axis oriented sapphire substrates by RF co-sputtering of ZnO target and Ga_2O_3 tablets in Ar/O_2 and Ar/N_2, respectively, by changing the number of Ga_2O_3 tablets (N_{Ga2O3}) placed on the ZnO target in the range of 0 to 16. They were subsequently annealed in N_2 at 800 °C and then, some of the samples formed by Ar/O_2-sputtering were subjected to NH_3 treatment at 650 °C for nitridation. XRD measurements revealed that the c-axis lattice parameter calculated from the ZnO (002) peak for GaZnON films on Si (100) was remarkably larger than for GaZnO films on Si (100). Moreover, ZnO (002) was observed up to N_{Ga2O3}=16 for GaZnON films formed on sapphire, while no XRD peaks were observed above N_{Ga2O3}=8 for GaZnON films on Si (100). Optical band-gap of GaZnO and GaZnON films became wider from 3.34 to 3.67 eV and from 3.21 to 3.40 eV, respectively, with increasing N_{Ga2O3} from 0 to 16. Photoluminescence spectra of GaZnO films showed band-to-band emission at 380 nm, while those of GaZnON films exhibited broad and weak peaks centered at 550 nm and 647 nm.

INTRODUCTION

ZnO is an interesting wurtzitic semiconducting material with a wide band-gap of 3.3 eV and it has a large exciton binding energy of 60 meV. ZnO-based materials are recently studied as phosphor-free white LED materials[1]. Since the white LED without the phosphor does not have the color conversion loss, high luminous efficiency can be expected. Crystal structure of GaN is the same as that of ZnO, which has hexagonal wurtzitic structure, closely matched lattice constant and nearly the same band-gap of about 3.3 eV.

It has been reported that the optical band-gap of $Zn_{1-x}Ga_xO$ thin films can be engineered from 3.3 to 4.9 eV by varying the Ga content[2]. Moreover, the reduction of the optical band-gap down to 2.4 eV has been observed from $(ZnO)_x(GaN)_{1-x}$ solid solution powders with x = 0.81[3]. According to another report, optical band-gap of ZnON decreased from 3.26 to 2.30 eV with increasing the N concentration[4]. In these previous reports, however, $(ZnO)_x(GaN)_{1-x}$ has been characterized as only powders, but it has not been characterized as thin films. Furthermore, there have been few observations on their luminescence properties.

From these results, we believe that the band-gap of GaZnON materials can be widely controllable from 2.3 to 4.9 eV. The purpose of this study is to produce GaZnON thin films for future phosphor-free white LED materials and to clarify their optical and structural properties. It should be noted that here we present the photoluminescence properties of GaZnO and GaZnON films and the dependence of the type of substrates on their optical and structural properties.

EXPERIMENTAL DETAILS

The substrates used in this study were p-Si (100) or sapphire substrates. GaZnO and GaZnON film were deposited on either substrate by RF co-sputtering of ZnO target (ϕ 4 inch, 99.99% in purity, Kojundokagaku Laboratory) and Ga_2O_3 tablets (ϕ 10 mm x t 5 mm, 99.9% in purity, Kojundokagaku Laboratory) at 1.0×10^{-2} Torr in constant Ar (8 sccm)/O_2 (2 sccm) flow and constant Ar (2 sccm)/N_2 (8 sccm) flow, respectively. The Ga_2O_3-to-ZnO ratio was controlled by varying the number of Ga_2O_3 tablets (N_{Ga2O3}) placed on the pure ZnO target in the range of 0 to 16. These samples were subsequently annealed at 800 °C for 60 min in N_2 with an infrared gold image furnace to improve the crystalline quality of films. After the N_2 annealing, some of the samples deposited by Ar/O_2-sputtering were subjected to NH_3 treatment at 650 °C for 20 min for nitridation.

These samples were examined by energy dispersive x-ray spectroscopy (EDS) (AZtec Advanced X-Max 20, Oxford Instruments), X-ray diffraction (XRD) (RINT-Ultima III, RIGAKU) and optical transmission (V670, JASCO) as well as room temperature photoluminescence (PL) (TRIAX320, HORIBA JOBIN YVON). CuKα radiation with a wavelength of 1.5443 A was used for XRD measurements. For PL measurements, He-Cd laser was used as an excitation source. Optical band-gap was determined from the optical transmission spectra using the following equation for direct band gap materials;

$$(\alpha h\nu) = A(h\nu - E_g)^{1/2} \tag{1}$$

where α is absorption coefficient, hν is photon energy, E_g is energy-gap and A is a constant.

DISCUSSION

Elemental composition measured by EDS

The elemental composition of films formed on Si (100) substrates by sputtering in Ar/O_2 and Ar/N_2 was measured by EDS and the results are shown in Figs. 1(a) and 1(b), respectively. One can see that in both figures, the amount of Ga and O atoms increases with increasing N_{Ga2O3}, while that of Zn atoms shows a decreasing tendency. We believe that the Ga atoms can replace Zn site and O atoms can bind to either Ga or Zn atoms. In Fig .1(b), the amount of N atoms

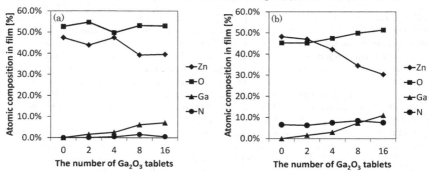

Figure 1. The dependence of N_{Ga2O3} on the atomic composition in GaZnO films formed on Si by Ar/O_2 sputtering (a), and GaZnON films formed on Si by Ar/N_2 sputtering (b).

involved in the films is independent of N_{Ga2O3} and it is almost constant about 7%.

X-ray diffractogram

The dependence of N_{Ga2O3} on x-ray diffractogram for GaZnO and GaZnON films formed on Si (100) with diamond structure substrates are shown in Figs. 2(a) and 2(b). In both figures, XRD peaks from ZnO (002) with hexagonal structure were observed up to $N_{Ga2O3}=4$ and they disappeared above $N_{Ga2O3}=8$, which seems to be due to amorphization of films by heavy Ga doping into the films. The crystallinity of GaZnO films is superior to that of GaZnON films because the full width at half maximum (FWHM) of ZnO (002) peak in Fig. 2(a) is narrower

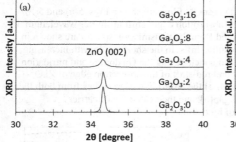 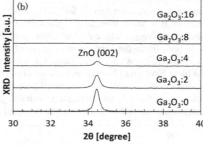

Figure 2. The dependence of N_{Ga2O3} on the x-ray diffractogram for GaZnO films formed on Si by Ar/O$_2$ sputtering (a), and GaZnON films formed on Si by Ar/N$_2$ sputtering (b).

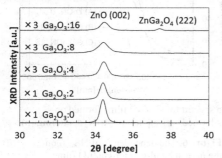 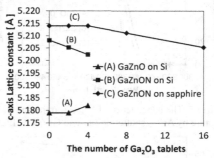

Figure 3. The dependence of N_{Ga2O3} on the x-ray diffractogram for GaZnON films formed on sapphire substrates by Ar/N$_2$ sputtering.

Figure 4. The dependence of N_{Ga2O3} on c-axis lattice constant calculated from the ZnO (002) peak for GaZnO films formed on sapphire substrates by Ar/O$_2$ sputtering (A), GaZnON films formed on Si substrates by Ar/N$_2$ sputtering (B), and GaZnON films formed on sapphire substrates by Ar/N$_2$ sputtering (C).

than that in Fig. 2(b). Next, the dependence of N_{Ga2O3} on x-ray diffractogram for GaZnON films formed on sapphire with hexagonal structure substrates are shown in Fig. 3. In Fig.3, the ZnO (002) peak was observed up to N_{Ga2O3}=16 and the additional weak peak presumably originating from $ZnGa_2O_4$ (222) with spinel structure appeared at N_{Ga2O3}=16. C-axis lattice parameters calculated from the ZnO (002) peak for GaZnO and GaZnON films formed on Si (100) and for GaZnON films formed on sapphire substrates are presented in Fig. 4. The lattice parameter of GaZnON films was relatively larger than that of GaZnO films on Si. The lattice parameter for GaZnON films on Si (100) became smaller with increasing N_{Ga2O3} up to 4, while that for GaZnON films on sapphire substrates started to decrease gradually above N_{Ga2O3}=4.

Optical transmission spectra

Optical transmission spectra for GaZnO and GaZnON films are shown Figs. 5(a) and 5(b), respectively. It is found from both figures that the absorption edges shift to shorter wavelength with increasing N_{Ga2O3}. Optical band-gap calculated from the transmission spectra are shown in Fig. 6. In both films, the absorption edge significantly shifts to the shorter side with increasing N_{Ga2O3} up to N_{Ga2O3}=8 and then it saturates. We can conclude that the Ga doping and nitridation in ZnO films have an influence on the blue-shift and red-shift of the absorption edge of ZnO films, respectively. The possible origin of blue-shift of the absorption edge comes from both the Burstein-Moss effect[5] and increasing the band gap[6], and the evaluation of carrier concentration is needed to determine the origin of blue-shift.

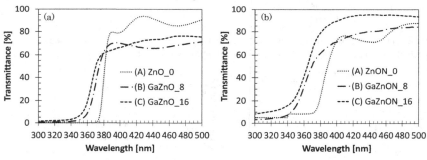

Figure 5. The dependence of N_{Ga2O3} on transmission spectra for GaZnO films formed on sapphire by Ar/O$_2$ sputtering (a), and GaZnON films formed on sapphire by Ar/N$_2$ sputtering (b).

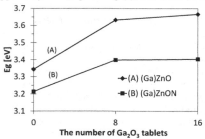

Figure 6. The dependence of N_{Ga2O3} on the optical band-gap for GaZnO films formed on sapphire by Ar/O$_2$ sputtering (A), and GaZnON films formed on sapphire by Ar/N$_2$ sputtering (B).

PL spectra

PL spectra for GaZnO and GaZnON films formed on Si (100) substrates are shown in Figs. 7 (a) and 7(b), respectively. In Fig. 7(a), the intensity of band-to-band emission observed at 380 nm from GaZnO films with $N_{Ga2O3}=2$ becomes stronger than that from undoped ZnO films and then it decreases with increasing N_{Ga2O3}. This observation is consistent with the report on Ga-doped ZnO films by Hong Quang Le et al.[7]. They reported that the PL intensity at around 380 nm becomes maximal at Ga content of 1.0% and then it decreases with increasing Ga content. On the other hand, in Fig. 7(b), the broad and weak peak centered at 647 nm was observed from undoped ZnON films. In N_{Ga2O3} range of 2 to 8, any peaks were hardly observed, but for GaZnON films with $N_{Ga2O3}=16$, the broad and weak peak centered at 550 nm appeared. The origin of a peak centered at 550 nm remains unclear but it could be due to the defects of ZnON, because it has not been observed for pure ZnO[8]. On the other hand, while the origin of a peak centered at 647 nm is assigned to be due to Ga vacancy of GaN[9]. It should be added that no noticeable band-to-band emission was observed from the whole GaZnON films. In order to examine if GaZnO films can be nitride after the deposition and subsequent N_2-annealing, GaZnO films with $N_{Ga2O3}=2$ formed on Si (100) substrates by Ar/O_2-sputtering and followed by N_2-annealing were subjected to NH$_3$ treatment at 650 °C for 20min. PL spectra for GaZnO films with $N_{Ga2O3}=2$ before NH$_3$ treatment and after NH$_3$ treatment are shown in Fig. 8. It was found that the PL intensity at 380 nm was enhanced presumably by improvement of the crystalline quality and annihilation of point defects.

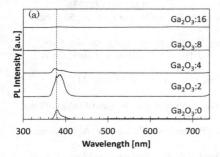 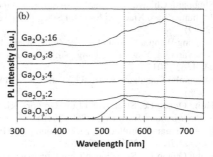

Figures 7. The dependence of N_{Ga2O3} on the PL spectra for GaZnO films formed on Si substrates by Ar/O_2 sputtering (a), and GaZnON films formed on Si substrates by Ar/N_2 sputtering (b).

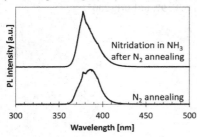

Figure 8. PL spectra for GaZnO films formed on Si substrates by Ar/O_2 sputtering using 2 pieces of Ga$_2$O$_3$ tablets. They were subsequently annealed in N_2 at 800 °C for 60 min (lower spectrum) and then subjected to nitridation in NH$_3$ at 650 °C for 20 min (upper spectrum).

CONCLUSIONS

Ga doping to ZnO showed widening the band-gap from 3.34 to 3.67 eV, which resulted in weakening the ZnO band-to-band PL emission intensity at 380nm and deteriorating crystalline quality at higher Ga doping. On the other hand, N doping to GaZnO caused narrowing the band-gap from 3.67 to 3.40 eV, which resulted in completely disappearing the ZnO band-to-band PL emission at 380nm, but instead extremely weak and broad PL emission was observed at 647 nm, which is assigned to be presumably due to Ga vacancy of GaN. C-axis lattice parameter calculated from the ZnO (002) XRD peak for GaZnON films was remarkably larger than for GaZnO films. The ZnO (002) peak was observed up to N_{Ga2O3}=16 for GaZnON films formed on sapphire substrates while no XRD peaks were observed above N_{Ga2O3}=8 for GaZnON films on Si substrates, which indicates the better crystalline quality for GaZnON films formed on sapphire substrates at higher Ga doping.

ACKNOWLEDGEMENTS

We are grateful for the personnel of collaborative innovation and incubation center of Meiji University, who support us to measure the EDS.

REFERENCES

1. Shabnam, Chhaya Ravi Kant, P. Arun, *Journal of Luminescence*, **132**, 1774 (2012).
2. Junliang Zhao, XiaoWei Sun and Swee Tiam Tan, *IEEE Trans Electron Devices*, **56**, 2995 (2009).
3. Junpeng Wang,a Baibiao Huang, Zeyan Wang, Peng Wang, Hefeng Cheng, Zhaoke Zheng, Xiaoyan Qin, Xiaoyang Zhang, Ying Daib and Myung-Hwan Whangboc, *J. Mater. Chem.*, **21**, 4562 (2011).
4. Masanobu Futsuhara, Katsuaki Yoshioka and Osamu Takai, *Thin Solid Films*, **317**, 322 (1998).
5. Hong Quang Le, Swee Kuan Lim, Gregory Kia Liang Goh, Soo Jin Chua and JunXiong Ong, *J. Electrochem. Soc.*, **157**, H769 (2010).
6. R. Al Asmar, S. Juillaguet, M. Ramonda, A. Giani, P. Combette, A. Khoury, A. Foucaran, *J. Crystal Growth*, **275**, 512 (2005).
7. Hong Quang Le, Swee Kuan Lim, Gregory Kia Liang Goh, Jin Chua and JunXiong Ong et al,. *J. Electrochem. Soc.*, **157**, 796 (2010).
8. M. D. McCluskey and S. J. Jokela, *J. Appl. Phys.*, **106**, 071101 (2009).
9. Michael. A. Reshchikov and Hadis Morkoç, *J. Appl. Phys.*, **97**, 061301 (2005).

Mater. Res. Soc. Symp. Proc. Vol. 1494 © 2012 Materials Research Society
DOI: 10.1557/opl.2012.1709

Band Structure and Effective Masses of $Zn_{1-x}Mg_xO$

Christian Franz, Marcel Giar, Markus Heinemann, Michael Czerner, and Christian Heiliger
I. Physikalisches Institut, Justus Liebig University, 35392 Giessen, Germany

ABSTRACT

We analyze the influence of the Mg concentration on several important properties of the band structure of $Zn_{1-x}Mg_xO$ alloys in wurtzite structure using *ab initio* calculations. For this purpose, the band structure for finite concentrations is defined in terms of the Bloch spectral density, which can be calculated within the coherent potential approximation. We investigate the concentration dependence of the band gap and the crystal-field splitting of the valence bands. The effective electron and hole masses are determined by extending the effective mass model to finite concentrations. We compare our results with experimental results and other calculations.

INTRODUCTION

Zinc oxide is a promising, sustainable material with many prospective applications, especially in opto-electronics. It is well known that the band gap and other properties can be tuned by adding magnesium. For Mg concentrations up to ca. 30%, the resulting $Zn_{1-x}Mg_xO$ alloy has wurtzite structure and a direct band gap [1]. This can be used in multilayer structures to form e.g. light-emitting diodes [2]. Recently, a two-dimensional electron gas with high charge carrier mobility was created in a ZnMgO-ZnO multilayer structure [3]. This paves the way to new fields of applications like high-frequency and high-power devices. Tsukazaki et al. were able to measure the integer [4] as well as the fractional quantum Hall effect [5] in ZnMgO-ZnO heterostructures. While both require a high degree of control of the material properties, the latter is of particular interest from a fundamental research point of view, since it arises from a strongly correlated state with extraordinary properties.

In order to advance these and other applications reliable numerical tools are of great value. Some of the necessary physical parameters like the valence band effective masses are still unknown for finite concentrations. Thus, recent calculations have to resort to linear interpolation between the pure components, or even use the ZnO value for all concentrations [6]. The most important part of the ZnO band structure, which is the bottom of the conduction band and the top of the valence bands in the vicinity of the Γ-point, can be well described within the effective mass approximation. We extend this approach to finite Mg concentrations using a Bloch spectral density [7] defined within the coherent potential approximation (CPA) [8]. Thereby, we provide the bang gap, the valence band splittings, and the electron and hole effective masses for concentrations up to 30%.

For pure ZnO the band gap and the valence band splittings are well established from experiments [9,10]. While the band gap is well described by modern *ab initio* methods, there are still open questions on how to compare the calculated masses to experimental results [11].

Ohtomo et al. were among the first to grow and investigate $Zn_{1-x}Mg_xO$ solid solution thin films with Mg concentrations up to 33% [1]. Among other things, they investigated the concentration dependence of the band gap. Later studies included the cubic phase at high Mg concentrations [12] and provided information on the exciton binding energies and valence band splittings [13,14]. The electron effective masses for finite concentrations were obtained indirectly by fitting a model equation to experimental data, and different studies obtain rather different concentration dependencies [15, 16].

Ab initio investigations include a comprehensive series of contributions on various properties of $Zn_{1-x}Mg_xO$ and $Zn_{1-x}Cd_xO$ by Schleife et al. They calculated the band structure of ZnO and MgO using a $HSE03+G_0W_0$ scheme [11]. They describe alloys using a cluster expansion, i.e. a supercell calculation with a subsequent statistical treatment employing various thermodynamic models [17,18]. Maznichenko et al. used the CPA to investigate the structural phase transitions and the fundamental band gaps of $Zn_{1-x}Mg_xO$ alloys [19].

In this contribution we provide results that are difficult to obtain by experiments or other methods including the concentration dependence of the effective masses and the valence band splittings.

THEORY

The results in this paper are obtained using a density functional theory method. We apply the local density approximation (LDA) for the exchange correlation functional. For the self-consistent density and band structure calculation we use an implementation of the KKR-method [20], which employs one-electron Green's functions expanded in spherical harmonics. The alloys are described using the CPA [8,20], which we recently implemented in our KKR-code. The CPA introduces a self-consistent effective medium, which restores the periodicity of a crystal. This allows us to calculate an averaged electron density in **k**-space called Bloch spectral density [7], which is closely related to the band structure. The CPA allows an accurate description of alloys at a relatively low computational effort. Popescu and Zunger proposed a different method to define an effective band structure of alloys [21]. They used a spectral decomposition to extract an alloy band structure from supercell calculations. This has several advantages but requires the calculation of very large supercells.

We consider ZnO and $Zn_{1-x}Mg_xO$ in wurtzite structure, which is the equilibrium structure of ZnO at ambient conditions. The calculations are performed for two sets of concentration-dependent lattice parameters: fully relaxed and c-plane growth (i.e. with a fixed a-parameter). The fully relaxed lattice parameters are taken from our earlier publication reference [22], where we used the ABINIT code to calculate the relaxed lattice structure of $Zn_{1-x}Mg_xO$ supercells for concentrations up to 31%. Note that we use the LDA results which are listed in Table I. For the c-plane growth we perform a similar calculation but keep the a-parameter fixed during the relaxation. The results are given in Table I, for the computational details see [22].

Table I. Mg concentration (x) dependent lattice parameters that are used in the calculations: the hexagonal lattice constant a, the axes ratio c/a and the parameter u.

	a(x)/Å	c(x)/a(x)	u(x)
Fully relaxed	3.2180+0.0354 x	1.6096-0.0473 x	0.3797+0.0097 x
c-Plane growth	3.22	1.6068-0.0135 x	0.3800+0.0040 x

For most applications the part of the band structure close to the band gap is most important. $Zn_{1-x}Mg_xO$ in wurtzite structure has a direct band gap at the Γ-point. In this energy range the bands are almost parabolic and can be described by the effective mass approximation

$$E_n(\mathbf{k}) \approx E_n(\mathbf{0}) + \frac{(\hbar\,\mathbf{k})^2}{2\,m_n^*}, \tag{1}$$

which approximates the band structure of the n-th band $E_n(\mathbf{k})$ by a parabolic band starting at $E_n(\mathbf{0})$ with an effective mass m_n^* for small $\mathbf{k}$. For the single conduction band with spherical symmetry this is a good approximation. The effective masses of the three valence bands differ significantly for directions of $\mathbf{k}$ along the k_z-axes ($\parallel$) and for directions in the k_x-k_y-plane ($\perp$). Hence, for the valence bands it is appropriate to introduce two masses

$$E_n(\mathbf{k}) \approx E_n(\mathbf{0}) + \frac{(\hbar\,k_\perp)^2}{2\,m_{n\perp}^*} + \frac{(\hbar\,k_\parallel)^2}{2\,m_{n\parallel}^*}. \tag{2}$$

The effective mass approximation is illustrated in Figure 1 (top).

The present calculations are carried out without relativistic effects like spin-orbit interaction. The latter introduces a small additional splitting of the valence bands, which results in three distinct bands at the Γ-point, which are usually labeled A, B and C. Without spin-orbit interaction the bands A and B are degenerate at the Γ-point ($E_A(\mathbf{0}) = E_B(\mathbf{0})$) and we obtain the crystal-field splitting $\Delta_{AC} = E_A(\mathbf{0}) - E_C(\mathbf{0})$ and the band gap $E_g = E_C(\mathbf{0}) - E_A(\mathbf{0})$. The Γ-A-line has the same symmetry and degeneracies as the Γ-point. In the k_x-k_y-plane the symmetry is lower and the degeneracy between the bands A and B is lifted. In that case these bands have a different dispersion. A has a large effective mass m_{hh} (heavy hole) and B a small effective mass m_{lh} (light hole). The C band is also referred to as crystal-field split-off band (m_C). In order to allow for a parabolic fit, the anticrossing of B and C is replaced by a crossing.

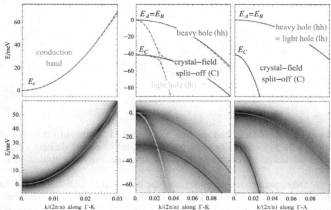

Figure 1. Calculated band structure of ZnO (top, dashed line) and the Bloch spectral density of $Zn_{0.85}Mg_{0.15}O$ (bottom, gray gradient), respectively, including the fitted effective mass approximation for the conduction (left) and valence bands (middle, right) for small $\mathbf{k}$.

For pure ZnO the effective mass approximation can be directly fitted to the calculated band structure (Figure 1, top). For finite concentrations of Mg the disorder leads to a broadening of the bands in the Bloch spectral density shown in Figure 1 (bottom). Despite this broadening, for the concentrations considered here (x = 0.0 – 0.3), it is possible to fit the effective mass approximation to the density. We consider an energy range of 26 meV for each band for the fitting.

RESULTS AND DISCUSSION

Table II. Some selected results on various properties of pure ZnO

E_g/eV	Δ_{AC}/meV	m_c/m_e	$m_{hh\perp}/m_e$	$m_{hh\parallel}/m_e$	$m_{lh\perp}/m_e$	$m_{C\perp}/m_e$	$m_{C\parallel}/m_e$
1.08	41.2	0.186	2.64	3.13	0.212	3.03	0.208

The results from the fits for pure ZnO are shown in Table II. We find that our results for ZnO are similar to other calculated results [10]. Some of the contributions in the literature include spin-orbit interaction in the calculation. While the electron mass is in rather good agreement with experimental values [9], some of the hole masses show a larger deviation. This is usually found in *ab initio* calculations (see e.g. reference [10,11,23]) and might be due to the fact that experiments can only measure some direction average of the hole masses [11]. The calculated band gap of 1.08 eV is too small, which is common for LDA calculations. More advanced *ab initio* methods are able to predict the band gap more accurately but give similar results for the masses (e.g. HSE03+G$_0$W$_0$ in reference [11]). On the other hand, the crystal-field splitting is in excellent agreement with experiments [9,10].

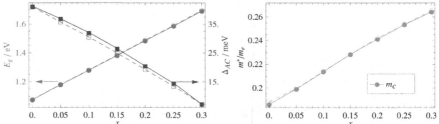

Figure 2. Band gap E_g and valence band splitting Δ_{AC} (left) and electron effective masses m_c (right) dependence on the Mg concentration x. (solid lines: fully relaxed, dashed: c-plane grown, the points show the calculated results, while the connecting lines are just a guide for the eye)

We analyze the concentration dependence of the band gap (Figure 2, left) and find a linear dependence: E_g/eV = 1.08 + 2.03 x. This linear slope is in outstanding agreement with experimental results. Ohtomo et al. obtained E_g/eV = 3.3 + 2.1 x from transmission spectra for the hexagonal phase and concentrations between 0 and 33% at room temperature [1]. Likewise, Chen et al. [12] found E_g/eV = 3.32 + 2.00 x at low temperature for concentrations between 0 and 33% and Wu et al. [24] obtained E_g/eV = 3.384 + 1.705 x for concentrations between 0 and 44% at room-temperature from optical absorption spectra. The latter two experiments included the cubic phase at high Mg concentrations and found a separate linear dependence with a different slope for this phase. Our calculated band gap agrees with the LDA result calculated by Maznichenko et al. [19]. The small concentration range considered here does not allow for a sound quadratic fit, but the data supports a small bowing in agreement with the cited experiments.

Further, we find that the crystal-field splitting Δ_{AC} between the valence bands, shown in Figure 2 (left), decreases approximately linearly over the considered range of concentrations and thus the order of the valence bands will eventually switch. A quadratic extrapolation yields a vanishing crystal-field splitting at ca. x = 0.35. The concentration dependence of the crystal-field splitting shows a moderate bowing which is notably reduced for the c-plane grown $Zn_{1-x}Mg_xO$. The reversal of the order of the valence bands for the pure components (without spin-orbit interaction) was also observed by Schleife et al. using HSE03+G_0W_0 [11], and by Xu et al. using the generalized gradient approximation [25]. Since the spin-orbit interaction introduces an additional splitting of the valence bands, the predicted concentration dependence of the crystal-field splitting cannot be directly observed in experiments. Nevertheless, since the additional spin-orbit splitting is much smaller than the crystal-field splitting, it can be considered a small perturbation. Thus, our result signifies a nonlinear behavior of the valence band splittings. Furthermore, this small perturbation can only have a very weak influence on the effective masses. The splittings are difficult to measure for finite concentrations and this has not yet been verified in experiments [13,14].

The electron effective mass is presented in Figure 2 (right) and we find an increase with the Mg concentration in accordance with experimental results [15,16]. A linear fit results in $m_c/m_e = 0.186 + 0.267$ x. Lu et al. [16] found a linear dependence with a larger slope, while the results by Cohen et al. [15] suggest a quadratic dependence. All our calculated hole effective masses shown in Figure 3 have a fair concentration dependence, which is quite different for each band. All hole masses increase for small concentrations x < 0.15 and some decrease after a maximum in the considered concentration range. A linear interpolation does not seem appropriate for any of the hole masses. We are not aware of any experimental or theoretical investigations of the hole effective masses for finite Mg concentrations. We find a very similar behavior of the effective masses for the two growing conditions considered.

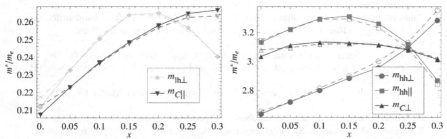

Figure 3. The dependence of the hole effective masses for the light holes (left) and heavy holes (right) on the Mg concentration x. (solid lines: fully relaxed, dashed: c-plane grown, the points show the calculated results, while the connecting lines are just a guide for the eye)

CONCLUSIONS

The influence of the Mg concentration on the band structure and in particular the effective masses of $Zn_{1-x}Mg_xO$ alloys is analyzed using *ab initio* calculations. We find that the Mg concentration has a strong influence on the band energies and a fair influence on the effective masses. While the band gap increases linearly, the crystal-field splitting decreases. The electron mass increases linearly and the hole masses show a rather complicated and unexpected concentration dependence. Our results are mostly in good agreement with experiments and can be helpful for the simulation of devices.

ACKNOWLEDGMENTS

We thank Prof. B. K. Meyer for helpful discussions. We acknowledge support from the German Science Foundation via grant HE 5922/1-1.

REFERENCES

1. A. Ohtomo, M. Kawasaki, T. Koida, K. Masubuchi, H. Koinuma, Y. Sakurai, Y. Yoshida, T. Yasuda, and Y. Segawa: Appl. Phys. Lett. **72** (1998) 2466.
2. Y. Choi, J. Kang, D. Hwang, and S. Park: IEEE Trans. Electron Devices **57** (2010) 26.
3. K. Koike, K. Hama, I. Nakashima, G. Takada, M. Ozaki, K. Ogata, S. Sasa, M. Inoue, and M. Yano: Jpn. J. Appl. Phys. **43** (2004) L1372.
4. A. Tsukazaki, A. Ohtomo, T. Kita, Y. Ohno, H. Ohno, and M. Kawasaki: Science **315** (2007) 1388.
5. A. Tsukazaki, S. Akasaka, K. Nakahara, Y. Ohno, H. Ohno, D. Maryenko, A. Ohtomo, and M. Kawasaki: Nat. Mater. **9** (2010) 889.
6. E. Furno, S. Chiaria, M. Penna, E. Bellotti, and M. Goano: J. Electron. Mater. **39** (2010) 936.
7. J. Faulkner and G. Stocks: Phys. Rev. B **21** (1980) 3222.
8. P. Soven: Phys. Rev. **156** (1967) 809.
9. Authors and editors of the volumes III/17B-22A-41B: Zinc oxide (ZnO) band structure. O. Madelung, U. Roessler, M. Schulz (ed.): The Landolt-Boernstein Database, www.springermaterials.com.
10. M. Goano, F. Bertazzi, M. Penna, and E. Bellotti: J. Appl. Phys. **102** (2007) 083709.
11. A. Schleife, F. Fuchs, C. Roedl, J. Furthmueller, and F. Bechstedt: Phys. Status Solidi B **246** (2009) 2150.
12. J. Chen, W. Shen, N. Chen, D. Qiu, and H. Wu: J. Phys.: Condens. Matter **15** (2003) L475.
13. R. Schmidt, B. Rheinlaender, M. Schubert, D. Spemann, T. Butz, J. Lenzner, E. Kaidashev, M. Lorenz, A. Rahm, H. Semmelhack, M. Grundmann: Appl. Phys. Lett. **82** (2003) 2260.
14. C. Teng, J. Muth, Ue. Oezguer, M. Bergmann, H. Everitt, A. Sharma, C. Jin, and J. Narayan: Appl. Phys. Lett. **76** (2000) 979.
15. D. Cohen, K. Ruthe, and S. Barnett: J. Appl. Phys. **96** (2004) 459.

16. J. Lu, S. Fujita, T. Kawaharamura, H. Nishinaka, Y. Kamada, and T. Ohshima: Appl. Phys. Lett. **89** (2006) 262107.
17. A. Schleife, M. Eisenacher, C. Roedl, F. Fuchs, J. Furthmueller, and F. Bechstedt: Phys. Rev. B **81** (2010) 245210.
18. A. Schleife, C. Roedl, J. Furthmueller, and F. Bechstedt: New J. Phys. **13** (2011) 085012.
19. I. V. Maznichenko, A. Ernst, M. Bouhassoune, J. Henk, M. Daene, M. Lueders, P. Bruno, W. Hergert, I. Mertig, Z. Szotek, W. M. Temmerman: Phys. Rev. B **80** (2009) 144101.
20. J. Zabloudil, R. Hammerling, L. Szunyogh, and P. Weinberger: *Electron scattering in Solid Matter: A Theoretical and Computational Treatise* (Springer, Berlin, 2005) Vol. 147 of *Springer Series in Solid-State Sciences*.
21. V. Popescu and A. Zunger: Phys. Rev. Lett. **104** (2010) 236403.
22. M. Heinemann, M. Giar, and C. Heiliger: MRS Symp. Proc. **1201** (2009) H05.
23. S. Karazhanov, P. Ravindran, A. Kjekhus, H. Fjellvag, U. Grossner, and B. Svensson: J. Cryst. Growth **287** (2006) 162.
24. C. Wu, Y. Lu, D. Shen, and X. Fan: Chin. Sci. Bull. **55** (2010) 90.
25. Q. Xu, X. Zhang, W. Fan, S. Li, and J. Xia: Comput. Mater. Sci. **44** (2008) 72.

Mater. Res. Soc. Symp. Proc. Vol. 1494 © 2012 Materials Research Society
DOI: 10.1557/opl.2012.1677

Characterization of Thin ZnO Films by Vacuum Ultra-Violet Reflectometry

T. Gumprecht[1,2], P. Petrik[1,3], G. Roeder[1], M. Schellenberger[1], L. Pfitzner[1], B. Pollakowski[4], B. Beckhoff[4]

[1]Fraunhofer Institute for Integrated Systems and Device Technology (IISB), Schottkystrasse 10, 91058 Erlangen, Germany

[2]Erlangen Graduate School in Advanced Optical Technologies (SAOT), Paul-Gordan-Strasse 9, 91052 Erlangen, Germany

[3]Institute for Technical Physics & Materials Science (MFA), Research Centre for Natural Sciences, Konkoly Thege u. 29-33, 1121 Budapest, Hungary

[4]Physikalisch-Technische Bundesanstalt (PTB), Abbestr. 2-12, 10587 Berlin, Germany

ABSTRACT

ZnO has a huge potential and is already a crucial material in a range of key technologies from photovoltaics to opto and printed electronics. ZnO is being characterized by versatile metrologies to reveal electrical, optical, structural and other parameters with the aim of process optimization for best device performance. The aim of the present work is to reveal the capabilities of vacuum ultra-violet (VUV) reflectometry for the characterization of ZnO films of nominally 50 nm, doped by Ga and In. Optical metrologies have already shown to be able to sensitively measure the gap energy, the exciton strength, the density, the surface nanoroughness and a range of technologically important structural and material parameters. It has also been shown that these optical properties closely correlate with the most important electrical properties like the carrier density and hence the specific resistance of the film. We show that VUV reflectometry is a highly sensitive optical method that is capable of the characterization of crucial film properties. Our results have been cross-checked by reference methods such as ellipsometry and X-ray fluorescence.

Key words:
Zink oxide, vacuum ultra-violet reflectometry, spectroscopic ellipsometry, atomic layer deposition

INTRODUCTION

ZnO is a key material in a range of optoelectronic applications, and has promising properties for numerous other applications. Therefore, it continues to be intensively studied in the recent years [1]. ZnO layers can be prepared by different methods including sputtering [2], atomic layer deposition (ALD) [3], pulsed laser deposition [4, 5, 6], spin coating from nanoparticulates [7], or spray pyrolysis [8]. Highly sensitive optical methods such as ellipsometry are frequently applied for the quick and non-destructive determination of a range of material and structural parameters such as thickness, surface nanoroughness, interface quality, density, homogeneity, band gap and exciton strength. These methods make indirectly also possible the determination of electrical properties.

The aim of the present work is the investigation of optical modeling of ZnO layers using a commercial vacuum ultra-violet (VUV) reflectometer, and the cross-checking of the result using reference metrologies. This study is part of a European project, intended to be a step towards the establishment of validated reference methodologies for a reliable characterization of key optoelectronic materials (IND07, "Metrology for the manufacturing of thin films") in the European Metrology Research Program of EURAMET. Furthermore, our investigations aim for the development of reference samples with controlled defect concentration and morphology or methods for elemental depth profiling [9].

EXPERIMENTAL DETAILS

GaInZnO (GIZO) samples with a nominal thickness of 50 nm were prepared by dual-target sputtering on oxidized (10 nm SiO_2) single-crystalline silicon wafers used for reflectometric measurements in comparison to ellipsometry and X-ray. A summary of the sample properties is shown in Table I. The Ga amount was measured by X-ray fluorescence (XRF) analysis.

Table I. Parameters of numbering and preparation of samples investigated in this study. The Ga content of sample VUV 3 is estimated by quadratic interpolation using the other three points measured by XPS.

Not annealed	Annealed	Ga (%)
VUV 1	VUV 9	0.00
VUV 3	VUV 11	0.09
VUV 5	VUV 13	0.19
VUV 7	VUV 15	0.37
Reference 10 nm SiO_2 layer, not annealed		

Reflectometric measurements were performed with a METROSOL VUV-7000 spectroscopic reflectometer (VUV-R) in the wavelength range of 120 nm to 800 nm with a spot size of 35 µm by 35 µm. A small spot size can be important to investigate lateral inhomogeneity. The reference measurements of a single point on the samples were performed using a SOPRA SE5 multi-channel rotating polarizer spectroscopic Ellipsometer (SE) in the wavelength range of 193 nm to 1690 nm using a spot size of about 4 mm.

The XRF measurements on the Ga- and In- doped Zinc oxides have been carried out at a four-crystal monochromator (FCM) beamline in the PTB laboratory at the synchrotron radiation facility BESSY II. This beamline provides monochromatic radiation from a bending magnet in the energy range between 1.75 eV to 11 keV.

For the measurement two different recipes are used. The first recipe was performed with a 30 point grid in a step size of 300 µm by 300 µm to measure the uniformity of the layer within a locally limited area. A repeated measurement of 30 points of a single location on the layer was used to get an information of a present surface contamination layer [10,11,12]. A surface contamination layer will affect the measurement results, furthermore the layer will change the optical properties and will lead to a misleading calculation.

RESULTS AND DISCUSSION

As mentioned above, a possible contamination layer would influence the reflectance significantly [10]. The spectra of each layer for the static measurement were analyzed at a wavelength of 130 nm. An absorbing contamination layer changes the intensity of the reflected light from the sample and this intensity behavior can be seen at the given wavelength. For the ZnO samples no significant changes in the intensities could be detected. This behavior is plotted in Fig. 1 in comparison with a thin SiO$_2$ layer as an example for the change in the intensity due to a VUV light contamination removal [10]. For the analysis of the ZnO samples we decided to use the third repeated measurement point.

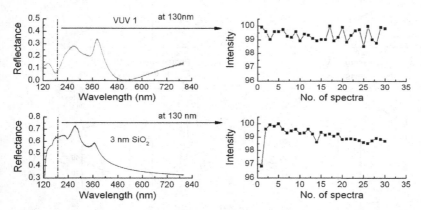

Figure 1. Selection of the measured spectra for analysis by means of a change in the intensity at a wavelength of 130 nm in comparison to a thin SiO$_2$ as an example for a visible influence of a contamination removed due to the VUV light exposure

The typical spectra measured by VUV reflectometry and ellipsometry on the ZnO samples are plotted in Fig. 2. Visible changes in the spectral region of the band gap can be seen for both measurement methods. Furthermore, there are also changes in the reflectometric spectra above 4.5 eV, which is not covered by standard spectroscopic hardware. The ellipsometric measurements were used to calculate the film thicknesses of the different layers, because using the reflectometric principle it is possible to calculate only one sample parameter (refractive index, extinction coefficient or thickness). The transparent and opaque photon energy ranges can clearly be distinguished by the interference oscillations characteristic to the transparent range [9] in the ellipsometric measurement. Hence, a simple Cauchy dispersion ($n=A+B/\lambda^2+C/\lambda^4$, where n denotes the refractive index, and A, B and C are the Cauchy parameters) using an optical model of c-Si/SiO$_2$/ZnO, whereas the refractive index of ZnO can be described by the Cauchy model, was used to calculate the layer thickness and surface nanoroughness from the ellipsometric data. The SiO$_2$ layer thickness was also measured on the reference sample by ellipsometry and a thickness of 11.5 nm was determined.

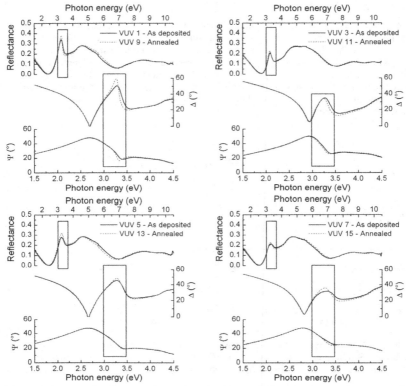

Figure 2. Measured reflectometric (static point 33) and ellipsometric Ψ and Δ spectra for the as deposited and annealed ZnO samples. The rectangles show the most relevant regions.

A summary of the sample properties measured by ellipsometry and reflectometry is compiled in Table II. The Cauchy parameters A, B and the layer thickness are used as start values for the regression to calculate the sample properties after the reflectometric measurements. The Cauchy parameter C was fixed to zero for a more stable analysis. A good correlation between the ellipsometric and reflectometric measurements was revealed. In Table III the resulting values are summarized. The thickness and nanoroughness of the ZnO layers determined by reflectometry are to a small extent but systematically thinner compared to the ellipsometer measurements. The optical properties show also a small deviation. A possible explanation could be an offset of the measurement location on the sample as well as the difference of the spot size between methods. The spot size of the reflectometric system is much smaller than the spot of the ellipsometer.

Table II. Reference data of the ZnO samples measured by spectroscopic ellipsometry, d_{ZnO} and d_r denote the thicknesses of the ZnO layer and the surface nanoroughness, respectively.

Nr.	A	B (μm^2)	n	d_{ZnO} (nm)	d_r (nm)
VUV1	1.934	0.038	2.030	50.7	9.2
VUV5	1.933	0.035	2.026	51.6	8.4
VUV7	1.932	0.038	2.033	46.6	8.9
VUV9	1.922	0.038	2.017	50.7	8.0
VUV13	1.916	0.040	2.018	51.3	8.0
VUV15	1.926	0.041	2.032	46.3	8.1

Table III. Data of the ZnO samples measured by VUV reflectometry, d_{ZnO} and d_r denote the thicknesses of the ZnO layer and the surface nanoroughness, respectively.

Nr.	A	B (μm^2)	n	d_{ZnO} (nm)	d_r (nm)
VUV1	1.897	0.040	1.941	49.1	8.4
VUV5	1.824	0.042	1.961	47.6	6.9
VUV7	1.901	0.036	1.997	43.6	5.6
VUV9	1.851	0.037	1.968	49.7	7.7
VUV13	1.869	0.045	1.943	49.3	7.6
VUV15	1.831	0.045	1.946	45.6	6.7

The last step of our investigations was the calculation of the gap energies in comparison of both methods. A Tauc-Lorentz model with one oscillator was used and the value of the layer thickness was fixed to the measured thickness for each method. The analysis of reflectometry was performed in the photon energy range from 2.5 to 3.1 eV, because the decaying excitonic line shape at higher photonic energies can not be described by the Tauc-Lorentz model. Both ellipsometry and reflectometry show a systematic increase of E_g with annealing, though with a slight offset (Table IV). The reason of the offset can partly be explained similar to that for the results of Tables II and III. Furthermore, the different implementation of the Tauc-Lorentz equation by two equipment manufacturers (Woollam and Metrosol) cannot be ruled out.

Table IV. Gap energies in eV calculated using the Tauc-Lorentz parameterization measured by spectroscopic ellipsometry and vacuum ultra-violet reflectometry.

	E_g	
	As deposited	Annealed
SE	2.95	3.02
	2.87	2.88
	2.82	2.85
VUV-R	2.63	2.75
	2.34	2.47
	2.41	2.59

CONCLUSIONS

In this study it has been shown that thin ZnO films can be characterized by vacuum ultra-violet reflectometry with results in agreement with ellipsometry. The effect of small changes in the optical film properties can be traced by optical methods. The results show that the effects of a surface nanoroughness must be taken into account during the optical modeling in order to determine correct layer thicknesses and optical properties. Hence, because of the smaller amount of directly measured information by reflectometry, the layer thickness must be measured with ellipsometry to have a starting value for the reflectometric analysis. Using this careful modeling approach, the reflectometric results show a good correlation with the ellipsometric ones. The band gap energy (which is consistent with decreasing specific resistance [13]) increase as a result of annealing and decrease with increasing Ga content. The refractive index decreases with annealing and increases with increasing Ga content. The surface nanoroughness was slightly but systematically smaller on the annealed samples. Finally, comparable physical properties between ellipsometry and reflectometry could be found.

ACKNOWLEDGEMENTS

Support from the European Community's Seventh Framework Program, European Metrology Research Program (EMRP), ERA-NET Plus, under Grant Agreement No. 217257 as well as from OTKA grant Nr. K81842 is greatly acknowledged.

REFERENCES

1. U. Özgür, Y. I. Alivov, C. Liu, A. Teke, M. A. Reshchikov, S. Dogan, V. Avrutin, S.-J. Cho, H. Morkoc, J. Appl. Phys. 98 (2005) 041301.
2. P. F. Carcia, R. S. McLean, M. H. Reilly, J. G. Nunes, Appl. Phys. Lett. 82 (2003) 1117.
3. P. F. Carcia, R. S. McLean, M. H. Reilly, Appl. Phys. Lett. 88 (2006) 123509.
4. V. Craciun, J. Elders, J. G. E. Gardeniers, I. W. Boyd, Appl. Phys.Lett. 65 (1994) 2963.
5. V. Craciun, S. Amirhaghi, D. Craciun, J. Elders, J. G. E. Gardeniers,I. W. Boyd, Appl. Phys. Lett. 65 (1994) 2963.
6. G. Socol, M. Socol, N. Stefan, E. Axente, G. Popescu-Pelin, D. C. L. Duta, C. N. Mihailescu, I. N. Mihailescu, A. Stanculescu, D. Visan, V. Sava, A. C. Galca, C. R. Luculescu, V. Craciun, J. Appl. Phys. 95 (2004) 4953.
7. M. Baum, S. Polster, M. Jank, I. Alexeev, L. F. M. Schmidt, Appl.Phys. A 107 (2012) 269.
8. S. A. Studenikin, N. Golego, M. Cocivera, J. Appl. Phys. 84 (1998) 2287.
9. P. Petrik, B. Pollakowski, S. Zakel, T. Gumprecht, B. Beckhoff, M. Lemberger, Z. Labadi, Zs. Baji, M. Jank, A. Nutsch, "Characterization of ZnO structures by optical and X-ray methods", accepted for publication in the Applied Surface Science.
10. T. Gumprecht, G. Roeder, P. Petrik, M. Schellenberger, L. Pfitzner, "Vacuum ultra-violet light induced surface cleaning and layer modification effects of thin dielectric films during reflectometric measurements", submitted for publication in the Applied Surface Science.
11. E. (Liz) Stein, D. Allred, Thin Solid Films 517 (2008) 1011–1015.
12. T. Gumprecht, G. Roeder, M. Schellenberger, L. Pfitzner, "Measurement strategy for dielectric ultra-thin film characterization by vacuum ultra-violet reflectometry", ASMC Conf. Proc. 2012.
13. C. Major, A. Nemeth, G. Radnoczi, Z. Czigany, M. Fried, Z. Labadi, I. Barsony, Appl. Surf. Sci. 255 (2009) 8907.

Mater. Res. Soc. Symp. Proc. Vol. 1494 © 2013 Materials Research Society
DOI: 10.1557/opl.2013.34

Sputter deposited ZnO porous films for sensing applications

Michał A. Borysiewicz[1], Elżbieta Dynowska[1,2], Valery Kolkovsky[2], Maciej Wielgus[1,3], Krystyna Gołaszewska[1], Eliana Kamińska[1], Marek Ekielski[1], Przemysław Struk[4], Tadeusz Pustelny[4] and Anna Piotrowska[1]

[1] *Institute of Electron Technology, Al. Lotników 32/46, 02-668 Warsaw, Poland*
[2] *Institute of Physics, PAS, Al. Lotników 32/46, 02-668 Warsaw, Poland*
[3] *Warsaw University of Technology, Pl. Politechniki 1, 00-661 Warsaw, Poland*
[4] *Silesian University of Technology, ul. Akademicka 2A, 44-100 Gliwice, Poland*

ABSTRACT

Nanoporous ZnO films are fabricated using a two step approach: sputter deposition of porous Zn followed by an ex-situ annealing in an oxygen flow at 400°C. The created structures have a porosity of 34% enabling use in surface-based absorption sensors. The films are used in a resistance-based sensor allowing easy discrimination between liquid methanol and ethanol by resistance measurements and also in a transmission-based sensor to detect NO_2, NH_3 and H_2 gases in concentration as low as 500 ppm.

INTRODUCTION

Every year, solid-state sensors gain widespread use due to their versatility, ease of tailoring by functionalization and cost-competitiveness. The heart of such a sensor is usually some volume of material to which the sensed species are absorbed. An enhanced surface to volume ratio in the sensing part of such a device can substantially improve their detection performance. The enhancement may be fabricated either through a top-down (dry or wet etching) or bottom-up (nanostructure growth) approaches, the latter of which includes the growth of a wide family of structures from the more common nanowires and nanorods [1] to more exotic flower-like structures [2]. In this report we focus on a cost-effective way of producing a highly porous dendrite-like ZnO film using a method enabling fast coating of large areas for potential large-scale applications, i.e. sputtering and apply it in two model sensor configurations: a resistance-based sensor and a transmission-based sensor. The first one is used to discriminate between two types of alcohols by measuring their resistivity after the application of a set volume to the porous ZnO film. The transmission-based sensor monitors changes in the transmission of the porous ZnO deposited on glass in a dry air environment after the application of set concentrations of contaminating gases such as NO_2, NH_3 and H_2.

EXPERIMENT

The porous ZnO films were prepared using a two-step method: first, a porous Zn film was grown via reactive magnetron sputtering of a 4N Zn target under 80W DC power in an argon-oxygen mixture with 17% oxygen content in a Surrey NanoSystems γ1000C sputtering system. The total pressure of the gas mixture during deposition was 1.5 mtorr and the base pressure prior to deposition was of the order of 10^{-7} torr. The second step comprised of ex-situ annealing in an oxygen flow at 400°C, performed in order to form ZnO. A detailed discussion of the growth procedure along with the influence of process parameters on the properties of the resulting films was previously reported by our group [3]. The films grown in this experiment were around 400 nm thick.

The substrates used, which were Si (100) and BK7 glass for the resistive and transmission sensors, respectively, were not intentionally heated during the deposition and were initially subjected to degreasing by boiling in trichloroethylene, acetone and isopropanol with subsequent deionised water rinse and nitrogen drying.

. The crystalline structure of the films was studied using X-ray diffraction (XRD) in the Bragg-Brentano mode on an upgraded DRON-1 diffractometer and their morphology was determined using Scanning Electron Microscopy (SEM) imaging on a Zeiss Auriga Neon 40 microscope. The porosity of the films was determined from the SEM images using image processing techniques.

The resistive sensors were prepared using nanoporous ZnO formed on Si (100) substrates cut into 10 mm x 5 mm bars, on the ends of which 0.5 mm wide ohmic contacts were deposited through a shadow mask. The contacts consisted of a 20 nm Ti/100 nm Al bilayer. I-V characteristics of pure and poisoned sensors were measured using a PC-controlled Keithley 2400 source-meter. 2 μl drops of 3N-pure methanol and ethanol were deposited on the sensors during poisoning. After drying the I-V characteristics were measured again.

The layers for the transmission sensors were formed on 25 mm x 25 mm BK7 glass slides with no further patterning. The measurement setup consisted of a transmission chamber with gas throughput, an Ocean Optics HR2000+ES spectrometer and an Ocean Optics DT-mini-2-GS light source, all PC-controlled. The ambient in the chamber was dry air and contaminating gases (NO_2, NH_3, H_2) were introduced in the chamber in a controlled manner using set concentrations of selected gas mixed with dry air for a given amount of time. The poisoning and purging times were set to be 20 minutes each and the operating temperature of the chamber was 120°C. The data was collected in 30 second intervals.

RESULTS AND DISCUSSION

As evidenced by XRD images presented in figure 1, the porous ZnO films are polycrystalline with no preferred orientation. Furthermore, XRD data shows that directly after deposition no ZnO fraction is present in the porous Zn film and that after the ex-situ annealing in oxygen no Zn crystallites are left.

From the SEM images it is seen that the morphology of the films is nanocrystalline with branching nanocrystallites similar to a coral reef. The dimensions of an individual crystallite are of the order of one hundred nanometers. Furthermore, the crystallites in the as-deposited Zn films are thinner than in the ZnO films formed after annealing.

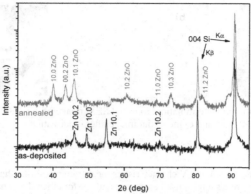

Figure 1. XRD images of the porous films after deposition and after subsequent annealing in oxygen at 400°C for 15 minutes. Please note, that as the diffractometer is not equipped with an incident beam filter, the presence of K_β line is evidenced by the doubling of the strongest peak, i.e. the 004 Si.

Figure 2. SEM images of the porous film after deposition (a) and subsequent annealing in an oxygen flow at 400°C (b).

As the samples of porous Zn and ZnO were respectively black and white, it was impossible to determine the porosity using the common ellipsometry measurements with subsequent effective medium modeling due to the strong absorbance and scattering of light from black and white surfaces, respectively. Therefore it was necessary to use digital processing of SEM images to achieve this goal. Intensity histograms were created for the images and a threshold was set for the differentiation between pixels belonging to a pore and the ones belonging to a crystallite (see figure 3.b). After a threshold was set a binarization of the image was performed, turning all pixels below threshold black and all above threshold white (see figure 3.c). The porosity was then calculated as the ratio of black to all pixels [4,5]. The porosity values obtained this way for the film after deposition and annealing were 28% and 34% respectively, meaning that although the crystallite size increased during ZnO formation (as evidenced by SEM imaging), the spaces between the crystallites increased more.

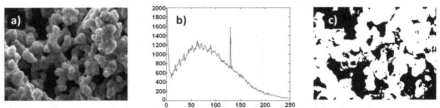

Figure 3. The steps of porosity determination: a) initial SEM image of the ZnO films, b) intensity histogram of the image and c) the initial image after binarization.

Resistance-based sensor

Figure 4 shows the results of I-V characteristics measurements for methanol and ethanol detection. There were separate samples used in the experiment for ethanol and methanol and the measured data were put into one image for brevity. The initial I-V characteristic of the ZnO porous film exhibited currents of the order of single nA at 3V bias. After the application of 2 μl of methanol and ethanol the current raised to 100 nA and 287 nA at 3V, respectively. Also for lower bias values the current recorded for ethanol is lower than for methanol, enabling easy discrimination between the two materials.

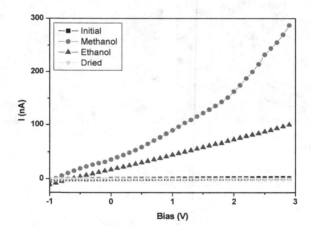

Figure 4. I-V characteristics of porous ZnO/Si resistance sensors upon the application of methanol and ethanol. The curves show data from two separate samples: one tested for methanol and the other for ethanol exclusively.

Transmission-based sensor

The shortwave region of the transmission spectra of the porous ZnO on BK7 glass used in the experiment before and after poisoning with selected gases is shown in figure 5. The (not shown) transmission in the visible up to 900 nm is constant at around 85% and exhibits no visible changes upon the addition of gases.

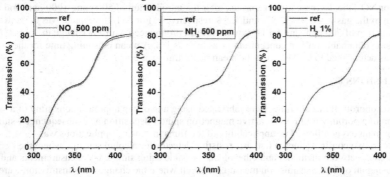

Figure 5. Transmission spectra near the absorption edge of the porous ZnO/BK7 glass used in the transmission sensor before and after poisoning with NO_2, NH_3 and H_2 gases at set concentrations.

The most pronounced changes in transmission for NO_2, NH_3 and H_2 were recorded at 410.10 nm, 323.15 nm and 390.22 nm, respectively. As it can be seen from the example in figure 6, after the introduction of the gas into the chamber, the transmission drops by a factor roughly proportional to the gas concentration. The most stable and strongest signal was obtained for NO_2 detection, while in the other cases the changes were less pronounced, the transmission signal significantly decreased in time and was more significantly affected by noise.

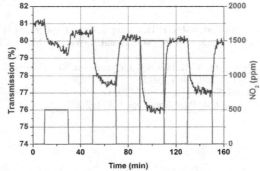

Figure 6. Changes in transmission values of the porous ZnO/BK7 glass as a function of time for subsequent purging and poisoning cycles using NO_2 at 410.10 nm.

The fall of the reference transmission value from 81% to around 80% in the first 80 minutes of the study is related to the fact, that no conditioning of the samples in the gas atmosphere though poisoning/purge cycling prior to the experiment was performed. It can be however seen that the reference transmission value stabilizes over time. Furthermore, the relative change in transmission for the addition of 1000 ppm NO_2 at 50 minutes and 130 minutes is the same at 2.8%, which shows proving the capability of reliable sensing.

For NO_2, the decrease in transmission after the introduction of 500 ppm, 1000 ppm and 1500 ppm of the gas was 1.3%, 2.8% and 4.2%, respectively. For NH_3 the change was 0.5% at the introduction of 500 ppm and for H_2 it was the least pronounced with 1% change in transmission obtainable only at concentrations as high as 1%. A mean poisoning time for the sensor was determined to be 240 s and the mean purge time was equal to it.

CONCLUSIONS

Nanoporous thin ZnO films were fabricated using a two-step approach consisting of a deposition of a porous Zn film via reactive magnetron sputter deposition and subsequent ex-situ annealing in an oxygen flow. The applicability of the films in sensing applications was confirmed in two configurations. First, as a resistive detector for alcohol sensing, where the discrimination between methanol and ethanol was possible using simple I-V measurements and second, using an optical transmission measurement cell where the changes in transmission were registered at different wavelengths upon the introduction of NH_3, NO_2 and H_2 gases into the cell, out of which NO_2 showed the most significant response of the detector. Work on further applications of the porous films, eg. in dye-sensitized solar cells, is under way.

ACKNOWLEDGMENTS

This study was partially supported by the European Union within European Regional Development Fund, through grant Innovative Economy (POIG.01.01.02-00-008/08 "Nanobiom").

REFERENCES

1. Xiang Liu, Xiaohua Wu, Hui Cao, and R. P. H. Chang, J. Appl. Phys. 95, 3141 (2004)
2. J. Shi, H. Hong, Y. Ding, Y. Yang, F. Wang, W. Cai and X. Wang, J. Mater. Chem. 21, 9000 (2011)
3. M.A. Borysiewicz, E. Dynowska, V. Kolkovsky, J. Dyczewski, M. Wielgus, E. Kamińska, A. Piotrowska, Phys. Stat. Sol. A 209, 2463 (2012)
4. M.F.M. Costa, J. Phys. Conf. Series 274, 012053 (2011)
5. Z.Y. Yang and M. Zhao, J. Opt. A: Pure Appl. Opt. 9 872 (2007)

Mater. Res. Soc. Symp. Proc. Vol. 1494 © 2013 Materials Research Society
DOI: 10.1557/opl.2013.4

Properties of Sputter Deposited ZnO Films Co-doped with Lithium and Phosphorus

T. N. Oder[1], A. Smith[1], M. Freeman[1], M. McMaster[1], B. Cai[2] and M. L. Nakarmi[2]
[1]Department of Physics and Astronomy, Youngstown State University, Youngstown, OH 44555, U.S.A.
[2] Department of Physics, Brooklyn College of the CUNY, Brooklyn, NY 11210, U.S.A.

ABSTRACT

Thin films of ZnO co-doped with lithium and phosphorus were deposited on sapphire substrates by RF magnetron sputtering. The films were sequentially deposited from ultra pure ZnO and Li_3PO_4 solid targets. Post deposition annealing was carried using a rapid thermal processor in O_2 and N_2 at temperatures ranging from 500 °C to 1000 °C for 3 min. Analyses performed using low temperature photoluminescence spectroscopy measurements reveal luminescence peaks at 3.359, 3.306, 3.245 eV for the co-doped samples. The x-ray diffraction 2θ-scans for all the films showed a single peak at about 34.4° with full width at half maximum of about 0.17°. Hall Effect measurements revealed conductivities that change from p-type to n-type over time.

INTRODUCTION

The properties of ZnO make it a very attractive material for fabricating optoelectronic devices for blue/UV applications. It is a wideband gap semiconductor with a direct energy band gap of 3.37 eV at 300 K or 3.437 at 4 K [1]. While it bears several similarities with gallium nitride (GaN) currently used for UV optoelectronic devices, ZnO has several advantages over GaN, making it a candidate for replacing or complimenting GaN [2]. For instance, it has a larger exciton binding energy (60 meV for ZnO, 24 meV for GaN) making it more attractive for fabricating efficient room-temperature LEDs and lasers. Successful development of ZnO-based devices, however, has been hampered by the lack of stable p-type materials. As-grown ZnO is n-type with usually very high concentration of electrons. Some of the reasons for the difficulty in p-type doping include the low solubility and self-compensation of the acceptor dopants by defects [3,4]. Native defects such as oxygen vacancies, zinc interstitials or zinc antisites were once thought to be the cause for the unintentional n-type conductivity in ZnO. However, while recent research have cast doubt on this concept, these defects nevertheless play a big role as compensating centers to p-type doping [5,6]. Furthermore, hydrogen-related defects have been suggested to be some of the hindrances to p-type doping in ZnO, especially when H substitutes for O in ZnO and act as a shallow donor [6]. Group V elements such as nitrogen (N), phosphorus (P), arsenic (As), and antimony (Sb) have been experimentally investigated for obtaining p-type ZnO by substituting oxygen. In theory, N substituting for O (N_O) was regarded as the most suitable dopants for producing p-type ZnO based on the strain effects, energy levels and the similarity of the atomic radius of N with O [6,7]. However, recently revisited theoretical work and experimental investigation showed that nitrogen in ZnO is a deep acceptor with large ionization energy of 1.3 eV [6].

From the theoretical standpoint, the shallow energy levels of group I elements should make them good candidates for producing p-type ZnO [8]. Among them, Li substituting for Zn (Li_{Zn}) should be the best candidate due to its shallower defect level of 0.09 eV and weaker repulsion between its substitution defects [4]. Further calculations showed that donor-like Li_I self-compensates the Li_{Zn} acceptors. Achievement of p-type conductivity in ZnO by mono-doping with both group V and group I acceptor elements have been reported [9,10]. The mechanism of p-type conductivity in X-doped (X = P, As, Sb) ZnO has been attributed to the formation of a $(X_{Zn}-2V_{Zn})$ complex [11,12]. Despite these reports, reliable and reproducible p-type conductivity in ZnO is still problematic due to the high background concentration of free electrons and compensating centers originating from defects. Single acceptor doping (mono-doping) with these atoms is always accompanied by compensating donor defects such as Li interstitials (Li_I) and N_2-on-O substitutions [$(N_2)O$] in N doped ZnO [7]. Due to the difficulty of reproducing the results of the reported p-type conductivity in ZnO materials, several reasons have been given to explain the supposed p-type conduction in ZnO. The main reasons seems to be incorrect assignment of p-type conduction from Hall measurements carried on inhomogeneous samples and interface or near surface states indicating the apparent p-type conductivity [13].

Co-doping was suggested as a possible solution to attain p-type ZnO materials. This is supposed to work by enhancing the solubility of acceptors and reducing the acceptor binding energy leading to highly doped p-type materials [14]. Successful realization of stable p-type ZnO by co-doping with Li and N was reported by Lu, et al [15]. Suggested mechanism for the p-type formation was given as due to the formation of complex acceptors such as $Li_{Zn}-N_O$ and $Li_{Zn}-N$ or that the co-doping minimizes the formation of donor-like compensating defects such as Li_I. [7].

In this work, we report the studies on sputter deposited ZnO films co-doped with Li and P. Post-deposition annealing using rapid thermal processor (RTP) was carried in O_2 or N_2 at 500 °C – 1000 °C for 3 min. The films obtained were analyzed by energy dispersive X-ray spectrometry (EDS), Photoluminescence (PL) spectroscopy, X-ray diffraction and Hall effect measurements. Our results indicate that samples annealed in N_2 show n-type conductivity, while those annealed in O_2 showed unstable conductivities that switch from p-type to n-type. Data from temperature dependent PL measurements indicate luminescence peak positions (at 12 K) at about 3.306 eV, 3.245 eV and 3.359 eV and the relative dominance of these peaks switch at about 50 K.

EXPERIMENT

The surface of the c-plane sapphire substrates used was prepared by cleaning using boiling acetone, alcohol and a 10 min dip in buffered HF acid prior to loading in a vacuum chamber which was then pumped to 2×10^{-7} Torr. The substrate was further heated at 900 °C for 30 min in oxygen atmosphere which was determined to be beneficial to achieving good quality ZnO films. To determine the optimum condition for a buffer layer, several undoped ZnO films were deposited at different substrate temperatures from 300 °C – 900 °C. The doped samples were prepared by first depositing a 1.0 μm-thick undoped ZnO buffer layer. The doped films consisted of several alternate layers of ZnO (~ 15 nm) and Li_3PO_4 (~5 nm) deposited from pure ceramic targets. Three doped samples were prepared, the first one (sample A) with 17 and the second one (sample B) with 60 alternate layers of ZnO and Li_3PO_4. The third sample (sample C) consisted of about 150 nm thick layer of Li_3PO_4 sandwiched between two 200 nm-thick layers of ZnO. The structures of these samples are illustrated in Fig. 1. All films were deposited using radio frequency (RF) magnetron sputtering at a substrate temperature of 700 °C and an RF power of 100 watts. The deposition gas used for all the films consisted of ultra-high purity (UHP) Ar and O_2 mixed to a ratio of 1:1 at a pressure of 10 mTorr and a flow rate of 20 standard cubic centimeter per minute at STP (sccm). Post deposition annealing was carried using a rapid thermal processor (RTP) in UHP O_2 or N_2 at 500 - 1000 °C for 3 min to improve the film quality.

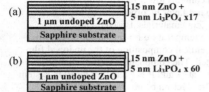

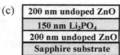

Figure 1. (a), (b) and (c) are diagrams illustrating the structures of samples A, B and C, respectively.

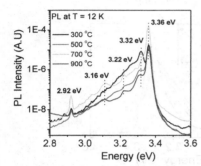

Figure 2. Photoluminescence spectra from undoped ZnO films deposited at different substrate temperatures.

Table I. Data from Hall effect measurements on undoped ZnO films deposited at different substrate temperatures.

Dep. Temp (°C)	n (cm^{-3})	Mobility (cm^2/V-s)	Resistivity (Ω-cm)
300	-6.3 x 10^{18}	28	0.037
500	-1.5 x 10^{17}	4.2	10
700	-3.4 x 10^{16}	4.0	47
900	-9.9 x 10^{15}	2.7	409

The optical properties of the films were obtained from photoluminescence (PL) spectroscopy measurements that were conducted using a 325 nm He-Cd laser source. The compositions of the films were determined using X-ray energy dispersive spectrometry (EDS) measurements using the EDAX Apollo XV with resolution of 128 eV (at Mn K). The film microstructures were investigated by X-ray diffraction measurements using the Bruker-Nonius D8 Advance Powder Diffractometer with a Cu K$_\alpha$ line of 1.54 Å. The electrical properties were investigated using Hall effect measurements.

DISCUSSION

Figure 2 shows photoluminescence spectra (at 12 K) from the undoped ZnO films deposited at different temperatures and annealed in O_2 900 °C for 3 min. The dominant band edge peak at 3.36 eV is identified as due to donor bound exciton (D°X) transition while the peak at 3.32 eV could be due to a lateral optical (LO) phonon replica of the free exciton (FX-1LO). The peak at 3.22 eV has been identified as due to transition from donor-acceptor pair (DAP) recombination. The 2θ X-ray diffraction scan (not shown) of the film deposited at 900 °C reveals a peak at 34.3°

that corresponds to the diffraction from the (0 0 2) plane of ZnO and indicates a strong c-axis orientation perpendicular to the surface at the sapphire substrate. Results from the Hall effect measurements of the undoped ZnO films are shown in Table I, which indicates that films deposited at higher temperatures are more resistive and contains lower background electron concentration. It is possible that higher temperatures do not favor incorporation of impurities (such as hydrogen) or formation of native defects in the ZnO materials. Based on this observation, we proceeded to use 900 °C as the substrate temperature for the buffer layers when depositing doped films. The EDS results indicate the chemical composition of the undoped film consists of Zn = 50.2 at % and O = 49.8 at %, as expected. The composition of the co-doped samples was found to comprise Zn = 39.0 at %, O = 59.8 at % and P = 1.2 at %. Our EDS system could not detect Li because of its low atomic weight. The electron beam used in the EDX measurements was accelerated at 15 kV, thus the oxygen quantization values could be due to the O in the sapphire substrate as well as that from the deposited ZnO films.

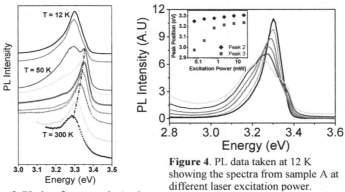

Figure 3. PL data from sample A taken at different temperatures.

Figure 4. PL data taken at 12 K showing the spectra from sample A at different laser excitation power.

The PL spectra from co-doped sample A, taken at different temperatures are shown in Figure 3. As can be seen from this figure and especially from the spectrum at 50 K, three main peaks can be resolved. At 50 K, these peaks appear at 3.357 eV (peak 1), 3.300 eV (peak 2) and at 3.206 eV (peak 3). The corresponding values at 12 K are 3.359 eV, 3.306 eV and 3.245 eV. At temperatures lower than 50 K, the intensity of peak 3 with the lower energy is higher than that of peak 1 with higher energy, and this is reversed for temperatures above 50 K. As stated for the case of the undoped sample, the peak at 3.359 eV (at 12 K) is usually a result of the transition from donor bound exciton (D°X). Noting that 50 K corresponds to a thermal energy of 4.31 meV which is close to the transition energy of a bound exciton (4.1 meV), this behavior could suggest that this bound exciton is more strongly excited as the temperature increases from 50 K. The peak at 3.306 eV could possibly be due to the transition of free electrons from conduction band to neutral acceptors (FA). On the other hand, the peak at 3.245 eV could be due to a donor to acceptor pair (DAP) transition [16].

To further determine the origins of the luminescence peaks, we collected PL spectra (at 12 K) from sample A at different laser excitation powers and the results are shown in Figure 4. At this temperature, the intensity of peak 1 (at 3.359 eV) is much smaller than for peak 2 (at 3.306 eV) and peak 3 (at 3.245 eV). For that reason, we resolved only peaks 2 and 3. The inset in Figure 4 shows the variation of the energy positions of these peaks as a function of excitation power. Peak 2 (at 3.306 eV) shows a very small change while peak 3 (at 3.245 eV) shows a much pronounced decrease in peak position as the laser power is decreased. From this, we can conclude that peak 3 is related to a donor-acceptor pair [16]. This result also means that peak 2 is not related to a DAP transition and could therefore be an FA transition, as has been observed on ZnO doped with phosphorus. XRD 2θ scans carried on co-doped samples A and B (not shown) reveals a peak at peak at 34.3° that corresponds to the diffraction from the (0 0 2) plane of ZnO and similarly indicates a strong c-axis orientation perpendicular to the surface at the sapphire substrate.

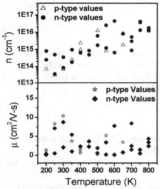

Figure 5. Data from Hall effect measurements on sample A showing random variation of carrier concentration (n) and mobility (μ).

Table II. Hall Effect data from samples A and C for different annealing temperatures.

Temp/Time	Sample A n (cm^{-3})	Sample C n (cm^{-3})
900C/3min	+2.31 x 10^{15}	-3.42 x 10^{11}
800C/3min	+2.05 x 10^{13}	-3.12 x 10^{13}
700C/3min	+1.44 x 10^{14}	-5.62 x 10^{12}
600C/3min	-2.89 x 10^{15}	-3.46 x 10^{12}
500C/3min	+1.77 x 10^{15}	

Table II shows data of Hall effect measurements from samples A and C annealed at different temperatures in O_2. These data were taken within two days after the samples were prepared, and the values are averages of at least three measurements from each sample analyzed. Samples in group C show n-type conductivity for all annealing temperatures. The samples from group A however show p-type conductivities for all temperatures (except 600 °C). Repeated Hall effect measurements were carried on one sample from group A that showed p-type conductivity (results not shown here). It was observed that over a three hour period, the conductivity switches randomly between p-type and n-type, the mobility vary from 0.5 – 7.8 cm^2/V-s, the carrier concentration decreases and the resistivity increases. The scatter in the results, as well as the inconsistencies in the carrier type, could be due to the small Hall voltages in the measurements which can be significantly impacted by small signal noise spikes during the measurements. Temperature dependent Hall effect measurements were conducted and Figure 5 shows the variation of the carrier concentration (n) and mobility (μ) as a function of temperature. These values change randomly and the conductivity also switches randomly. However, the resistivity shows an exponential decrease with increasing temperature. From the resistivity versus temperature plots (not shown), we obtained two values of activation energies, one of 76 meV (for T = 200 – 400 K) and the second one of 376 meV (for T = 450 – 700 K). These could be related to defect states in the films that are activated by different amounts of energy.

CONCLUSIONS

Co-doped ZnO films deposited using RF magnetron sputtering were analyzed. A buffer layer deposited at 900 °C was used to reduce background electron concentrations in the doped films. The p-type conductivity obtained in samples annealed in O_2 was found to be unstable. Resistivity measurements carried up to 800 K showed two activation energies at 76 meV and 376 meV, which could be related to defects in the films. Temperature dependent and power dependent PL measurements resolved three peaks from D°X, FA, DAP transitions.

ACKNOWLEDGMENTS

The authors wish to gratefully acknowledge funds from the National Science Foundation (#DMR-1006083) which supported this work. Assistances by Dr. Matthias Zeller (for the XRD measurements) and Dr. Dingqiang Li, (for EDS measurements) from the Department of Chemistry, Youngstown State University are also gratefully acknowledged.

REFERENCES

1. M. R. Wagner, U. Haboeck, P. Zimmer, A. Hoffmann, S. Lautenschläger, C. Neumann, J. Sann and B. K. Meyer, *Proc. of SPIE* **6474**, 64740X-1 (2007).
2. Ü. Özgür, Ya. I. Alivov, C. Liu, A. Teke, M. A. Reshchikov, S. Doğan, V. Avrutin, S.-J. Cho and H. Morkoç, *J. Appl. Phys.* **98**, 041301 (2005).
3. V. Avrutin, D. J. Silversmith, and H. Morkoç, *Proc IEEE Inst. Electr. Electron. Eng.* **98**(7), 1269-1280 (2010).
4. C. H. Park, S. B. Zhang, and Su-Huai Wei, *Phys. Rev. B* **66**, 073202 (2002).
5. L. S. Vlasenko and G. D. Watkins, *Phys. Rev. B* **72**, 035203 (2005).
6. A. Janotti and C. G. Van deWalle, *Phys. Rev. B* **75**, 165202 (2007).
7. X. Y. Duan, R. H. Yao and Y. J. Zhao. *Appl. Phys. A* **91**, 467–472 (2008)
8. C.H. Park, S.B. Zhang and S.H.Wei, *Phys. Rev. B* **66**, 073 202 (2002).
9. D. C. Look, R. L. Jones, J. R. Sizelove, N. Y. Garces, N. C. Giles, and L. E. Halliburton, *Phys. Stat. Sol. a* **195**, 171 (2003).
10. X. M. Fan, J. S. Lian and Z. X. Guo, *Appl. Surf. Sci.* **239**, 176 (2005).
11. S. T. Tan, B. J. Chen, X. W. Sun, W. J. Fan, H. S. Kwok, X. H. Zhang and S. J. Chua, *J. Appl. Phys.* **98**, 013505 (2005).
12. S. J. Kang, H.-H. Shin and Y.-S. Yoon, *Journal of the Korean Physical Society* **51**(1), 183-188 (2007).
13. M. D. McCluskey and S. J. Jokela *J. Appl. Phys.* **106**, 071101(2009).
14. T. Yamamoto. *Phys. Stat. Sol. (a)* **193**(3), 423–433 (2002).
15. J. G. Lu, Y.Z. Zhang, Z.Z. Ye, L.P. Zhu, L. Wang, B.H. Zhao and Q.L. Liang, *Appl. Phys. Lett.* **88**, 222 114 (2006).
16. D. Yu, L. Hu, S. Qian, H. Zhang, S. A. Len, L. K. Len, Q. Fu, X. Chen and K. Sun, *J. Phys. D: Appl. Phys.* **42**, 055110 (2009).

Mater. Res. Soc. Symp. Proc. Vol. 1494 © 2013 Materials Research Society
DOI: 10.1557/opl.2013.409

Control of the Formation of $Zn_{1-x}Mg_xO$ Films by Zinc Sulfate Concentration

Hiroki Ishizaki[1] and Seishiro Ito[2]
[1] Department of Electronic System Engineering, Tokyo University of Science Suwa,
5000-1 Toyohira, Chino-shi, Nagano 391-0292, Japan
[2] Faculty of Science and Engineering, Kinki University,
4-1 Kowakae 3-chome, Higashiosaka, Osaka 577-8502, Japan

ABSTRACT

Magnesium doped ZnO films were electrochemically grown on the NESA conductive glass substrate from the magnesium nitrate aqueous solution with zinc sulfate, kept at 323K and the cathodic potential of -0.9V vs. Ag/AgCl. The Mg/(Mg+Zn) atomic ratio of $Zn_{1-x}Mg_xO$ films increased with the decrease in the zinc sulfate concentration. The optical band gap energy of these $Zn_{1-x}Mg_xO$ films decreased with increasing content of zinc sulfate. Thus, the optical band gap energy and Mg/(Mg+Zn) atomic ratio of $Zn_{1-x}Mg_xO$ films would depend on the zinc sulfate concentration.

INTRODUCTION

Recently, $Zn_{1-x}Mg_xO$ films with wide band gap energy were paid much attention for many applications such as optic device, electric luminescence device and transparent conductive oxide of solar cell [1]. For the doping of magnesium atoms into $Zn_{1-x}Mg_xO$ films, this $Zn_{1-x}Mg_xO$ film had the wider band gap energy than that of ZnO film. The band gap energy of the $Zn_{1-x}Mg_xO$ films would be easily controlled by the magnesium content of this $Zn_{1-x}Mg_xO$ film. The $Zn_{1-x}Mg_xO$ films with the dopant of magnesium atoms present interesting electrical and optical properties, which find the wide applications in the fields of optoelectronic and transparent conductive oxide [2].

Other authors reported that $Zn_{1-x}Mg_xO$ films were deposited on the substrates by physical vapor deposition such as RF-magnetron sputtering, molecular beam epitaxy, metal organic chemical vapor deposition and pulsed laser deposition. On the other hand, the electrochemical preparation of oxide films from aqueous solutions presents several advantages over these techniques mentioned above; (1) the thickness and morphology of film can be controlled by electrochemical parameters, (2) relatively uniform films can be obtained on the substrates with complex shape, (3) films can be obtained on substrates with melting point below 373K such as polymer, (4) the technique is less hazardous and more environmentally friendly and (5) the equipment is not expensive.

In this paper, the influence of zinc sulfate concentration on electrochemical growth of $Zn_{1-x}Mg_xO$ films will be discussed in detail.

EXPERIMENT

$Zn_{1-x}Mg_xO$ films were electrochemically grown on the conductive NESA glass (NESA glass, approximately $12\Omega/\square$, Asahi glass Co., Ltd.) substrate from 0.1 mol/L $Mg(NO_3)_2$ aqueous solutions containing zinc sulfate ranging of 0.0 mmol/L to 5 mmol/L, at the deposition temperature of 323K, cathodic potentials of -0.9V. A Pt/Ti sheet (99.999% purity) was used as an active anode. An Ag/AgCl electrode was used as a reference electrode. The electrolysis was potentiostatically carried out by using a potentio/galvanostat without stirring.

The structural properties of these films were characterized by x-ray diffraction measurements. This x-ray diffraction measurement was performed using X-ray diffraction meter (Rigaku Instrument Rint 2000) with using monochromatic Cu Kα radiation operated at 40 kV and 30 mA. The compositions of the films were evaluated with an induction coupled plasma atomic emission spectrometer (ICP-AES, Seiko Instruments SPS7700). In order to clarify the dependence of Mg/(Mg+Zn) atomic ratio on the lattice parameters of the films, the lattice parameters and the crystal structure of these films are determined by the Rietveld analysis[3]. The surface morphology and the cross section morphology were observed by using scanning electron microscopy (FE-SEM, Hitachi S4800). The optical properties of these thin films were measured by using UV-VIS-NIR scanning spectrophotometer (UV-VIS-NIR, Shimadzu, MPC3100).

DISCUSSION

Figure 1 shows the observed and the calculated x-ray diffraction spectra for $Zn_{1-x}Mg_xO$ film. For $Zn_{1-x}Mg_xO$ film electrochemically grown on NESA glass substrate from a magnesium nitrate aqueous solution containing zinc sulfate, the wurtzite structure with the space group of C6mc is determined by Rietveld analysis, regardless of the cathodic potential. The degree of fitness is quite satisfactory with Rwp=10.86, Rp=8.43 and Rf=2.04. The solid lines are

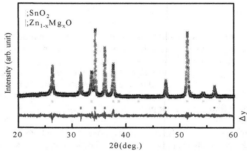

Figure 1 The observed and the calculated x-ray diffraction spectra for $Zn_{1-x}Mg_xO$ film
The observed XRD data are indicated by the plus signs, the calculated profile by the continuous line overlaying them, and the difference between observed intensity and calculated intensity by the lower curve.

calculated intensities, the crosses overlying them are observed intensities, and Δy is the difference between observed and calculated intensities. Since Δy is very small, the agreement between the calculated and observed intensities is very satisfactory.

Figure 2 shows the influence of zinc sulfate concentration on the x-ray diffraction patterns of films electrochemically grown at cathodic potential of –0.9V. For the zinc sulfate

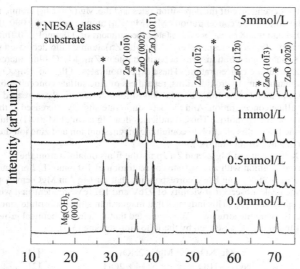

Figure 2 The x-ray diffraction patterns of $Zn_{1-x}Mg_xO$ films grown by electrochemical techniques

concentration above the 0.3 mmol/L, all diffraction lines are assigned to those of SnO_2 substrate and ZnO with wurtzite structure. Any diffraction lines attributed to other magnesium compounds and zinc hydroxide cannot be observed. For the film obtained from the magnesium nitrate aqueous solution, all diffraction lines are identified with SnO_2 and $Mg(OH)_2$ with the hexagonal structure. It indicated that adding zinc sulfate into the magnesium nitrate aqueous solution gave rise to electrochemical growth of $Zn_{1-x}Mg_xO$ with the wurtzite structure on the conductive NESA substrate.

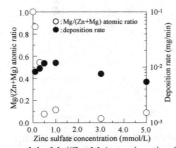

Figure 3 The deposition rates and the Mg/(Zn+Mg) atomic ratio of $Zn_{1-x}Mg_xO$ film vs. zinc sulfate concentration

Figure 3 shows the dependence of the deposition rates and the Mg/(Zn+Mg) atomic ratio of $Zn_{1-x}Mg_xO$ film on the zinc sulfate concentration. Mg/(Mg+Zn) atomic ratios of 1.000, 0.866, 0.544, 0.077, 0.115, 0.055 and 0.008 occur for zinc sulfate concentrations of 0.0mmol/L, 0.1mmol/L, 0.3mmol/L, 0.5mmol/L and 1.0mmol/L respectively. Mg/(Mg+Zn) atomic ratio decreased with the increase of the zinc sulfate concentration. The deposition rate of $Zn_{1-x}Mg_xO$ films increase with an increase of the zinc sulfate concentration. These deposition rates of the $Zn_{1-x}Mg_xO$ films were higher than that of magnesium hydroxide films, regardless of zinc sulfate concentration. For the zinc sulfate concentration below 1mmol/L, the deposition rate sensitively increased with an increase of the zinc sulfate concentration. And the deposition rate slightly decreased at the zinc sulfate concentration above 1mmol/L. Thus, it indicates that Mg atom of $Mg(OH)_2$ cluster will be substituted for zinc ion of this electrolyte containing magnesium ion and zinc ion during electrochemical growth of $Zn_{1-x}Mg_xO$ film[4-5].

Figure 4 shows XPS spectra of Mg 2s and Zn 2p for the films obtained from the magnesium nitrate aqueous solution with zinc sulfate concentration of 1.0 mmol/L. For figure 4-a, the peak of Mg 2s is observed at about 88eV corresponding that for Mg^{2+} in MgO envelope [6]. For figure 4-a and 4-b, the peaks observed at vicinity of 92eV and 1021eV are identified with Zn 3s and Zn 2p of ZnO, respectively [7]. This indicates that magnesium atoms substitute zinc atoms in ZnO films with the wurtzite structure. We suggested that the electrochemical growth reaction of $Zn_{1-x}Mg_xO$ film would be described by the following scheme [3-4].

$$Mg(NO_3)_2 \rightarrow Mg^{2+} + 2NO_3^- \qquad [1]$$
$$NO_3^- + H_2O + 2e^- \rightarrow NO_2^- + 2OH^- \qquad [2]$$
$$Mg^{2+} + 2OH^- \rightarrow Mg(OH)_2 \qquad [3]$$
$$ZnSO_4 \rightarrow Zn^{2+} + SO_4^{2-} \qquad [4]$$
$$Mg(OH)_2 + (1-x)Zn^{2+} \rightarrow Zn_{1-x}Mg_xO + 1-xMg^{2+} + H_2O \qquad [5]$$

The NO_3^-/NO_2^- reduced reaction will play an important role to electrochemically grow

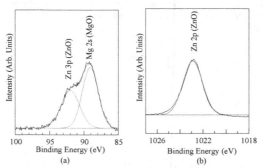

Figure 4 XPS spectra of Mg 2s and Zn 2p for the films obtained from the magnesium nitrate aqueous solution with zinc sulfate concentration of 1.0 mmol/L
(a) Mg 2s XPS spectrum and (b) Zn 2p XPS spectrum

Mg(OH)$_2$ cluster. Thus, the reduced reaction of NO$_3^-$ ion and the substitution reaction mentioned at equation (5), will play an important role to electrochemically grow Zn$_{1-x}$Mg$_x$O film with the wurtzite structure.

Mentioned above, the substitution of Mg atom into the zinc site of these ZnO films electrochemically grown on the conductive NESA glasses will give the increase of band gap energy for ZnO films. Figure 5 shows the optical properties of Zn$_{1-x}$Mg$_x$O films. The transmission data and reflection data measurements at the wavelength ranging from 300nm to 800nm are performed by the using UV-VIS-NIR scanning spectrophotometer at the room temperature. The absorption function α of the films is calculated by the following the expression [8].

$$T = (1-R)^2 \exp(-\alpha d) \qquad [6]$$

Where R is the reflectance, α is the absorption function, d is the thickness of film, T is the transmitance. With the increase in the zinc sulfate concentration, $(\alpha h\upsilon)^2$-hυ curves were shifted at the high energy side. And the optical band gap energy of Zn$_{1-x}$Mg$_x$O films was obtained from the intersection point between the hυ axis and the line extrapolated the linear portion of $(\alpha h\upsilon)^2$-hυ curve. The optical band gap energy of Zn$_{1-x}$Mg$_x$O films increased with the decrease in zinc sulfate concentration. Thus, it indicated that the increase of Mg/(Zn+Mg) atomic ratio of Zn$_{1-x}$Mg$_x$O film gave the increase of the band gap energy of Zn$_{1-x}$Mg$_x$O films.

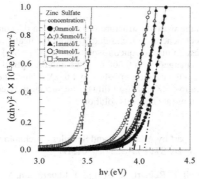

Figure 5 The optical properties of Zn$_{1-x}$Mg$_x$O films

Figure 6 shows the surface morphology of these films electrochemically grown from a 0.1mol/L Mg(NO$_3$)$_2$ aqueous solution containing zinc sulfate ranging of 0.0mmol/L to 5mmol/L. For zinc sulfate concentration of 0.0mmol/L, MgO films are composed of aggregates of hexagonal grains. And any defects such as pores and cracks cannot be observed in MgO film. For zinc sulfate concentration above the 0.5mmol/L, Zn$_{1-x}$Mg$_x$O films are composed of aggregates of hexagonal grains. At the zinc sulfate concentration ranging of 0.5mmol/L to 1.0mmol/L, the roughness and the grain size of Zn$_{1-x}$Mg$_x$O films decrease with an increase in the zinc sulfate concentration. However, the roughness and the grain size of Zn$_{1-x}$Mg$_x$O films

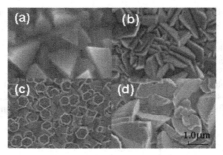

Figure 6 The surface morphology of these films electrochemically grown from a Mg(NO₃)₂ aqueous solution containing zinc sulfate
(a) zinc sulfate concentration of 0.0mmol/L, (b) 0.5mmol/L, (c) 1.0mmol/L and (d) 5.0mmol/L

increase at zinc sulfate concentration above the 1.0mmol/L. Thus, it indicated that grain size of $Zn_{1-x}Mg_xO$ films depended on the deposition rate, referring to figure 3.

CONCLUSIONS

The $Zn_{1-x}Mg_xO$ films with the wurtzite structure were grown on the conductive NESA glass substrates at 323K by the electrochemical techniques without the heat treatment. By the adding the zinc sulfate into magnesium nitrate aqueous solution, Mg/(Zn+Mg) atomic ratio of $Zn_{1-x}Mg_xO$ films were sensitively changed from 0.008 to 0.866 and the optical band gap energy of $Zn_{1-x}Mg_xO$ films were changed in the ranging of 3.4eV to 4.05eV. Thus, the optical band gap energy and Mg/(Zn+Mg) atomic ratio of $Zn_{1-x}Mg_xO$ films were successfully controlled by the adding zinc sulfate into magnesium nitrate aqueous solution.

ACKNOWLEDGMENTS
This work was supported in part by Adaptable and Seamless Technology Transfer Program through target-driven R&D (A-STEP) AS231Z03075B

REFERENCES
1. F. Erfurth, B. Husmann, A. Scholl, F. Reinert, A. Grimm, I. Lauermann, M. Bar, Th. Niesen, J. Palm, S. Visbeck, L. Weinhardt, and E. Umbach, *Appl. Phys. Lett.*, **95**, 122104 (2009)
2. K. Matsubara, H. Tampo, H. Shibata, A. Yamada, P. Fons, K. Iwata and S. Niki, *Appl. Phys. Lett.*, **85**, 1374 (2004)
3. F. Izumi and T. Ikeda, *Mater. Sci. Forum*, **198**, 321 (2000)
4. M. Pourbaix, *"Atlas of electrochemical Equilibria in Aqueous Solution"*, p.139, Pergamon Press, Paris, (1966)
5. M. Pourbaix, *"Atlas of electrochemical Equilibria in Aqueous Solution"*, p.406, Pergamon Press, Paris, (1966)
6. J. Chastain, *"Handbook of X-ray Photoelectron Spectroscopy"*, p.52, Perkin-Elmer Corporation Physical Electronics Division, America (1992)

7. J. Chastain,"*Handbook of X-ray Photoelectron Spectroscopy*", p.88, Perkin-Elmer Corporation Physical Electronics Division, America (1992)

8. J. M. Dona, J. Electrochem. Soc., **141**, 205 (1994)

Mater. Res. Soc. Symp. Proc. Vol. 1494 © 2013 Materials Research Society
DOI: 10.1557/opl.2012.1743

Substrate Temperature Effects of the ZnO:AlF₃ Transparent Conductive Oxide

Tien-Chai Lin[1], Wen-Chang Huang[2*], Chin-Hung Liu[1] and Shang-Chou Chang [1]
[1] Department of Electrical Engineering, Kun Shan University, No. 949, Da Wan Road, Yung-Kang District, Tainan, 710, Taiwan, ROC
[2] Department of Electro-Optical Engineering, Kun Shan University, No. 949, Da-Wan Road, Yung-Kang District, Tainan, 710, Taiwan, ROC
*Corresponding author: email: wchuang@mail.ksu.edu.tw

ABSTRACT

Thermal effects on the crystal structure, electrical and optical characteristics of the Al and F co-doped ZnO films (ZnO:AlF₃) are discussed in the paper. The ZnO:AlF₃ thin films are prepared by RF sputtering with a constant power (ZnO/AlF₃=100W/75W) toward the ZnO and AlF₃ targets. The substrate temperature varied from room temperature to 250 °C with a step of 50 °C during thin film deposition. The crystalline quality of the ZnO:AlF₃ film improved as the substrate temperature increased, with a corresponding increase in grain size. The improvement of the film quality leads to a higher electron mobility, with electron mobility of 0.85 cm²/V-s for the film deposited at the substrate temperature of 250 °C. The doping effect of fluorine in ZnO, and hence carrier concentration, was reduced at high temperature due to the vaporization of fluorine. This led to a reduction of carrier concentration with increase of temperature from 25 to 200°C. The corresponding resistivity increased from 3.60×10^{-2} to 6.0×10^{-2} Ω-cm. While for a further increase in substrate temperature, the doping of Al to the ZnO film was increased and resulted in an increase in carrier concentration.

INTRODUCTION

The transparent conducting film, Indium tin oxide (ITO) has been widely used because of its good electrical and optical properties. However, it shows the disadvantages of high cost, toxicity and low stability to H_2 plasma. On the other hand, zinc oxide (ZnO) films have the advantages of low cost, non-toxicity and high stability in H_2 plasma atmosphere with good electrical and optical properties [1]. Furthermore, the electrical properties of ZnO film can be modified by appropriate doping process [2].

Al-doped ZnO thin films (AZO) have been obtained by several techniques [3]. The resistivity values of these films are in the range of 10^{-4}–10^{-2} Ω-cm depending on the preparation technique used. One approach to further increase conductivity of AZO is the addition of another dopant element to the thin film. Fluorine is an adequate anion doping candidate, due to its similar ionic radius relative to oxygen. F-doped ZnO thin films (FZO) have also been deposited by many methods [4,5]. Lately, ZnO thin films doped with Al and F have been deposited by radio frequency (RF) magnetron co-sputtering of a ZnO target containing Al_2O_3 and a ZnO target containing ZnF_2 [5],with notable effect on the resistivity and carrier mobility. It was found that F dopants improved crystallization of ZnO films, and with additional post-deposition vacuum annealing desorption of oxygen at the grain boundaries led to reduced grain boundary scattering. So, co-doping of ZnO film with Al and F is an effective method to improve the conductivity of the transparent conductive film.

For a sputter system, the aid of thermal energy during deposition is a key parameter towards obtaining a high quality film. The thermal effects passivate stress at the film/substrate interface and repair defects in the thin $ZnO:AlF_3$ film[6]. In this paper, the temperature effect on the deposition of thin film was investigated. The rf power applied to ZnO and AlF_3 targets were fixed, while the substrate temperature was varied in order to evaluate the thermal effect of the thin film deposition.

EXPERIMENTAL DETAILS

ZnO thin films were prepared by RF magnetron co-sputtering equipped with a ZnO target (99.9%, 7.62 cm) and a AlF_3 target (99.99%, 7.62 cm). Microscope slide with area of 2.5×2.5 cm^2 was used to be substrate for thin film deposition. The base pressure was under 3×10^{-5} torr and the working pressure was maintained at $5 \sim 20 \times 10^{-3}$ torr. The RF power applied to the ZnO target was 100 W and to the AlF_3 target was 75W with deposition time of 1 hour. In order to investigate the thermal effects on the $ZnO:AlF_3$ film, the substrate temperature was varied between room temperature, 150°C, 200°C and 250°C, respectively. The phase and crystallinity of the films were investigated by X-ray diffraction (XRD). The energy dispersive spectrum (EDS) was used to analyze the elements of the thin film. The sheet resistances of the films were measured by a four-point probe and the Hall measurements were used to evaluate the resistivity, carrier concentration and mobility. The surface morphology of the film was observed by the field emission scanning electron microscopy (FE-SEM).

DISCUSSION

Figure 1 shows the XRD spectrum of the $ZnO:AlF_3$ films with different substrate temperature. All the films showed strong ZnO (002) peaks, indicating that the ZnO thin films were in a preferential c-axis orientation, due to the (002) plane having the lowest surface free energy. At higher substrate temperature the sputtered particles gained sufficient energy to diffuse and migrate on the sample surface leading to better crystalline quality in the films. The increase of substrate temperature also reduced the crystal defects and stress effects which resulted from the lattice mismatch between thin film and substrate[6]. Figure 2 shows the full width half maximum (FWHM) and grain size of ZnO(002) of the $ZnO:AlF_3$ films at different substrate temperature. It shows that the grain size increased with substrate temperature during thin film deposition. The grain size of ZnO(002) was observed to be 38.85 nm at the substrate temperature of 250 °C.

Figure 3 shows the plot of angle shift and space distance of ZnO (002) with respect to substrate temperature. It shows that the diffraction angle of ZnO (002) shifted from $2\theta=34.27$ to 34.40° as substrate increased from 25 to 250°C. This corresponds to the decrease of space distance of ZnO (002) at the increase of substrate temperature. The phenomena was due to the doping process of the $ZnO:AlF_3$ film, where partial replacement of Zn by Al and O by F occurred. This is due to the ion radius of Al^{3+} (0.53Å) and F^{-1}(1.33 Å) being less than that of Zn^{2+}

(0.74 Å) and O^{-2} (1.40 Å), respectively, therefore resulting in a decrease of the average space charge distance and change in the lattice constant. The corresponding change in the lattice led to the observed angle shift toward higher diffraction angle[7].

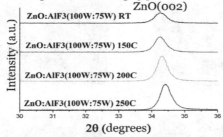

Figure. 1 The XRD spectrums of the ZnO:AlF₃ films with different substrate temperature.

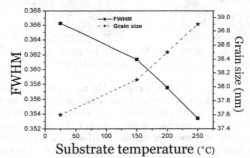

Figure. 2 The FWHM and grain size of ZnO (002) of ZnO:AlF₃ films at different substrate temperature.

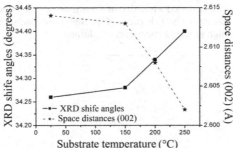

Figure. 3 The plot of angle shift and space distance of ZnO (002) vs. substrate temperature.

The surface morphology of the ZnO:AlF₃ films that were grown under various substrate temperature are shown in Figs 4 (a)~(d). The film shows a denser surface and larger grains at the sample deposited at higher substrate temperature. At higher substrate temperatures the particles gained more energy to diffuse producing a denser surface structure and larger grains growth.

(a)   (b)

(c)   (d)

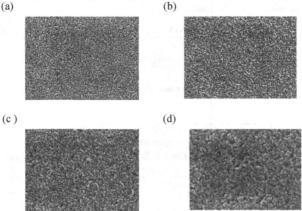

Figure. 4 The surface morphology of the ZnO:AlF$_3$ thin films under different substrate temperature: (a)room temperature, (b)150˚C, (c)200˚C, and (d) 250˚C.

Figure 5 shows the EDS analysis of the samples. The atomic percentage of F was 1.68 for the sample deposited at room temperature. As the substrate temperature was increased the percentage of F decreased in the ZnO:AlF$_3$ film. No F was detected in the film deposited at the substrate temperature above 200˚C due to the volatilization of F at the high temperature condition [8]. On the other hand, the atomic percentage of Al was increased in the film as the substrate temperature was increased. This gives evidence, supporting the XRD analysis, that the decrease of the space distance of ZnO (002) as the substrate temperature was increased primarily came from the substitutional effect of Al at the Zn sites. For the elements of Zn and O, it shows the atomic percentage of O was increased while that of Zn was decreased in the films as the substrate temperature was increased. This is because of the high vapor pressure of Zn, causing some Zn to evaporate at the high process temperature conditions.

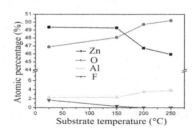

Figure. 5 The EDS analysis of the ZnO:AlF$_3$ films with respect to substrate temperature.

The carrier concentrations, carrier mobility and resistivity of the films with respect to various substrate temperatures are shown in Fig. 6 and Fig. 7, respectively. The carrier concentration was 4.2×10^{20} cm^{-3} in the room temperature prepared ZnO:AlF$_3$ film. As the substrate temperature was increased, the carrier concentration of the film decreased. This is because the evaporation of F dopant from the film at higher substrate temperatures, as discussed in the EDS analysis. The carrier concentration reached its minimum value at the temperature of 200°C. With a further increase of the substrate temperature, the carrier concentration of the film increased. EDS measurements showed the atomic percentage of Zn decreased from a deposition temperature of 150°C relative to 250°C, while that of Al increased. This indicates that the increase of carrier concentration at the higher temperature region was due to the doping effect of Al to the ZnO film. The carrier mobility of the film increased with temperature. The resistivity of the room temperature processed film was 3.60×10^{-2} Ω-cm, as the process temperature was increased to 200°C, the corresponding resistivity was increased to 6.0×10^{-2} Ω-cm. The resistivity decreased with a further increase in temperature, due to the increased incorporation of Al into the ZnO:AlF$_3$ film[9,10].

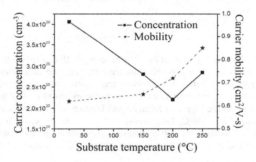

Figure. 6 The carrier concentration and mobility of the ZnO:AlF$_3$ films with respect to various substrate temperature.

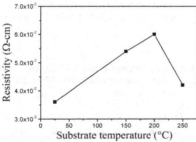

Figure. 7 The resistivity of the ZnO:AlF$_3$ films with respect to various substrate temperature.

Figure 8 shows the transmittance of the various substrate-temperature deposited ZnO:AlF$_3$ films with respect to wavelength of incidence. The average transmittance of the films was 90%. It showed no obvious variation of the optical characteristics at the ZnO:AlF$_3$ films that prepared at different substrate temperature.

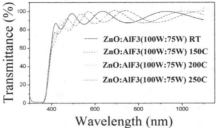

Wavelength (nm)

Figure. 8 The transmittance of the various substrate temperature deposited ZnO:AlF$_3$ films with respect to wavelength of incidence

CONCLUSION

The ZnO:AlF$_3$ thin films were prepared by RF sputtering with a constant power, ZnO(100 W) and AlF$_3$(75 W) for one hour at various substrate temperature. The films showed improved crystal quality and larger grain growth as the substrate temperature was increased. The carrier mobility also increased with substrate temperature. Due to the evaporation of F at high temperature, the doping of F to the film was reduced and led to a reduction of carrier concentration in the region of 25~200°C. The corresponding resistivity was increased from 3.60×10^{-2} to 6.0×10^{-2} Ω-cm. While for a further increase of substrate temperature, the doping of Al to the ZnO film was increased and resulted in the increase of carrier concentration.

ACKNOWLEDGEMENTS

The authors would like to give thank to National Science Council (NSC) for their financial support of this work under the project of contract no. 101-2918-I-168-001.

REFERENCES

1. S. M. Hyun, K. Hong, B. H. Kim, J. Korean Ceram. Soc. **33**, 149–154 (1996).
2. J.H. Lee, K.H. Ko, B.O. Park, J. Cryst. Growth **247**, 119–125 (2003).
3. R.J. Hong, X. Jiang, B. Szyszka, V. Sittinger, A. Pflug, Appl. Surf. Sci. **207**, 341–350(2003).
4. M.L. Olvera, A. Maldonado, R. Asomoza, Sol Energy Mater Sol Cells **73**, 425–433 (2002).
5. B. G. Choi, I. H. Kim, D. H. Kim, K. S. Lee, T. S. Lee, B. Cheong, J. Eur. Ceram. Soc. **25**, 2161–2165 (2005).
6. K. Ellmer, K. Diesner, R. Wendt, S. Fiechter, Solid State Phenomena, **51-52**, 541 (1996).
7. W.F. Wu, B.S. Chiou, S.T. Hsien, Semicond. Sci Technol. **9**, 1242 (1994).
8. K. Kawamata , T. Shouzu, N. Mitamura, Vacuum, **51**, 559-564 (1998).
9. G. Fang, D. Li, Vacuum **68**, 363-372 (2003).

10. Y.M. Hu , C.W. Lin, Thin Solid Films **497**, 130-134 (2006).

Mater. Res. Soc. Symp. Proc. Vol. 1494 © 2013 Materials Research Society
DOI: 10.1557/opl.2013.373

Preparation of ZnO:Mo Thin Films by RF Sputtering

Tien-Chai Lin[1], Shang-Chou Chang[1], Wen-Chang Huang*[2] and Wen-Feng Huang [1]
[1]Department of Electrical Engineering, Kun Shan University, No. 949, Da Wan Road, Yung-Kang District, Tainan, 71003, Taiwan, ROC
[2]Department of Electro-Optical Engineering, Kun Shan University, No. 949, Da Wan Road, Yung-Kang District, Tainan, 71003, Taiwan, ROC
*Corresponding author: email: wchuang@mail.ksu.edu.tw

ABSTRACT

Based on the electron configurations of Mo and Zn, the valence electron difference between Mo^{6+} and Zn^{2+} is 4. Therefore, a small amount of Mo doping can produce sufficient free carriers to reduce the ion scattering effects. The Mo doped ZnO (MZO) thin film prepared by RF sputtering was studied in this research. Structural, electrical, and optical characteristics of the films were discussed. The MZO film shows a resistivity of 1.1×10^{-2} Ω·cm, a carrier concentration of 2.2×10^{21} cm^{-3}, a mobility of 0.63 cm^2/V·s, and average transparency of 81.0% at both the powers of 20 W to the Mo target and of 125 W to the ZnO target. The MZO film becomes a stable p-type semiconductor at high power process toward Mo target. The film preserves its p-type characteristics after exposure to air for one and a half months. The crystal structure of the p-ZnO films is amorphous with an average transparency of 34.5%.

INTRODUCTION

Zinc oxide (ZnO) is a wide bandgap semiconductor material with attractive properties such as high piezoelectric constant, high optical transparency in the range of 0.4-2 μm optical wavelength, and large exciton binding energy [1]. In general, for the development of ZnO-based optical devices, it is necessary to obtain n-type and p-type ZnO films. Owing to the low solubility of the dopants and the high self-compensating processes due to doping, it is difficult to obtain good quality and reproducible ZnO p-type materials [2]. ZnO occurs naturally as an n-type semiconductor owing to its intrinsic donor defects (e.g., zinc interstitials or oxygen vacancies), which can be enhanced by B, Al, Ga, or In donor dopants [3-6]. The low solubility and ionic scattering effects [7] of the dopant elements, B, Al, Ga, or In, limit the production of high free-carrier concentrations in ZnO. The carrier concentrations obtained from the above scheme are in the range of $1.5 - 2.0 \times 10^{20}$ cm^{-3}. However, the element Mo is also a donor element in the doping process of ZnO films. Based on the electron configurations of Mo and Zn, the valence electron difference between Mo^{6+} and Zn^{2+} is 4. Therefore, a small amount of Mo doping can produce sufficient free carriers to reduce the ion scattering effects [7]. In addition, Mo is a high thermally stable metal, ZnO based transparent conductive oxide (TCO) thin films show great potential in numerous applications as it was doped with Mo. Only few reports of Mo doped ZnO are found in the literature. Xiu et al. [8] reported the MZO films which were deposited onto glass substrates by radio frequency (RF) magnetron sputtering system using a sintered oxide target ZnO/MoO$_3$ and obtained a MZO film with a low resistivity 9.2×10^{-4} Ω cm, and its visible light transmittance exceeds 84%. Lin et al. [9] prepared a MZO thin film on a glass substrate using pulsed DC magnetron sputtering. The source target combined Mo metallic pieces on a ZnO target material. A MZO thin film with a low resistivity of approximately 8.9×10^{-4} Ω cm and a visible light transitivity of 80% was reported.

In our research, the MZO films on glass substrates were prepared by RF magnetron co-sputtering equipped with both a ZnO target and a Mo target. The RF power applied to the ZnO target was constant at 125 W while the power applied to the Mo target was varied. The structural, electrical, and optical effects of the MZO films are discussed in this work.

EXPERIMENTAL DETAILS

The MZO thin films were prepared by RF magnetron co-sputtering equipped with both a ZnO target (99.9%, 7.62 cm) and a Mo target (99.99%, 7.62 cm). Glass substrates were used, and they were cleaned with a standard cleaning procedure. The MZO thin films were prepared under Ar (99.99 %) as an ambient gas and its flow rate was 3 sccm. The base pressure was under 8×10^{-3} Pa, and the working pressure was maintained at 0.67 Pa. In order to investigate the effects of the RF power of the Mo target on the properties of ZnO films, the RF power applied to the ZnO target was fixed at 125 W, and the RF power of the Mo target varied from 0 to 40 W. The substrate temperature was constant at 100 °C, and the deposition time was 45 min. The deposition rate and thickness of the MZO thin films were measured using an alpha-step profilometer. The phase and crystallinity of the films were investigated by X-ray diffraction (XRD) using a Ni-filtered Cu Kα source. Film resistivity, carrier concentration, and mobility were measured by Hall-effect measurements. The atomic percentages of each element of the MZO film were evaluated by Energy Dispersive Spectroscopy (EDS).

DISCUSSION

Figure 1 shows the XRD results of the MZO thin films. The RF power applied to the ZnO target was constant at 125 W, and the RF powers applied to the Mo target were varied from 0 to 40 W. The films showed obvious ZnO (002) peaks of preferred orientation when the power applied to the Mo target was in the range of 0 to 30 W. This indicates that the ZnO thin films were in a preferential c-axis orientation, due to the lowest surface free energy of the ZnO(002) plane. At the same time the peak of Mo was not observed in the spectrum. This is because some of the Zn atoms in the ZnO lattice were replaced by Mo atoms. The diffraction peak of ZnO disappeared as the RF power of Mo was increased in the range of 35~40 W. The film crystallinity became worse at a higher (>35 W) deposition power of Mo target. The collision from higher energy plasma spaces destroyed the structure, leading to the possible deterioration of the microstructure of the thin film. Similar results were also discussed by Malinovska et al. [10] on a RF sputtered AZO film. Lin et al. [9] prepared MZO thin film by sputtering, showing the crystallinity of the MZO film was degraded as the sputtering power was too high. These results match the experimental result of this experiment.

The peaks of ZnO (002) were shifted as the power of Mo was varied in the range of 20–30 W. The peaks began to shift toward higher diffraction angles and then shifted toward lower diffraction angles as the power was increased. The positive angle shifting was due to the fact that the ion radius of Mo (Mo^{3+} 0.069 nm, Mo^{4+} 0.065 nm, Mo^{5+} 0.061 nm and Mo^{6+} 0.059 nm) is less than that of Zn (0.074 nm). Consequently, the space distance of the lattice decreased as the substitution of Mo into the ZnO lattice. According to Bragg's law [11], this substitution should shift the diffraction peak toward a higher diffraction angle. However, as the RF power increased in the range of 25–30 W, the peak diffraction angle shifted toward a lower angle, which is inconsistent with Bragg's law. This is due to the excess doping of Mo in the ZnO lattice. The excess Mo sited at the interstitials and resulted in an increase of lattice plane spacing. The lattice constants increased from 0.525 to 0.542 nm as the RF powers toward Mo target increased from 20 to 30 W, respectively. Thus, the diffraction peak shifted toward a lower diffraction angle.

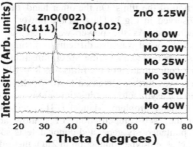

Figure. 1 XRD spectrum of MZO films.

The atomic ratio of the MZO films with respect to the RF power of the Mo target is shown in Fig. 2. Results showed that the atomic ratio of Mo increased from 0.3% to 7.8% as the RF power of the Mo target increased from 20 to 40 W. This means that the incorporation of Mo into the ZnO film was dependent on the RF power of the Mo target. The atomic ratio of Zn increased from 10.5% to 36.4% as the RF power of the Mo target increased from 20 to 25 W. A better crystallinity of the MZO film at the power of 25 W was observed in the XRD analysis. As the RF power further increased, the atomic ratio of Zn was decreased. This is due to the increase of the atomic ratio of Mo element. The high atomic ratio of oxygen at low power (20 W) came from the signal of the glass substrate.

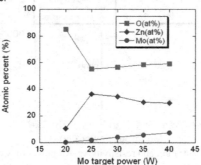

Figure. 2 Atomic ratio of MZO films with respect to RF power of Mo target.

Both Figs. 3 and 4 show the plot of carrier concentration, resistivity and mobility of the MZO films as a function of the RF power of the Mo target. A very high carrier concentration of 2.5×10^{21} cm^{-3} was obtained at the MZO film with a power of 20 W. The corresponding resistivity was 1.1×10^{-2} $\Omega \cdot$cm, and the electron mobility was 0.63 cm^2/V·s. The carrier concentration of the MZO films decreased as the power increased from 20 to 30 W. It was found that the electrons recombined, and that the electron concentration decreased as the power increased. This may be due to the low solubility of Mo in the ZnO film. The excess Mo did not increase the electron concentration of the films. On the other hand, the excess Mo provide hole carriers to the ZnO film as observed in Fig.3. Thus, as the power was increased, the carrier concentration of electrons decreased. As the power was increased to 30 W, the carrier concentration was not only too low, but it was also difficult to distinguish between the types of conductivity. As the power was increased to 35 W, the MZO films transformed from n-type to p-type semiconductors. The carrier concentration of the films was 8.5×10^{13} cm^{-3}, and the resistivity was 300 $\Omega \cdot$cm. When the power was further increased to 40 W, the carrier concentration only showed a slight increase. Due to the asymmetric doping limitations in ZnO [12], the lack of good and reliable p-ZnO has been a major problem. Mandel [13] suggested that the resistance of forming p-ZnO is because of the "self-compensation" of shallow acceptors resulting from various naturally occurring or spontaneously generated donor defects such as oxygen vacancies or zinc interstitials. In order to achieve p-type conduction in ZnO, these donor-like native defects have to be suppressed or even eliminated. Here, the formation of p-MZO was caused by the over-doping effect of Mo into ZnO. The excess Mo twisted the lattice and transformed it into an amorphous structure. For n-type semiconductors, the transport carriers are electrons, and the electrons came from the oxygen vacancy of MZO. In the case of the Mo over-doped MZO films, some of the Zn-O binding was replaced by the metallic Mo-Zn binding. This resulted in a change of the original crystal structure of MZO. The hole carriers mainly came from the metallic vacancy of Mo-O and/or Zn-O bindings, and as the power was increased up to 35 W, the MZO films showed p-type semiconductor characteristics.

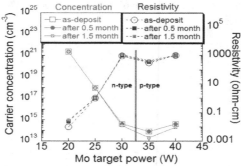

Figure. 3 Plots of carrier concentration and resistivity vs. RF power of Mo target of MZO films

It is known that p-ZnO is an unstable film [14]. The p-ZnO films reacted with oxygen in the air, and they were transformed into n-type thin films. Figs. 3 and 4 also show the test of stability of the MZO films that were exposed to air for a period of time. Results showed that the carrier concentration and resistivity did not exhibit an obvious change after half a month of exposure to air. The electrical characteristics of the samples with applied powers of 20 W and 25 W were unchanged even after one and a half months of testing. The sample with an applied power of 30 W was transformed from an uncertain type to an n-type conductor after exposure to air for one and a half month. This is attributed to the presence of donor-like Zn interstitial defects in MZO, and these defects make the MZO films self-compensating, thereby transforming them into n-type conductors [15]. For the samples with applied powers of 35 W and 40 W, the MZO films still maintained their p-type characteristics, and did not show obvious self-compensating effects. This is due to the formation of the stable Zn-Mo binding. They showed a slight decrease of carrier concentration, and a slightly increase of resistivity after one and a half months of exposure to air. Therefore, they were very stable p-type ZnO thin films.

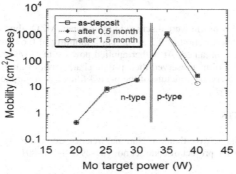

Figure. 4 Plots of mobility vs. RF power of Mo target of MZO films

Figure 5 shows the transparency of the MZO films obtained at various Mo RF powers. The MZO film of 20 W RF power shows cutoff wavelength of 380 nm which is due to good crystallinity of MZO films with optical bandgap like ZnO film.

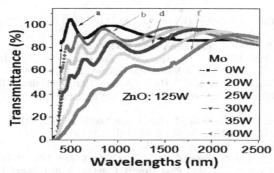

Figure. 5 Transparency of MZO films obtained at various Mo target RF powers.

As the RF power increased the carrier concentration decreased. The absorption edge red shifted slightly in the range of 20-30 W, and more obviously at 35-40W. The transparency of the MZO films in visible region decreased with increasing power. Because the atomic radius of Mo is different from that of Zn, the doping of Mo produced both lattice defects and twisting effects. These effects resulted in a decrease in transparency. The transparency for short wave lengths decreased more than it did for longer wave lengths. A possible reason may be that the excess Mo creates energy levels within the band gap of ZnO. These energy levels might be near the band edge so the absorption effect is more obvious in the shorter wavelength region. The average transparency (in the range of 400–800 nm) decreased from 81.8% to 26.19% as the power increased from 0 to 40 W.

CONCLUSION

ZnO thin films doped with Mo were discussed in this paper. MZO films with a resistivity of $1.1 \times 10^{-2} \,\Omega\cdot cm$, carrier concentration of $2.2 \times 10^{21} \, cm^{-3}$, mobility of 0.63 $cm^2/V\cdot s$, and average transparency of 81.0% were obtained. At higher applied powers, the MZO films became stable p-type semiconductors. The films preserved their p-type characteristics after exposure to air for one and a half months. The crystal structure of the p-ZnO films was amorphous with an average transparency of 34.5%.

ACKNOWLEDGEMENTS

The authors would like to give thank to National Science Council (NSC) for their financial support of this work under the project of contract no. 101-2918-I-168-001.

REFERENCES

[1] T. Minami, Mater. Res. Bull. **25,** 38 (2000).
[2] A. Kobayashi, O.F. Sankey, J.D. Dow, Phys. Rev. **B 28**, 946 (1983) 946.
[3] B.J. Lokhande, P.S. Patil, M.D. Uplane, Phys. **B 59**, 306 (2001).
[4] H. Kim, C.M. Gilmore, J.S. Horwitz, A. Pique, H. Murata, G.P. Kushto, Appl. Phys. Lett. **76**, 259 (2000).
[5] D.J. Kwak, K. Park, B.S. Kim, S.H. Lee, S.J. Lee, D.G. Lim, J. Korean Phys. Soc. **45**, 206 (2004).
[6] J.Y. Hwang, C.R. Cho, J. Korean Phys. Soc. **47**, S288 (2005).
[7] H.K. Kim, S.H. Huh, J.W. Park, J.W. Jeong, G.H. Lee, Chem. Phys. Lett. **354**, 165 (2002).
[8] X. Xiu, Y. Cao, Z. Pang, S. Han, J. Mater. Sci. Technol., **25**, 785 (2009).
[9] Y.C. Lin, B.L. Wang, W.T. Yen, C.T. Ha, Chris Peng, Thin Solid Films **518**, 4928 (2010).
[10] D. Dimova-Malinovska, N. Tzenov, M. Tzolov, L. Vassilev, Mater. Sci. Eng. **B52**,59 (1998).
[11] Y. Sun, G. Guo, D. Tao, Z. Wang, J. Phys. Chem. Solids, **68**, 373(2007).
[12] T. Yamamoto, H.K. Yoshida, Physica **B 302/303**, 155 (2001).
[13] G. Mandel, Phys. Rev. **A 134**, 1037(1964).
[14] C. Wang, Z. Ji, K. Liu, Y. Xiang, Z. Ye, J. Cryst. Growth **259**, 279 (2003).
[15] C.H. Park, S.B. Zhang, S.H. Wei, Phys. Rev. B: Condens. Matter. Mater. Phys. **66**, 073202(2002).

Mater. Res. Soc. Symp. Proc. Vol. 1494 © 2013 Materials Research Society
DOI: 10.1557/opl.2013.178

The Electrical Characteristics of ZnO :Ga/p-Si Junction Diode

Tien-Chai Lin[2], Wen-Chang Huang[1]*, Chia-Tsung Horng[1], Shu-Hui Yang[1]
[1]Department of Electro-Optical Engineering, Kun Shan University, No. 949, Da-Wan Road, Yong-Kang district, Tainan, 71003, Taiwan, ROC
[2]Department of Electrical Engineering, Kun Shan University, No. 949, Da-Wan Road, Yong-Kang district, Tainan, 71003, Taiwan, ROC
*Corresponding author: e-mail:wchuang@mail.ksu.edu.tw

ABSTRACT

The junction characteristics between ZnO:Ga (GZO) film and p-Si substrate are discussed in the research. For the transparent semiconductor ZnO, the element Ga is chosen to be the dopant source to produce a high quality n-type ZnO thin film. The ZnO:Ga (GZO) film shows a average transmittance is 84.7% (above 400 nm), a bandgap energy of 3.37 eV, a carrier concentration of 7.29×10^{13} cm^{-3} and a resistivity of 118 Ω-cm. For the GZO/p-Si junction, it shows a junction barrier height of 0.54 eV with an ideality factor of 1.24. The capacitance-voltage measurement shows that it has a uniform reverse bias depletion layer. The Cheung function is also brought to discussion the diode characteristics.

INTRODUCTION

Zinc oxide (ZnO) is a wide band gap semiconductor exhibiting many interesting properties making it promising for electro-optical devices [1]. ZnO is a direct band gap material with a bandgap of 3.37 eV and shows more resistance to radiation damage than Si and GaN. Moreover, ZnO has a high exciton binding energy (about 60 meV) making the excitons thermally stable at room temperature. It is an n-type semiconductor due to the non-stoichiometry existing in the excess of Zn [2].While, ZnO can be doped with the appropriate metal atoms, such as Al, Sn, Cd, Ga, In, etc. to increase its conductivity. The Ga element was chose to be the dopant source to produce highly conductive as well as high quality film.

The structure of n-ZnO/p-Si junction is based on the application of the n-ZnO film as a practical antireflecting photon-window for Si photodiodes. The thin film, ZnO also is used to be a semiconductor layer and produce a built-in potential barrier at the n-ZnO/p-Si junction interface. Because of the wide bandgap of ZnO, low energy photons in the visible range may be collected mainly at the depletion region of the p-Si after they are transmitted through the ZnO. It is interesting that such a simple photodiode structure as the n-ZnO/p-Si has not been widely reported except by a small number of studies to date [3-8]. In order to observe a good photoelectric effects from the photodiode, a good pn junction interface quality for a low leakage current and a good film quality for the light transmission and carrier transport is very important.

In the present work, we fabricated the n-ZnO/p-Si heterojunction. The Ga element was chose to be the dopant source to produce highly conductive as well as high quality film for ZnO. The junction reliability such as barrier height and ideality factor was discussed. The temperature dependent current-voltage (I-V-T) measurement was used to evaluate the carrier transport mechanism and barrier height homogeneity.

EXPERIMENTAL DETAILS

The heterojunction structure GZO/p-Si was discussed in the paper. A p-type (5 Ω-cm) Si (100) wafer was used to be the substrate for the diodes. Both front and backside use metal aluminum to be the ohmic contact metals. For backside substrate ohmic contact, a 2000Å thickness of Al was evaporated on the backside of silicon substrate through thermal evaporation system. Then, the wafer was annealed in a furnace at 450 °C for 5 min to form ohmic contact. Prior to the deposition of GZO, the samples were ultrasonically cleaned with acetone, methanol, and de-ionized water for 10 min. Then, the wafers were dipped for 30 sec into buffered oxide etchant (BOE) to remove native oxides. Finally, the wafers were blown dry with nitrogen. The metal mask was used to define the pattern of the diode. GZO films were deposited in a RF magnetron sputtering system using a zinc oxide target with 2 inches diameter. The chamber was evacuated to a base pressure of $1=10^{-6}$ Torr. The ZnO films were deposited on Si substrates at a substrate temperature of 300 °C with a working pressure of 10 mTorr. The chosen RF power and the deposition time were 120 W and 1 h, respectively. The thickness of the GZO film was 3000Å. The metal, Al was then evaporated through the metal mask to be the contact electrode of GZO and its thickness was 2000 Å. The diode was then annealed in a rapid thermal annealing (RTA) system at 600 °C for 50 sec to obtain a high quality junction diode. The current-voltage characteristic (I-V) of the diode was performed at a semiconductor parameter analyzer, HP-4145B. The film quality of GZO was also evaluated.

DISCUSSION

The X-ray diffraction pattern of the GZO film on the silicon substrate is shown in Fig. 1. The diffraction angle, 2θ, in the region between 20 to 50° has been applied to observe the phases of the deposited film. The XRD results indicate that the GZO film has a strong ZnO (002) peaks of preferred orientation, indicating that GZO thin films is preferential c-axis orientation, due to the lowest surface free energy of ZnO (002) plane. The peak centered at 34.58 ° and the full width at half maximum (FWHM) of the ZnO (002) is 0.34 °. From the Scherrer equation $L_{hkl}=$ $0.89\lambda/B \cos\theta$ (L_{hkl}: size of crystal grain vertical to diffraction surface, λ: wavelength of X-ray, B: FWHM, θ: Bragg angle), the average grain size of the ZnO (002) is estimated to be 24.5 nm. The space distance is 0.52 nm.

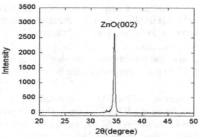

Figure. 1 XRD spectrum of the GZO thin films prepared on silicon substrate.

Fig. 2 shows the transmittance spectra of the GZO film measured at room temperature by a conventional ultraviolet–visible spectrometer. The GZO film shows a high transmittance (84.7%) in the visible region. The transmittance spectra show a strong decrease after 400 nm. This decrease is related to the optical transitions occurring in the optical band gap. We figured out that the GZO film have a band gap of 3.37 eV. The electrical characteristics of GZO film are evaluated through Hall measurement. The GZO film shows a resistivity of 118 Ω-cm, a carrier concentration of 7.29×10^{13} cm^{-3} and a carrier mobility of 7.26 cm^2/V-s.

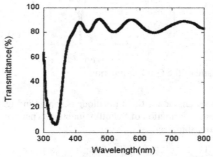

Figure. 2 The transmittance spectra of the GZO film measured at room temperature by a conventional ultraviolet–visible spectrometer.

The current–voltage characteristics of ZnO/p-Si junctions could be described by the thermionic emission model. According to thermionic emission theory, the current in such a device can be expressed as

$$I = I_s \left[\exp\left(\frac{q(V - IR_s)}{nkT} \right) - 1 \right] \qquad (1)$$

where $I_s = AA^*T^2 \exp[-q\phi_b / kT]$ (2)

where R_s is the series resistance of the diode, V is the applied voltage, q is the electronic charge, k is the Boltzmann constant, T is the absolute temperature, A is the diode contact area, A^* is the effective Richardson constant, ϕ_b is the barrier height at zero bias, and n is the ideality factor. Theoretical $A*$ value of 32 Acm^{-2}K^{-2} is used for Si. The saturation current density J_s was obtained by extrapolating the linear region of the forward-bias semi-log $I-V$ curves to the zero applied voltage and the ϕ_b values were calculated from Eq. (2). The values of n were determined from the slope of the linear region of the forward bias semi-log $I-V$ characteristics using the relation: $n = q / kT [\partial V / \partial (\ln J)]$. Note that n depends on the current flow at the interface and is equal to 1 for an ideal diode.

Fig. 3 shows the I-V characteristics of the 600 °C-annealed ZnO/p-Si junction diode. The saturation current density J_s of the diode is 1.92×10^{-3} A-cm^{-2}. The effective junction barrier height, which is derived from the TE model, is 0.54 eV, and the ideality factor is 1.24. The barrier height and the ideality factor values of the diode is derived from the intercept and slopes of the forward-bias current at each temperature, respectively. The value of n is obtained by linear fitting of the forward bias region of the $I-V$ plot, as the series resistance effect is negligible in this region.

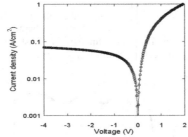

Figure 3. The I-V characteristic of the GZO/Si junction

The capacitance-voltage characteristic of the diode is shown in Fig. 4. The plots of $1/C^2$ vs. V is linear which indicates the formation of Schottky junction. Therefore, it follows a standard Mott–Schottky relationship:

$$\frac{1}{C^2} = \frac{2[V_{bi} - V - kT/q]}{q \varepsilon_o \varepsilon_s A^2 N_A} \quad (3)$$

where C is the diode capacitance, V_{bi} is the built in voltage, ε_s is the semiconductor dielectric constant, ε_o is the permittivity in vacuum, V is the applied voltage, q is the charge, A is the diode active area, kT/q is the thermal voltage at 300 K, and N_A is the charge carrier concentration. The value of V_p [$V_p = (kT/q)\ln(N_V/N_A)$] is the energy difference between the valence band and the Fermi level. The value of the ϕ_{cb} can be obtained by the relation: $q\phi_{cb} = V_{bi} + V_p$. The charge carrier concentration can be determined from the slope of $1/C^2$ vs. V plots. From the extrapolated intercept on voltage axis, V_{bi} can be estimated. Fig. 4 shows the $1/C^2$-V plot of the diode that was measured at the frequency of 100 K Hz. It shows that the built in voltage is $V_{bi} = 0.6$ V, the effective carrier concentration is 3.44×10^{16} cm^{-3}, the value of V_p is 0.17 V, and the effective barrier height is 0.77 eV. The effective carrier concentration (3.44×10^{16} cm^{-3}) of the silicon substrate was evaluated from the slope of the $1/C^2$-V plot. However from the resistivity (5 Ω-cm) of p-Si, the doping concentration is in the order of 10^{15} cm^{-3}. The difference of the carrier concentration may arise from the underestimation of GZO's carrier concentration (7.29×10^{13} cm^{-3}) in Hall effect measurement.

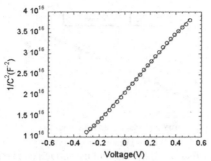

Figure 4. The C-V characteristic of the GZO/Si junction

The *I–V* characteristics of the metal/semiconductor contacts are linear and at a low forward bias voltage, but deviate considerably from linearity due to the some factors at large voltages of a semi-log scale I-V plot. The series resistance (R_s) is one of the factors of the deviation from the linearity. According to Cheung and Cheung [9,10], the forward biased *I–V* characteristics for a large applied voltage ($V>3kT/q$) from the TE model of a Schottky diode with a series resistance can be expressed as Eq. (1). The IR_s term indicates the voltage drops across the series resistance of the diode described in the equation. The values of the series resistance can be determined from the following equations:

$$\frac{dV}{d(\ln I)} = \frac{nkT}{q} + IR_s \qquad (4)$$

$$H(I) = V - \left(\frac{nkT}{q}\right)\ln\left(\frac{I}{AA^*T^2}\right) \qquad (5)$$

where *H(I)* is given by

$$H(I) = n\phi_b + IR_s \qquad (6)$$

The series resistance can be found from the slope of the *dV/d(lnI)* vs. *J* plots, and nkT/q is obtained at the interception of the y-axis, as indicated by Eq.(4). Fig. 5 shows the relations between *dV/d(lnJ)* vs. *J* of the GZO/Si junction diode. The values of *n* and R_s were calculated, yielding values of $n = 2.29$ and $R_s =351$ Ω. The value of ideality is slightly larger than that of the forward-bias *ln I* vs. *V* plot. Thus, it can clearly be seen that there is relatively difference between the values of n obtained from the downward curvature regions of forward bias lnI vs.V plots (Cheung's function) and from the linear regions of the same characteristics. The reason of this difference can be attributed to the existence of the series resistance and the bias dependence barrier heights according to the voltage drop across the interfacial layer and change of the interface states with bias in the concave region of the current–voltage characteristics. The plot of *H(J)* vs. *J* of the GZO/Si junction diode is also shown in Fig. 5. Using the value of the *n* obtained from Eq. (4), the value of the effective barrier height is obtained from the y-axis intercept. The value of the Schottky barrier height of the diode is 0.54 eV. The series resistance obtained by the Cheung's function is 360 Ω at the diode.

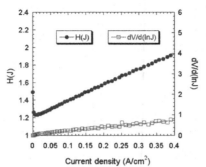

Figure 5. The plots of *dV/d(lnJ)* vs. *J* and *H(J)* vs. *J* of the GZO/Si junction diode.

CONCLUSION

In the present work, we fabricated the n-ZnO/p-Si hetero-junction. Gallium was chosen to be the dopant source to produce highly conductive as well as high quality film for ZnO. For the GZO/p-Si junction, it shows a junction barrier height of 0.54 eV with an ideality factor of 1.24. The capacitance-voltage measurement shows it has a uniform reverse bias depletion layer. The series resistance of the diode was found to be 351 Ω. And a consist effective barrier of 0.54 eV was also found by Cheung's function.

ACKNOWLEDGEMENTS
The authors wish to thank the Kun Shan University for their financial support of this work under contract 100-N-274-KSU-M-005.

REFERENCES
[1] C. Jagadish, S.J. Pearton, Zinc Oxide Bulk, Thin Films and Nanostructures, Elsevier, Amsterdam, 2006.
[2] T. Okamura, Y. Seki, S. Nagakari, H. Okushi, Japan. J. Appl. Phys. 31 (1992) L762.
[3] J.Y. Lee, Y.S. Choi, W.H. Choi, H.W. Yeom, Y.K. Yoon, J.H. Kim, S. Im, Thin Solid Films **420/421**,112 (2002) 112
[4] C.H. Park, J.Y. Lee, S. Im, T.G. Kim, *Nucl. Instru. Meth.* **B206**, 432 (2003).
[5] L. Luo,Y. Zhang, S.S. Mao, L. Lin, Sensors and Actuators A 2006;127:201–6.
[6] H.Y. Kim, J.H. Kim, M.O. Park, S. Im, Thin Solid Films, **93**, 398(2001).
[7] S.J. Young, L.W. Ji, R.W. Chuang, S.J. Chang, X.L. Du, Semi. Sci. Tech. **21**,1507(2006).
[8] L.C. Chen, C.N. Pan, The Open Crystallography Journal, **1**, 10(2008).
[9] A. Turut, M. Saglam, H. Efeoglu, N. Yalcin, M. Yildirim, B. Abay, Phys. B **205**, 41 (1995).
[10] S.K. Cheung, N.W. Cheung, Appl. Phys. Lett. **49**, 85 (1986).

Mater. Res. Soc. Symp. Proc. Vol. 1494 © 2012 Materials Research Society
DOI: 10.1557/opl.2012.1649

Oxidation state of tungsten oxide thin films used as gate dielectric for zinc oxide based transistors

Michael Lorenz[1], Marius Grundmann[1], Sandra Wickert[2] and Reinhard Denecke[2]

[1]Institut für experimentelle Physik II, Fakultät für Physik und Geowissenschaften, Universität Leipzig, Linnéstrasse 5, 04103 Leipzig, Germany

[2]Wilhelm-Ostwald-Institut für Physikalische und Theoretische Chemie, Fakultät für Chemie und Mineralogie, Universität Leipzig, Linnéstrasse 2, 04103 Leipzig, Germany

ABSTRACT

We present an investigation of the degree of oxidization of tungsten oxide (WO_x) thin films used as gate dielectric for metal-insulator-semiconductor field-effect transistors (MISFET). By means of X-ray photoelectron spectroscopy WO_x thin films grown by pulsed-laser deposition at room temperature were investigated. The electrical and optical properties depend significantly on the oxygen pressure during deposition and are affected by the stoichiometric ratio of oxygen and tungsten.

INTRODUCTION

Tungsten trioxide (WO_3) thin films have recently been demonstrated as high-κ gate dielectric for highly transparent and temperature stable thin-film transistors based on zinc oxide channel material [1,2]. WO_x can also be used as an n-type semiconductor [3] or electrochromic material [4]. However the optical and electrical properties of the WO_x thin films depend significantly on the oxygen pressure during deposition [1,5,6]. For pressures $p(O_2) > 0.1$ mbar during PLD at room temperature highly insulating tungsten oxide is obtained, whereas $p(O_2) < 0.03$ mbar gives rise to metallic like properties of the resulting thin film samples due to oxygen vacancies resulting in a high electron concentration. As gate dielectric, a highly insulating WO_x layer with a stoichiometric ratio close to $x = 3$ is necessary.

EXPERIMENT

Tungsten oxide thin films have been prepared by pulsed-laser deposition (PLD) on a-oriented sapphire substrates using a 248 nm KrF excimer laser. The substrate was kept at room temperature. The ablation target was pressed from 99.9% WO_3 powder and sintered for several hours to obtain a ceramic target. Two samples, deposited at an oxygen pressure of $p(O_2) = 0.002$ mbar (sample A) and $p(O_2) = 0.2$ mbar (sample B) were grown (thickness about 250 nm). To investigate the effect of the PLD on the stoichiometric ratio O:W, a reference sample was prepared using WO_3 powder on carbon tape. By means of an ESCALAB 220iXL XPS/AES device with a spectral resolution below 1.5 eV, using Al K_α radiation (energy of 1486.6 eV) generated by an Al/Mg dual mode X-ray anode, X-ray photoelectron spectroscopy (XPS) was performed ex situ. Calibration of the spectrometer was performed according to ISO 15472. The quantification of the contribution of the different oxidized tungsten states of the samples was performed considering a product of the mean free path, the transmission factor and the ionization cross section.

DISCUSSION

Figs. 1(a) and (b) depict the relative intensity of the W4f and O1s signal, respectively. The peak positions of the raw data show a shift which is attributed to the charging of the thin film due to the insulating substrate. Accordingly, the data was calibrated with the energetic maximum position of the O1s peak from the reference sample with a binding energy E_B=530.5 eV which is in good agreement with the literature [5]. Also note the broadening of the high energetic flank of the O1s signal which is attributed to adsorbed H_2O and OH in the thin films [5].
The binding energy of the W4f core level of the WO_3 reference sample is E_B=35.6 eV and agrees well with literature values [7].
Comparing the XPS measurement of the W4f core level for each sample, a small shift of the peak position of the thin films compared to the WO_3 reference is observed. This shift is attributed to contributions from lower oxidized WO_2/$WO_{2.5}$ within the thin films. From the spectra depicted in Fig.1 (a) the relative contribution of the different oxidized states to the overall XPS signal were fitted to the data using the *UNIFIT* software using Voigt profiles (convolution of a Gaussian and Lorentzian profile). The results are depicted in Fig.1(c)-(e) for the reference sample and for the thin films A and B. For the reference sample, only a signal attributed to the W^{6+} core level is obtained, corresponding to a ratio O:W = 3:1, which was calculated by the ratio of the area below the O1s and W4f signal. For sample B, a signal which has been attributed to W^{5+} is needed for a satisfactory fit [5,7]. The contribution of the W^{5+} to the overall W4f signal is only 10%. The lower oxidized W^{4+} state was not detected. Since the films are highly insulating ($\rho > 10^9$ Ωcm) at this deposition pressure, it can be concluded that only a negligible lowering of the ratio O:W is introduced with x being close to 3. The sample A shows an increased W^{5+} signal and also a contribution from W^{4+} at E_B=32.4 eV which is due to WO_2 [5,8]. The contributions of W^{6+}:W^{5+}:W^{4+} to the overall W4f signal are 65%:17%:18%, respectively.The low oxygen pressure during deposition gives rise to a lowered ratio O:W. In consequence, due to oxygen vacancies the electron concentration is increased, significantly lowering the resistivity of the tungsten oxide with decreasing deposition pressure as reported in [1] ($\rho > 10^{-4}$ Ωcm for sample A). Stolze *et al.* reported a ratio of O:W < 2.5 for a similar spectrum as depicted in Fig.1(e). For O:W > 2.5 a much lower W^{4+} signal is reported, hence a ratio O:W well below 3 can safely be assumed here.

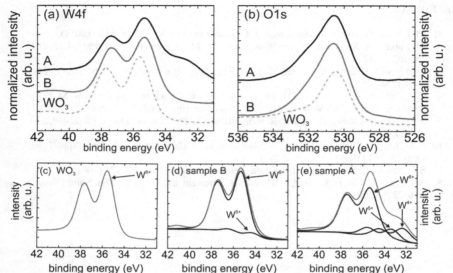

Figure 1: (a) W4f and (b) O1s XPS spectra. (c) XPS spectrum of W4f from the reference sample. For the reference, the fit equals the data which is due to the sole contribution of W^{6+}. XPS spectra in (d) and (e) depict the contributions due to W^{6+}, W^{5+} and W^{4+} state (black) fitted to the experimental spectra (grey) of samples A and B. Note the doublet splitting of the tungsten spectra due to $W4f_{7/2}$ and $W4f_{5/2}$.

CONCLUSIONS

Tungsten oxide thin films grown at different oxygen pressures by PLD at room temperature show a significant dependence of the stoichiometric ratio O:W. The reference sample, consisting of WO_3 powder only shows a signal attributed to W^{6+} without the presence of lower oxidized tungsten states. However, even at high growth pressures during PLD a small fraction of W^{5+} is incorporated within the thin film, probably originating from the dissociation of WO_3 due to the high energetic laser fluence or due to different bonding geometries (e.g. at the surface). Lowering the pressure gives rise to a significant contribution of WO_2 which decreases the overall stoichiometric ratio below x well below 3. This in turn explains the different optical and electrical properties of the WO_x thin films.

ACKNOWLEDGMENTS

We gratefully acknowledge financial support by Deutsche Forschungsgemeinschaft in the framework of Sonderforschungsbereich 762 "Functionality of Oxide Interfaces", M.L. was also supported by the Leipzig School of Natural Sciences "Building with Molecules and Nano-objects (BuildMoNa)" (GS 185).

REFERENCES

1. M. Lorenz, H. von Wenckstern and M. Grundmann, Adv. Mater. **23**, 5383 (2011).
2. M. Lorenz, A. Reinhardt, H. von Wenckstern and M. Grundmann, Appl. Phys. Lett. **101**, 183502 (2012)
3. J. Meyer, S. Hamwi, M. Kröger, W. Kowalsky, T. Riedl and A. Kahn, Adv. Mater. **24**, 5408 (2012)
4. C.G. Granqvist, A. Azens, A. Hjelm, L. Kullman, G.A. Niklasson, D. Rönnow, M. Strømme Mattson, M. Veszelei and G. Vaivars, Sol. Energy **63**, 199 (1998)
5. M. Stolze, B. Camin, F. Galbert, R. Reinholz and L. Thomas, Thin Solid Films **409**, 254 (2002)
6. S.-H. Lee, H. Cheong, C. Tracy, A. Mascarenhas, A. Czanderna and S. Deb, Appl. Phys. Lett. **75**, 154 (1999)
7. S.C. Moulzolf, S.-a. Ding and R.J. Lad, Sensor Actuat B-chem **77**, 375 (2001)
8. S. Penner, X. Liu, B. Klötzer, F. Klauser, B. Jenewein and E. Bertel, Thin Solid Films **516**, 2829 (2008)

Mater. Res. Soc. Symp. Proc. Vol. 1494 © 2013 Materials Research Society
DOI: 10.1557/opl.2013.260

Room Temperature Ferromagnetism and Band Gap Investigations in Mg Doped ZnO RF/DC Sputtered Films

Sreekanth K. Mahadeva[1,2], Zhi-Yong Quan[1,3], J. C. Fan[1], Hasan B Albargi[4], Gillian A Gehring[4], Anastasia Riazanova[1], L. Belova[1], K. V. Rao[1]

1. Department of Materials Science, Royal Institute of Technology, Stockholm, SE100 44 Sweden
2. Department of Physics, Amrita Vishwa Vidyapeetham University, Amritapuri Campus, Kollam 690 525, Kerala, India
3. Key Laboratory of Magnetic Molecules and Magnetic Information Materials of Ministry of Education, Shanxi Normal University, Linfen 041004, China
4. Department of Physics and Astronomy, University of Sheffield, Sheffield S3 7RH, UK

ABSTRACT

Mg@ZnO thin films were prepared by DC/RF magnetron co-sputtering in (N_2+O_2) ambient conditions using metallic Mg and Zn targets. We present a comprehensive study of the effects of film thickness, variation of O_2 content in the working gas and annealing temperature on the structural, optical and magnetic properties. The band gap energy of the films is found to increase from 4.1 to 4.24 eV with the increase of O_2 partial pressures from 5 to 20 % in the working gas. The films are found to be ferromagnetic at room temperature and the saturation magnetization increases initially with the film's thickness reaching a maximum value of 14.6 emu/cm^3 and then decreases to finally become diamagnetic beyond 95 nm thickness. Intrinsic strain seems to play an important role in the observed structural and magnetic properties of the Mg@ZnO films. On annealing, the as-obtained 'mostly amorphous' films in the temperature range 600 to 800°C become more crystalline and consequently the saturation magnetization values reduce.

INTRODUCTION

Dilute Magnetic semiconductors obtained by doping magnetic impurities into host semiconductors, mostly II - VI and III - V compounds, are the key materials for developing magneto-optic and spin electronics devices [1,2]. With wide direct band gap ($Eg = 3.37$ eV) and large exciton binding energy (~60 MeV) at room temperature (RT), ZnO thin films were predicted to be promising host materials to achieve room temperature ferromagnetism (RTFM) [3-5]. Extensive studies show that defects and non-magnetic impurities like Li, play an important role in inducing RTFM in ZnO [6, 7]. Various types of intrinsic and extrinsic defects in non-transition metal doped and un-doped ZnO have been attributed to give rise to RTFM. However, the experimental results on obtaining RTFM by different approaches have raised more questions because the properties of the materials obtained are highly dependent on the complexities of sample preparation and processing conditions. For example, P. Zhan *et al.* observed the FM of un-doped ZnO was induced by the singly occupied oxygen vacancies and provided a way to further enhance its ferromagnetic property [8]. Also, RTFM has been reported in both pristine and doped MgO films [9, 10]. The wide tunability of Eg in Mg incorporated ZnO films open the door for tailoring novel optoelectronic devices especially short wavelength light emitters and photo detectors. Tunability of Eg from 3.3 eV to 7.8 eV covers the UV regions [11]. Recently,

Li *et al.* reported that tailoring the *Eg* and engineering the defects were effective in tuning the RTFM in Mg@ZnO alloy [12].

We report a comprehensive study of RTFM in RF/DC sputtered thin films of Mg@ZnO. We find that the saturation magnetization (M_S) increases with film thickness up to a value of about 14.6 emu/cm^3 at about 52 nm exhibiting robust FM and then decreases with increasing film thickness finally exhibiting diamagnetic behavior characteristic of the bulk material above about 95 nm thick film. The experimental investigations show that intrinsic strain in the films deposited on silicon plays an important role in the observed RTFM. Furthermore, we find that the *Eg* and RTFM in Mg@ZnO thin films is tunable by varying the film thickness and oxygen partial pressure (P_{O_2}). On annealing the films in air the M_S values drops monotonically with increasing temperatures and above 800°C the films are no magnetic.

EXPERIMENTAL DETAILS

Mg@ZnO films were deposited on Si and glass substrates by co-sputtering pure Mg (99.99%) and Zn (99.99%) targets in the Leybold-Heraeus sputtering system with direct current (DC) power of 10W on Mg target and the radio frequency (RF) power of 50W on Zn target at RT. The vacuum chamber was evacuated to ~10^{-6} mbar. The film depositions on glass substrates were made at a fixed oxygen partial pressure $P_{O_2} = 1.5 \times 10^{-4}$ mbar in a mixture of N_2 and O_2 with total pressure of 1.5×10^{-3} mbar, for different deposition times. We also deposited films on Si and glass substrates for a fixed time of 60 min but varying P_{O_2} from 5 to 20%. We have also investigated the effects of thermal treatment of the film at 600, and 800°C in air for 1h for the sample deposited at 60 min and P_{O_2} 10%. The crystal structure of the films was characterized by X-ray diffraction (XRD). The surface morphologies and the thickness of the thin films were analyzed by using a SEM/FIB system. The RT optical absorption measurements were performed using a UV-visible-near infrared spectrophotometer. The magnetic properties of the films were measured at RT by means of SQUID magnetometer. In the evaluation of the magnetic raw data the diamagnetic response arising from the respective Si substrates was carefully accounted for.

RESULTS AND DISCUSSION

The cross-section of Mg@ZnO films deposited on Si substrate was analyzed by using Focused Ion Beam (FIB) technique in our Nova 600 Nanolab facility. Figure 1(a) shows the typical cross-section of Mg@ZnO films deposited for 1h in P_{O_2} 10% of the working gas. Similar uniform thickness of ~52 nm was found at different sections of the film. The deposition rate of the film was determined to be ~0.86 nm/min. From this value we calculated the film thickness for films deposited for different time scales to be ~26, 52, 65, 77 and 104 nm respectively. Figure 1(b), & (c) show the typical EDS spectrum of as-grown Mg@ZnO films deposited for 1h in P_{O_2} of 5 and 10% respectively. Besides Si from the substrate only the elements Mg, Zn and O were detected, indicating that there is no other possible form of metallic contamination in the films within the limits of detection. The inset of figure 1(b, & c) shows the contents of Mg, Zn and O. From EDS results, our analyses reveals that the content of Mg varies from ~10 to ~16 at.% with the increase of P_{O_2} from 5 to 10%. The Mg content slightly increases up to ~18 at.% in the film deposited in P_{O_2} of 20%.

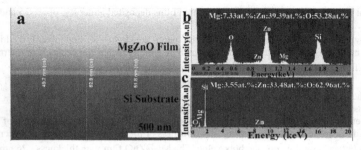

Figure 1. *(a) The typical FIB cross-section of the Mg@ZnO film deposited for 1h in P_{O_2} of 10%. The EDS spectrum of the Mg@ZnO film deposited for 1h (b) in P_{O_2} - 10% and (c) with P_{O_2} 5% resp.*

The XRD patterns of Mg@ZnO films deposited at different thicknesses are shown in figure 2. All of the Mg@ZnO films show the *c*-axis orientation of ZnO structure with (002) peak. Figure 3 shows the XRD result of Mg@ZnO films deposited at 1h for different P_{O_2}. A strong peak with 2θ value at 33.97 degrees corresponding to the (002) crystalline plane of ZnO is present in all the four XRD patterns. In addition to this peak, the film deposited in P_{O_2} content of 5% in the working gas exhibits the peaks at 2θ angles of 30.86 and 62.07degrees corresponding to ZnO planes (100) and (103), respectively, indicating the polycrystalline nature of the film.

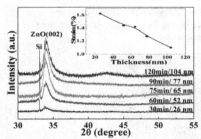

Figure 2. *XRD patterns of Mg@ZnO thin films deposited on Si for 30, 60, 75, 90 and 120 min respectively. The inset: the corresponding thickness dependence of the lattice strain along the c-axis.*

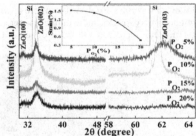

Figure 3. *XRD patterns of Mg@ZnO thin films deposited under different P_{O_2} on Si substrates. The inset: the effect of P_{O_2} on the lattice strain along the c-axis.*

The XRD spectra for the films with different thickness contain virtually only one peak of hcp ZnO (002). As the film thickness increases the intensity of the (002) peak increases indicating that the films with thickness higher than 26 nm are strongly textured. Besides thickness of the films, the partial oxygen pressure P_{O_2} in which the films are deposited has a significant effect on the film texture. In figure 3, the film deposited under P_{O_2} of 5% shows the presence of (100), (002) and (103) peaks of ZnO indicating the random orientation. That is, the sample has a weak texture comparable to other three films deposited under P_{O_2} of 10, 15 and 20%.

From figures 2 and 3, the ZnO (002) diffraction peaks are slightly shifted towards the higher angles with increasing film thickness and P_{O_2}, which essentially means the decrease of internal strain in the films [13]. We calculated the strain by using the following equation,

$$\varepsilon = (c\text{-}c_o)/c_o \times 100\% \qquad (1)$$

where c is the lattice parameter of the strained ZnO films calculated from XRD data, and c_o=5.207Å, is the value of the unstrained lattice parameter of ZnO [14].

The insets of figures 2 and 3 illustrate the variations of the c-axis strain as a function of thickness and P_{O_2} respectively. Increasing film thickness or increasing P_{O_2} resulted in a decrease in the c-axis lattice strain towards the bulk value. The Mg@ZnO film with thickness ~26 nm possesses the largest strain of ~1.62 %. As the film thickness increases, the c-axis strain decreases and the strain is only ~1.1 % when the film is ~104 nm thick. In case of films of different thickness strain may originate from the lattice mismatch of 40% at the interface between ZnO films and the substrates, which may be related to the presence of cation and anion defects. Similarly, with the increase of O_2 content in working gas from 5% to 20%, the strain is relaxed from ~1.53% to ~0.48%. Since the films were deposited on the same batch of substrate, it can be assumed that the strain due to lattice mismatch should be at the same level for all the films. Thus the variation of strain among the films would be primarily due to the variation of intrinsic defects concentrations. That is, oxygen vacancy (V_O) in the films decreases with an increase of P_{O_2}.

The SEM images of as-grown and annealed Mg@ZnO film deposited for 1h in P_{O_2} content of 10% on Si substrate are shown in figure 4, which exhibits a smooth and dense structure without any micro cracks, indicating homogeneous surface of the films. After annealing, the size of the grains has increased and the structure of the films improves.

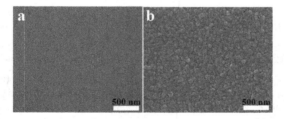

Figure 4. The typical SEM images of the Mg@ZnO film deposited for 1h in P_{O_2} of 10% as-grown (a), and (b) annealed at 700^0C for 1h in air.

The Eg of the Mg@ZnO films were evaluated using the relation:

$$(\alpha h\nu) = \sqrt{(h\nu - E_g)} \qquad (2)$$

where α is the absorption coefficient and $h\nu$ is the photon energy. Figure 5 shows $\alpha h\nu^2$ plot of Mg@ZnO thin films deposited on glass substrate under different P_{O_2} as a function of photon energy $h\nu$. For the films grown under different P_{O_2}, the Eg increases from 4.10 to 4.24 eV attributed to the increase of Mg content (as shown in figure 1(b) and (c)) in Mg@ZnO films caused by the sputtering conditions. It is clear that the E_g of the Mg@ZnO can be tuned by changing P_{O_2} ratio of the N_2+O_2 ambient.

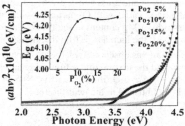

Figure 5: *Optical absorption spectra of Mg@ZnO thin films deposited under different P_{O_2} on glass substrates and the inset shows corresponding optical Eg as a function of P_{O_2}.*

The values for the saturation magnetism, Ms of Mg@ZnO as a function of film thickness is shown in figure 6. The M_S value of films is found to vary in the range of 0.87-14.6 emu/cm^3 with a maximum value for the film of 52 nm thickness. Films of thickness greater than 104 nm show only diamagnetism (although not shown in figure 6) with susceptibility values close to what is known for bulk ZnO. Also, we find that the decrease in Ms values for films of thickness >52 nm is found to have the same functional form as that of the decreasing intrinsic strain in the films with increasing film thickness. We thus have the following picture for the observed magnetic behavior of Mg doped ZnO: Initially, on doping Mg in films Mg produces cation vacancies and the polarized oxygen atoms around these defect sites induce magnetism. The effective magnetization increases to become a maximum when some sort of equilibrium is reached between the cation concentration, the strain due to induced magnetism and the total oxygen content which appears to occur around 52nm in the present case. Above such optimized thickness in the film the density of the relative cation induced defect concentration begins to decrease with increasing oxygen content that in turn reduces the strain to eventually reach the bulk diamagnetic value of ZnO. The inset of figure 6 shows M_S as a function of P_{O_2}. Figure 7 shows the M-H curves at RT for as-grown and annealed Mg@ZnO films at 600, 700 and 800°C in air. Annealing results in a reduction of the M_S values from 14.6 to 0.36 emu/cm^3 due to the reduction of V_O after annealing.

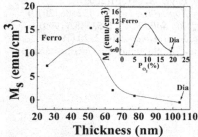

Figure 6. *M_S as a function of film thickness. (inset: M_S as a function of different P_{O_2} in working gas).*

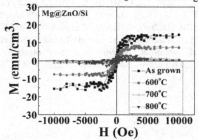

Figure 7. *The M-H loops at RT for as-grown, and films annealed at various temperatures in air for 60 min.*

CONCLUSIONS

Mg@ZnO films were prepared on Si and glass substrates by co-sputtering Mg and Zn targets. We observed the Eg of the films varies with O_2 partial pressure in the working gas, which may be ascribed to the variation of Mg content in the films. The RTFM was observed in Mg@ZnO films prepared on Si. With the increase in film thickness and P_{O_2}, a transition from ferromagnetic to diamagnetic behavior was observed which is attributed to the cation defect induced strain. The intrinsic strain and the texture in the films play an important role in structural and magnetic properties in Mg@ZnO films. Extensive XPS, NEXAS, and XMCD investigations on these films will be published in the near future.

ACKNOWLEDGMENTS

The work was supported by the Swedish funding Agencies, VINNOVA, Hero-M Centre of Excellence at KTH. Sreekanth K Mahadeva acknowledges a graduate study fellowship funded by the India 4EU- Erasmus Mundus External Cooperation Window program. Jin-Cheng Fan and Zhi-Yong Quan acknowledge the Carl Trygg's Foundation in Sweden for post-doctoral Scholarships.

REFERENCES

1. J. K. Furdyna, *J. Appl. Phys.* **64,** R29 (1988).
2. S. D. Sarma, *American Scientist* **89,** 516 (2001).
3. T. Dietl, H. Ohno, F. Matsukura, J. Cibert, and D. Ferrand, *Science* **287,** 1019 (2000)
4. J.M.D. Coey, A.P. Douvalis, and C.B. Fitzgerald, *Nature Mater.* **4,** 173 (2005).
5. M. Kapilashrami, J. Xu, V. Ström, K.V. Rao, and L. Belova, *Appl. Phys. Lett.* **95,** 033104 (2009).
6. G. Xing, D. Wang, J. Yi, L. Yang, M. Gao *et al., Appl. Phys. Lett.* **96,** 112511 (2010).
7. J. B. Yi, C. C. Lim, G. Z. Xing, H. M. Fan *et al. Phys. Rev. Lett.* **104,** 137201 (2010).
8. P. Zhan, W. Wang, C. Liu, Y. Hu, Z. Li, Z. Zhang *et al., J. Appl. Phys.* **111,** 033501 (2012).
9. C. M. Araujo, M. Kapilashrami, X. Jun *et al., Appl. Phys. Lett.* **96**, 232505 (2010).
10. S. Nagar, O.D Jayakumar, L. Belova, and K.V. Rao, *Materials Express* **2,** 233(2012).
11. Ü. Özgür, Ya. I. Alivov, C. Liu, A. Teke *et al., J. Appl. Phys.* **98,** 041301(2005).
12. Y. Li, R. Deng, B. Yao, G. Xing, D. Wang, and T. Wu, *Appl. Phys. Lett.* **97,** 102506 (2010).
13. E.M. Bachari, G. Baud, S. Ben Amor, and M. Jacquet, *Thin Solid Films* **165,** 348, (1999).
14. M. K. Puchert, P. Y. Timbrell, and R. N. Lamb, *J. Vac. Sci. Technol. A* **14**, 2220 (1996).
15. B. Straumal, A. Mazilkin, S. Protasova, A. Myatiev, P. Straumal, E. Goering, and B. Baretzky, *Phys. Status Solidi B* **248,** 1581 (2011).

Mater. Res. Soc. Symp. Proc. Vol. 1494 © 2013 Materials Research Society
DOI: 10.1557/opl.2013.261

Grazing Incidence X-ray Topographic Studies of Threading Dislocations in Hydrothermal Grown ZnO Single Crystal Substrates

Tianyi Zhou[1], Balaji Raghothamachar[1], Fangzhen Wu[1], and Michael Dudley[1,a]

[1]Department of Materials Science and Engineering, Stony Brook University, Stony Brook, New York, 11794, USA
[a] mdudley@notes.cc.sunysb.edu

ABSTRACT

ZnO single crystal substrates grown by the hydrothermal method have been characterized by grazing incidence X-ray topography using both monochromatic and white synchrotron X-ray beams. $11\bar{2}4$ reflection was recorded from the (0001) wafers and the different contrast patterns produced by different threading defects were noted. To uniquely identify the Burgers vectors of these threading dislocation defects, we use ray tracing simulation to compare with observed defect contrast. Our studies showed that threading screw dislocations are not commonly observed. Most threading edge dislocations have the Burgers vector of $1/3\,[2\bar{1}\bar{1}0]$ or $1/3\,[1\bar{2}10]$ and a density of $2.88\times10^4/\mathrm{cm}^2$.

INTRODUCTION

Zinc oxide is a promising semiconductor with many desirable properties, such as wide band gap (3.37 eV) and a large exciton banding energy (60 meV) at room temperature [1,2], which make it highly suitable for application in emitter devices in the blue to ultraviolet region and as a substrate material for GaN-based devices. The hydrothermal method is the prevalent technique for growing wurtzite ZnO single crystals [3,4]. The presence of structural defects will strongly influence the performance, lifetime and reliability of devices grown on these substrates. Different dislocation defect types will affect devices to different extents [5] and therefore it is necessary to carry out detailed characterization of all dislocation types. X-ray topography is a powerful technique to image defects in single crystalline materials of low defect densities ($<10^6/\mathrm{cm}^2$). Back reflection and more specifically grazing geometries, where the penetration depth is low, are used to image the threading dislocations. Comparison of recorded contrast details with the ray tracing simulations of expected defect images can be used to uniquely identify dislocation types. The ray tracing model has been successfully used in interpreting various dislocations in various geometries for 4H-SiC [6,7].

EXPERIMENT

Commercial c-plane ZnO wafers grown by the hydrothermal method were initially imaged in the transmission geometry to characterize defects in the bulk. $11\bar{2}4$ reflection was recorded in grazing geometry with 2° incident angle to image threading defects. Synchrotron white beam X-ray topography experiments were carried out at the Stony Brook Synchrotron Topography Station, Beamline X-19C, at the National Synchrotron Light Source while synchrotron monochromatic X-ray topography experiments were carried out at the Advanced Photon Source, Beamline XOR-33BM/UNI-CAT. All images were recorded on Agfa Structurix D3-SC films. Ray tracing simulation work is carried out using Wolfram Mathematica 8.0 software.

THEORY

There are mainly two contrast mechanisms associated with dislocation image formation on X-ray topographs: orientation contrast and extinction contrast. Under the low absorption conditions used in our study, orientation contrast dominates contrast contributions to the image. Therefore ray tracing simulation, which is based on orientation contrast, is an excellent model to interpret the observed dislocations in ZnO. The diffracting plane is distorted by stress fields associated with dislocations and therefore, the vector of the diffracted beam varies. The contrast on the simulated image is determined by the superposition or separation of beams diffracted from the tiny squares on sample surface. The relationship between plane normal and stress field has been proven to be [8]:

$$\vec{n}(x,y,z) = \vec{n}_o(x,y,z) - \nabla[\vec{n}_o(x,y,z) \cdot \vec{u}(x,y,z)] \dots\dots\dots\dots\dots\dots\dots\dots\dots(1)$$

where $\vec{n}_o(x,y,z)$ is the plane normal for perfect crystal, $\vec{n}(x,y,z)$ is the plane normal after distortion. $\vec{u}(x,y,z)$ is the displacement vector due to the stress field associated with dislocations. Assume the threading dislocation line is on the z-axis, the Burgers vector is on the basal plane, which is composed by x-axis and y-axis. Stress field for threading edge dislocation is [9]:

$$u_x = \frac{b_1}{2\pi}\left[\tan^{-1}\frac{y}{x} + \frac{xy}{2(1-v)(x^2+y^2)}\right], \quad u_y = -\frac{b_1}{2\pi}\left[\frac{1-2v}{4(1-v)}\ln(x^2+y^2) + \frac{x^2-y^2}{4(1-v)(x^2+y^2)}\right] \dots\dots\dots\dots (2)$$

where b_1 is the Burgers vector of threading edge dislocations (TEDs), which is approximately 3.426Å, v is the Poisson's ratio of ZnO (0.351 without porosity). If the Burgers vector of TED is at angle θ to the x-axis (assume the clock-wise direction is the reference direction), the displacement field of TED is:

$$u'_x = u_x\cos\theta - u_y\sin\theta; \quad u'_y = u_x\sin\theta + u_y\cos\theta,$$

For threading screw dislocations (TSDs), the strain components perpendicular to the crystal surface has to be zero in order to satisfy the free surface condition. Surface relaxation effect [10] has to be taken into consideration. Therefore, the stress field for TSD is given by:

$$u_\theta = \frac{-b_2}{2\pi}\sum_{n=0}^{\infty}(-1)^n\left\{\frac{\sqrt{x^2+y^2}}{(2n+1)t-(z+t)+\sqrt{[(2n+1)t-z]^2+(x^2+y^2)}} - \frac{\sqrt{x^2+y^2}}{(2n+1)t+(z+t)+\sqrt{[(2n+1)t+z]^2+(x^2+y^2)}}\right\}\dots\dots(3)$$

$$u_x = u_\theta\frac{-y}{\sqrt{x^2+y^2}}, \quad u_y = u_\theta\frac{x}{\sqrt{x^2+y^2}}, \quad u_z = \frac{b_2}{2\pi}\tan^{-1}\frac{y}{x}\dots\dots\dots\dots\dots\dots\dots\dots (4)$$

where b_2 is the Burgers vector of TSD (approximately 5.1948 Å), t is the half thickness of the sample. For $c+a$ dislocation, it can be considered that the stress field is the sum of a-component and c-component [6]. Once the plane normal $\vec{n}$ after distortion is known, the vector of diffracted beam could be calculated using the following equation:

$$\vec{s_0} \times \vec{n} = -\vec{n} \times \vec{s} \dots\dots\dots\dots\dots\dots\dots\dots\dots\dots\dots\dots\dots\dots\dots\dots\dots\dots\dots\dots\dots\dots(5)$$

$\vec{s_0}$ is the vector of incident beam; $\vec{s}$ is the vector of diffracted beam. For each square on the film, the number of diffracted beams from the sample which could hit this area is counted. Then, a mapping of the diffracted beam density is simulated, which is the dislocation image.

RESULTS

For the basal plane cut wurtzite ZnO crystal, there are 2 possible TSDs, 6 possible TEDs and 12 possible $c+a$ dislocations. The diffraction vector used for the simulated images of all these dislocations is $\langle 11\bar{2}4 \rangle$, pointing upwards as shown in Fig.1.

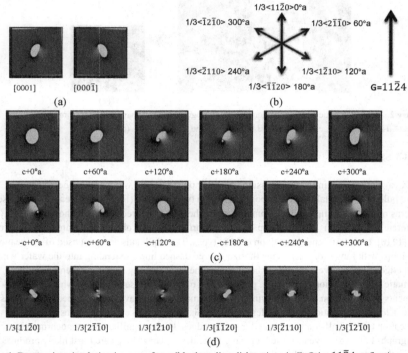

Figure 1. Ray tracing simulation images of possible threading dislocations in ZnO in $11\bar{2}4$ reflection topograph (a) TSDs; (b) Directions of Burgers vectors of a-components in $c+a$ dislocations and g vector is $11\bar{2}4$; (c) TEDs (d) $c+a$ dislocations.

Here, $c+0°a$ means, the c-component's Burgers vector is [0001] and the a-component's Burgers vector is $1/3\langle11\bar{2}0\rangle$. Detailed studies of grazing incidence monochromatic X-ray topographs from multiple ZnO wafers reveal no contrast features matched to TSD or $c+a$ simulated images. This means, few c-component dislocations are present in this sample. Most dislocation features we observed in this study are TEDs with Burgers vector of either $1/3\,[2\bar{1}\bar{1}0]$ or $1/3\,[1\bar{2}10]$, while very few other type TEDs could be found from the topographic images (see Fig. 2)

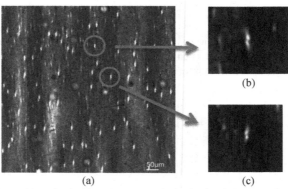

(a) (b) (c)

Figure 2. Monochromatic grazing incidence X-ray topography (a) showing contrast from chiefly two types of TEDs with Burgers vectors $1/3\,[1\bar{2}10]$ (b) and $1/3\,[2\bar{1}\bar{1}0]$ (c).

DISCUSSION

X-ray topography in the transmission geometry on these ZnO wafers reveal defects in the entire bulk of the crystal wafers. Topograph from a typical wafer (Fig. 3(a)) reveals criss-crossed patterns of prismatic slip bands (S) running along the $\langle11\bar{2}0\rangle$ directions and inhomogeneously scattered inclusions (I). Prismatic slip on all three prism planes is observed but predominantly of the $[1\bar{2}10]$ on $(10\bar{1}0)$ and $[2\bar{1}\bar{1}0]$ on $(01\bar{1}0)$ types. The slip bands are composed of dislocation half-loops with long screw segments (horizontal red dashed lines) extending into the wafer with short edge segments (vertical red dashed lines) at the ends (see Fig. 4(a) for schematic). The prismatic slip dislocations exhibit the phenomenon of cross-slip (see Fig.3(b)) where the screw segments of dislocation half-loops lying in the prismatic planes have cross-slipped onto the basal plane. Closer examination of the wafer edges reveals a dense concentration of basal plane dislocation (BPD) half-loops nucleated from the edges. High magnification monochromatic topograph (Fig. 3(d)) reveals these loops in greater detail. These loops are most likely produced during the cutting process, while the prismatic loops are mainly post-growth thermal-stress induced dislocations. The monochromatic topograph also reveals that the ends of the prismatic slip bands are associated with a TED contrast (Fig. 3(d)).

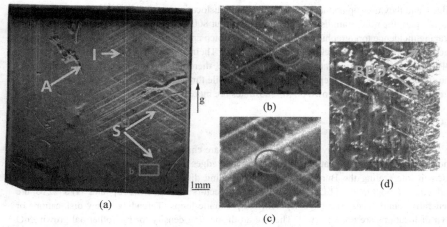

Figure.3 (a) White beam transmission topograph of a typical ZnO wafer showing prismatic slip bands (S), inclusions (I) and surface artifacts (A). (b) & (c) High magnification topographs showing the cross-slip of screw segments on to the basal plane (inside circle). (d) Monochromatic topograph showing TED images at the end of prismatic slip bands (inside circle) and BPD loops nucleated from wafer edge.

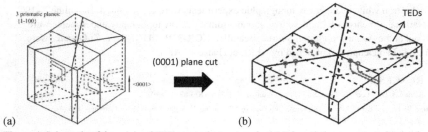

Figure 4. Schematic of the origin of TEDs on c-plane cut wafer (a) Prismatic loops generated in boule due to thermal gradient induced stresses. (b) Wafer slicing perpendicular to edge segments resulting in TEDs on surface.

From these observations, a clear picture of the origin of TED contrast is obtained. When the c-plane wafer is cut from the boule, the predominantly $[1\bar{2}10]$ and $[2\bar{1}\bar{1}0]$ prismatic dislocation loops can be cut in the middle perpendicular to the edge segments. Therefore, the edge segments intersect the surface nearly along c-axis and appear as TEDs on the intersection plane which is the surface plane in this study (see Fig. 4 for schematic).

The dislocation density of these two TED types is approximately $2.88 \times 10^4/cm^2$ near wafer edges, and they take up about 50% of all threading dislocations in the $11\bar{2}4$ reflection. In the central part, the minimum dislocation density is about 800/cm^2. The remaining features on the grazing incidence topographic images are BPDs, which appear as a light straight white line, and inclusions which appear as circles of white contrast. Their density is 4.89×10^4/cm^2. Note that the BPDs usually do not terminate on the crystal surface, therefore, they will not strongly influence the structural quality of the epilayer grown on it; while threading dislocations which will propagate into the device grown on the substrates are more damaging.

CONCLUSIONS

ZnO substrates grown by hydrothermal method are characterized by prismatic slip bands and inclusions with basal dislocations concentrated near edges. Ray tracing simulation is successfully used in interpreting the Burgers vectors of threading dislocations in $11\bar{2}4$ grazing incidence reflection topographs. TEDs whose Burgers vectors are $1/3\,[2\bar{1}\bar{1}0]$ and $1/3\,[1\bar{2}10]$ are generally found. These are edge segments of prismatic loops. Threading screw dislocations or $c+a$ dislocations are not observed. The overall dislocation density for hydrothermal grown ZnO is 5000/cm^2. To minimize threading dislocations in the sample, prismatic slip must be eliminated by minimizing thermal gradient stresses.

ACKNOWLEDGEMENT

Synchrotron white beam X-ray topographic experiments were carried out at Beamline X-19C, in the National Synchrotron Light Source; monochromatic X-ray topography experiments were carried out at the Advanced Photon Source, Beamline XOR-33BM/UNI-CAT. ZnO wafer were grown by Tokyo Denpa and provided by AFRL at Hanscom AFB.

REFERENCES
[1] D.C. Reynolds, D.C. Look, B. Jogai, Solid State Commun. **99** (12), pp.873-875 (1996).

[2] P. Zu, Z.K. Tang, G.K.L. Wong, M. Kawasaki, A. Ohtomo,H. Koinuma, Y. Segawa, Solid State Commun. **103** (8), pp.459-463 (1997).

[3] Katsumi Maeda, Mitsuru Sato, Ikuo Niikura and Tsuguo Fukuda, Semicond. Sci. Technol. **20** (4), pp.49-54 (2005).

[4] Nathan Johann Nicholas, George V. Franks and William A. Ducker, CrystEngComm, **14**, pp.1232–1240 (2012).

[5] C.Y. Lin, W.-R. Liu, C.S. Chang, C.-H. Hsu, W.F. Hsieh and F.S.-S. Chien, J. Electrochem. Soc. **157** (3), pp.H268-H271 (2010).

[6] F.Z. Wu, S.Byrappa, H.H. Wang, Y.Chen, B. Raghothamachar, M. Dudley, E.K. Sanchez, G. Chung, D. Hansen, S.G. Mueller and M.J. Loboda, Mater. Res. Soc. Symp. Proc. **1433** (2012).

[7] Y. Chen, M. Dudley, E.K. Sanchez and M.F. MacMillan, J. Electron. Mater., **37**, pp.713-720 (2008).

[8] D.K. Bowen and B. K. Tanner, "High Resolution X-Ray Diffractometry and Topography", Taylor & Francis, p.189 (1998).

[9] J.P. Hirth and J. Lothe, "Theory of Dislocations", 2nd Edition, John Wiley& Sons, pp.59-79 (1982).

[10] J.D. Eshelby and A. N. Stroh, Phil. Mag. **42**, p.1401 (1951).

Mater. Res. Soc. Symp. Proc. Vol. 1494 © 2013 Materials Research Society
DOI: 10.1557/opl.2013.240

Electrical properties of zinc oxide thin films deposited using high-energy H₂O generated from a catalytic reaction on platinum nanoparticles

Kanji Yasui, Naoya Yamaguchi, Eichi Nagatomi, Souichi Satomoto, and Takahiro Kato
Nagaoka University of Technology, 1603-1 Kamitomioka, Nagaoka, Niigata 940-2188, Japan

ABSTRACT

Zinc oxide (ZnO) with excellent crystallinity and large electron mobility was grown on a-plane (11-20) sapphire (a-Al$_2$O$_3$) substrates by a new chemical vapor deposition method via the reaction between dimethylzinc (DMZn) and high-energy H$_2$O produced by a Pt-catalyzed H$_2$-O$_2$ reaction. The electron mobility at room temperature increased from 30 cm^2/Vs to 189 cm^2/Vs with increasing film thickness from 0.1 μm to approximately 3 μm. Electron mobility increased significantly with decreasing temperature to approximately 110 – 150 K, but decreased at temperatures less than 100 K for films greater than 500 nm in thickness. On the other hand, the mobility hardly changed with temperature for films lesser than 500 nm in thickness. Based on the dependence of the electrical properties on the film thickness, the ZnO films grown on a-Al$_2$O$_3$ substrates are considered to consist of an interfacial layer with a high defect density (degenerate layer) generated due to a large lattice mismatch between ZnO and Al$_2$O$_3$ substrates and an upper layer with a low defect density.

INTRODUCTION

Zinc oxide (ZnO) is a useful material for many applications such as surface acoustic wave devices [1], gas sensors [2], photoconductive devices [3], and transparent electrodes [4]. Due to its large bandgap (3.37 eV at RT) and large exciton binding energy (60 meV) [5], its application to optoelectronic devices such as light emitting diodes and laser diodes operating in the ultraviolet region has been intensively investigated [6-12]. Many growth techniques, including molecular beam epitaxy (MBE) [8, 9], pulsed laser deposition (PLD) [7, 10], laser MBE (LMBE) [6], and metal-organic chemical vapor deposition (MOCVD) [11, 12], have been used to prepare ZnO thin films. Although MOCVD has many advantages for industrial applications, such as a high growth rate on large surface substrates and a wide selection of metalorganic and oxygen source gases, ZnO film growth by conventional MOCVD requires high electric power to react the source gases and raise the substrate temperature. To overcome this, a more efficient means for reacting oxygen and metalorganic source gases is needed. In addition to the low reaction efficiency, conventional CVD methods yield low-quality ZnO films (exhibiting small electron mobility) compared to those prepared by MBE and PLD, due to incomplete reaction between metalorganic and oxygen source gases in the gas phase. However, if thermally excited water is used to hydrolyze the metalorganic source gases, high-energy ZnO precursors are produced in the gas phase, thus allowing the growth of ZnO films in a manner similar to PLD and MBE. In a previous paper [13], we reported a new growth method for ZnO films using the reaction between dimethylzinc (DMZn) and high-energy H$_2$O produced by a Pt-catalyzed H$_2$ - O$_2$ reaction, and also reported the excellent electrical and optical properties of the ZnO films grown on a-plane (11-20) sapphire (a-Al$_2$O$_3$) substrates. Based on the dependence of the electrical properties on the film thickness, however, the ZnO films grown on a-Al$_2$O$_3$ substrates are considered to consist of an interfacial layer with a high defect density (degenerate layer) generated due to a large lattice mismatch between ZnO and Al_2O_3 substrates and an upper layer with a low defect density.

In this paper, variations in electrical properties of the ZnO films as functions of film thickness and temperature are reported.

EXPERIMENT

CVD apparatus used in this study is the same as that shown in a previous paper [13]. H_2 and O_2 gases were introduced into a catalyst cell containing a Pt-dispersed ZrO_2 catalyst. The temperature of the catalyst cell increased rapidly to over 1000°C after introduction of the H_2-O_2 gas supply due to the exothermic reaction of H_2 and O_2 on the catalyst, as shown in Fig. 1. The catalytic cell temperature becomes stable within five minutes after the introduction of the H_2-O_2 gas supply. Then a shutter placed between a skimmer cone and a substrate holder was opened 10 min after the H_2-O_2 gas supply. The resulting thermally excited H_2O molecules were ejected from a fine nozzle into the reaction zone and allowed to collide with DMZn ejected from another fine nozzle. The distance between the H_2O nozzle and the substrate was 50 – 60 mm. The skimmer cone was placed between the H_2O nozzle and the substrate to select only high-velocity H_2O molecules and direct them to the substrate. ZnO films with various film thicknesses were grown directly on a-Al_2O_3 substrates at substrate temperatures of 773 K for 2 – 60 min with no buffer layer. Although the H_2 and O_2 gas flow ratios varied somewhat from deposition to deposition, typical values were 200 and 40 sccm, respectively. The reaction gas pressure in the chamber during the deposition was 0.4 – 1.0 Pa. This variability was due to the fact that the burning temperature in the catalytic cell was set to the maximum (over 1000°C) for each deposition experiment. The sapphire substrates were first degreased by methanol and acetone, etched with a H_2SO_4+H_3PO_4 solution, then rinsed with ultrapure water, and finally set on a substrate holder in the CVD chamber.

The Hall mobility and residual carrier concentration were measured using the van der Pauw method (ECOPIA, HMS-5000) with a magnetic field of 0.57 T at room temperature. All films showed a n-type character. The temperature dependence of the Hall mobility and residual carrier concentration were also measured from 80 to 330 K.

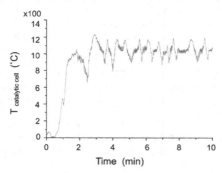

Fig. 1 Catalytic cell temperature during catalytic reaction between H_2 and O_2 on Pt nanoparticles

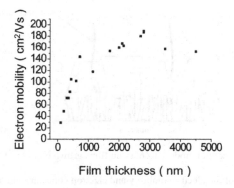

Fig. 2 Variation in electron mobility as a function of film thickness

RESULTS AND DISCUSSION

Figure 2 shows the variation in electron mobility as a function of film thickness measured at room temperature (290 K). The electron mobility of the ZnO films 100 – 400 nm in thickness increases significantly with increasing film thickness. The increase in the mobility becomes slower when the film thickness becomes greater than 800 nm, and the mobility reaches the saturation point for thicknesses greater than 2000 nm. Compared to the Hall mobility of the ZnO films grown by MOCVD reported by Dr. K. T. Roro [14], the maximum mobility was much grater than that grown by the MOCVD (120 cm^2/V for the 4.4 μm thick sample). On the other hand, the electron concentration of the ZnO film 100 – 400 nm in thickness is on the order of 10^{18} cm^{-3}, and decreases to the order of 10^{17} cm^{-3} with increasing film thickness at greater than 500 nm, as shown in Fig. 3. The electron concentration is less than 3×10^{17} cm^{-3} for the films greater than 1500 nm in thickness.

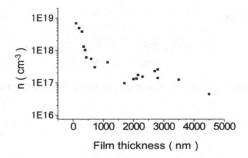

Fig. 3 Variation in electron concentration as a function of film thickness

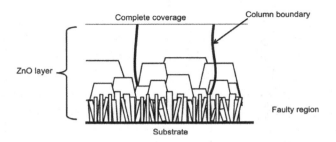

Fig. 4 Structural model of ZnO thin films grown on a-Al$_2$O$_3$

Based on dependence of the electron mobility and electron concentration on the film thickness, the ZnO films grown on a-Al$_2$O$_3$ substrates are estimated to consist of an interfacial degenerate layer with a high-defect density (degenerate layer) and an upper layer with a low-defect density [15], as shown in Fig. 4. If most of the ZnO film consists of the degenerate layer, the residual carrier concentration is high, and the electron mobility and the electron concentration hardly depend on temperature. As the upper layer with a low-defect density grows, temperature dependence of the electron mobility and electron concentration becomes large, and the electron mobility increases and electron concentration decreases with decreasing temperature (semiconductor-like dependence).

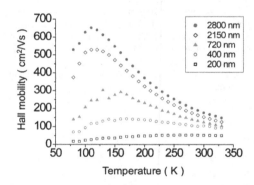

Fig. 5 Temperature dependence of electron mobility for various film thicknesses

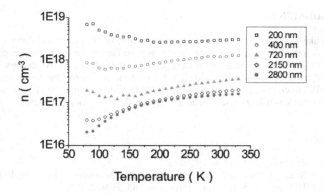

Fig. 6 Temperature dependence of electron concentration for various film thicknesses

From the temperature dependence of electron mobility as shown in Fig. 5, the mobility increases significantly with decreasing temperature to approximately 110 – 150 K, but decreases at temperatures less than 100 K for films greater than 500 nm in thickness. The electron mobility of the thickest film (189 cm^2/Vs at room temperature, 290 K) increases to 660 cm^2/Vs at 100 K, but decreases at lower than 100 K. On the other hand, the mobility hardly changes with temperature for the ZnO films lesser than 500 nm in thickness. In the case of the film 200 nm in thickness, the electron mobility monotonically decreases from 50 to 17 cm^2/Vs with decreasing temperature from 330 K to 80 K.

The electron concentration significantly decreases from 1.6×10^{17} to 2.0×10^{16} cm^{-3} for the thickest ZnO film with decreasing temperature from 330 K to 80 K, as shown in Fig. 6. On the other hand, in the case of the film 200 nm in thickness, the electron concentration slightly increases from 3×10^{18} cm^{-3} to 7×10^{18} cm^{-3} with decreasing temperature from 330 K to 80 K. These results would reflect the difference in the influence of the interfacial layer on the electrical properties for the ZnO films with various thicknesses.

CONCLUSIONS

In conclusion, electrical properties of the ZnO films grown by a CVD method using high-energy H_2O generated by a catalytic reaction of H_2-O_2 on Pt-nanoparticles were investigated. The electron mobility at room temperature increased from 30 cm^2/Vs to 189 cm^2/Vs with increasing film thickness from 0.1 μm to approximately 3 μm. The electron concentration at room temperature decreased from 7.0×10^{18} cm^{-3} to 4.7×10^{16} cm^{-3} with increasing film thickness from 0.1 μm to 4.5 μm. From the temperature dependence of the electron mobility and electron concentration on the thickness and temperature, the structure of the ZnO films grown on a-A_2O_3 was estimated to consist of an interfacial degenerate layer with a high-defect density and an upper layer with a low-defect density.

ACKNOWLEDGMENTS

We would like to express our sincere gratitude to Professor N. Tsuboi and Associate Professor H. Shimizu of Niigata University for offering the facility for the measurement of Hall effect. We would also like to express our sincere gratitude to Professor M. Aketagawa of Nagaoka University of Technology for allowing us to use the corresponding equipment for measuring the film thickness. This work was supported in part by a Grant-in-Aid for Scientific Research (B) (No. 24360014) from the Japan Society for the Promotion of Science.

REFERENCES

1. F. S. Hickernell, *Proc. IEEE*, **64**, 631 (1976).
2. S. Pizzini, N. Butta, D. Narducci, and M. Palladino, *J. Electrochem. Soc.*, **136**, 1945 (1989).
3. I. S. Jeong, J. H. Kim, and S. Im, *Appl. Phys. Lett.*, **83**, 2946 (2003).
4. T. Minami, *Semicond. Sci. Technol.*, **20**, S35 (2005).
5. B. K. Meyer, H. Alves, D. M. Hofmann, W. Kriegseis, D. Forster, F. Bertram, J. Christein, A. Hoffmann, M. Straßburg, M. Dworzak, U. Haboeck, and A. V. Rodina, *Phys. Stat. Sol.* (b), **241**, 231 (2004).
6. A. Tsukazaki, A. Ohtomo, T. Onuma, M. Ohtani, T. Makino, M. Sumiya, K. Ohtani, S. Chichibu, S. Fuke, Y. Segawa, H. Ohno, H. Koinuma, and M. Kawasaki, *Nature Materials*, **4**, 42 (2005).
7. E. M. Kaidashev, M. Lorenz, H. von Wenckstern, A. Rahm, H. C. Semmelhack, K. -H. Han, G. Benndorf, C. Bundesmann, H. Hochmuth, and M. Grundmann, *Appl. Phys. Lett.*, **82**, 3901 (2003).
8. P. Fons, K. Iwata, S. Niki, A. Yamada, and K. Matsubara, *J. Cryst. Growth*, **201-202**, 627 (1999).
9. K. Miyamoto, M. Sano, H. Kato, and T. Yao, *J. Cryst. Growth*, **265,** 34 (2004).
10. A. Ohtomo and A. Tsukazaki, *Semicond. Sci. Technol.*, **20**, S1 (2005).
11. C. K. Lau, S. K. Tiku, and K. M. Lakin, *J. Electrochem. Soc.*, **127**, 1843 (1980).
12. J. Dai, F. Jiang, Y. Pu, L. Wang, W. Fang, and F. Li, *Appl. Phys. A*, **89,** 645 (2007).
13. K. Yasui, H. Miura, and H. Nishiyama, *MRS Symp. Proc.,* **1315**, 21 (2011).
14. K. T. Roro, G. H. Kassier, J. K. Dangbegnon, S. Sivaraya, J. E. Westraadt, J. H. Neethling, A. W. R. Leitch, and J. R. Botha, S*emicond. Sci. Technol.*, **23**, 055012 (2008).
15. D. C. Look and R. J. Molnar, *Appl. Phys. Lett.*, **70**, 3377 (1997).

Mater. Res. Soc. Symp. Proc. Vol. 1494 © 2012 Materials Research Society
DOI: 10.1557/opl.2012.1579

Crystal Polarity and Electrical Properties of Heavily Doped ZnO Films

Yutaka Adachi, Naoki Ohashi, Isao Sakaguchi and Hajime Haneda

National Institute for Materials Science, 1-1 Namiki, Tsukuba, Ibaraki, 305-0044, Japan

ABSTRACT

In this study, ZnO films heavily doped with Al or Ga were grown on a polarity-controlled buffer layer using pulsed laser deposition. The films prepared using a 1 mol% Al-doped target with the buffer layer grown at 700 °C had the c(+)-face, whereas the films with the buffer layer grown at 400 °C had the c(-)-face, which means that the polarity control can be successfully carried out using the buffer layer. However, the films prepared using targets doped with more than 1 mol% Al or Ga had the c(+)-face regardless of the polarity of the buffer layer. The 1 mol% Al-doped ZnO film with the c(+)-face had lower electron concentration and higher growth rate than the film with the c(-)-face. This result indicates that the Al content in the film with the c(-)-face was larger than that in the film with the c(+)-face.

INTRODUCTION

ZnO has attracted considerable attention for its application to transparent electrodes for solar cells owing to its transparency in the range of visible light and its low resistivity [1]. ZnO has the wurtzite structure, which has no inversion symmetry; thus, it has the spontaneous polarization along the c-axis, which results in two types of surface, the c(+) (Zn-terminated) face and the c(-) (O-terminated) face. It is known that the polarity affects the properties of ZnO, for example, its electronic structure [2], chemical stability [3] and interfacial properties [4]. ZnO films tend to have the c-axis orientation. Therefore, for device applications it is important to investigate how the polarity affects the properties of ZnO films.

Recently, we have found that Al doping into ZnO films results in the inversion of polarity [5]. ZnO films grown on sapphire using pulsed laser deposition (PLD) with a nominally undoped ZnO target exhibited c(-)-polarity, while films deposited with a 1 mol% Al-doped ZnO target exhibited c(+)-polarity. This inversion of polarity occurs not only on sapphire but also on SiO_2 glass substrates [6]. These results indicate that impurity doping affects the polarity of a ZnO film. Doping efficiency is also affected by the polarity of ZnO. Maki et al. reported that nitrogen doping is favored by the c(+)-face of ZnO [7]. Recently, we found that the Mg content in PLD-grown MgZnO films depends on the film polarity [8]. The Mg content in MgZnO films grown on the c(-)-face of a ZnO single crystal was larger than that in films grown on the c(+)-face of a single crystal, despite the same growth conditions used.

For the application to transparent electrodes for solar cells, Al or Ga, which both act as a donor in ZnO, is heavily doped into ZnO films. As described above, the polarity of ZnO is expected to affect the incorporation of impurities, which would affect the electric properties, especially the electron concentration, of the films. In this study, we investigated the polarity and electric properties of Al- or Ga-doped ZnO films.

EXPERIMENTAL DETAILS

The deposition of ZnO films was carried out by PLD using the fourth-harmonic generation of a neodymium-doped yttrium gallium garnet (YAG:Nd) laser (=266 nm) with a pulse width of 5 ns, a repetition rate of 5 Hz and an average fluence of about 1 J/cm^2. The target used was an Al or Ga doped ZnO ceramic prepared by an ordinary ceramics process. ($11\bar{2}0$) sapphire with a mirror like polished surface was used as a substrate. The substrates were ultrasonically cleaned in acetone and ethanol before deposition. The base pressure of the growth chamber prior to deposition was 4×10^{-7} Pa and the films were prepared in an O_2 partial pressure of 2×10^{-3} Pa.

To obtain films heavily doped with Al- or Ga with the c(+)- or c(-)-face, a polarity-controlled Al-doped ZnO buffer layer was inserted between the heavily doped films and substrates. The structure of the films prepared in this study is shown in figure 1. It has been reported that the films prepared at high temperatures, such as 700 °C, using PLD with a 1 mol% Al-doped ZnO target have the c(+)-face, while the films prepared at low temperatures, such as 400 °C, have the c(-)-face [5]. To obtain heavily doped films with the c(+)-face, an 80-nm-thick Al-doped film was prepared at 700 °C on sapphire as a buffer layer and then a heavily doped film with a thickness of approximately 1 μm was deposited at 700 °C. To obtain films with the c(-)-face, a buffer layer was prepared at 400 °C and then a heavily doped film was deposited at 200 °C. The polarity of the buffer layers was confirmed by photoluminescence (PL) measurements as described below, which showed that the high- and low-temperature-grown buffer layers had the c(+)- and c(-)-polarity, respectively.

The crystallinity of the PLD-grown films was analyzed using an X-ray diffraction (XRD) (PANalytical X'Pert Pro MRD) equipped with a hybrid 2-bounce asymmetric Ge (220) monochromator and a Cu Ka source. Hall measurement was performed using the Van der Pauw geometry to determine the electron concentration (n) and mobility (μ).

The crystal polarity of the films was determined from the results of PL measurements. The peak position of the near-band-edge emission (NBE) from MgZnO films is proportional to the Mg content of the films [9]. The Mg content in the films with the c(-)-face grown using PLD is larger than that in the films with the c(+)-face [8]. Therefore, the peak position of the NBE from the MgZnO films with the c(-)-face is at a shorter wavelength than that for the films with

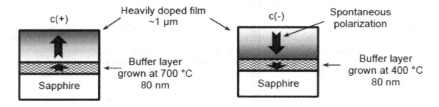

Figure 1. Schematic diagrams of heavily doped films with a 1 mol% Al-doped buffer layer grown at 700 °C and 400 °C.

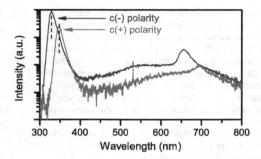

Figure 2. PL spectra of (Mg, Zn)O films grown on ZnO single crystals with the c(+)- and c(-)-faces. The films were prepared at 500 °C using PLD with a $(Mg_{0.1}, Zn_{0.9})O$ target. The peaks at approximately 650-700 nm are due to second-order diffraction of the NBE.

the c(+)-face when the films were prepared using the same growth conditions except the film polarity, as shown in figure 2. In this study, to determine the polarity of the films, MgZnO films were grown on the Al- or Ga-doped ZnO films at 500 °C using PLD with a 10 mol% Mg doped ZnO target and then PL measurements were carried out. The films with the NBE peak position at about 325 nm were determined to have the c(-)-polarity, and the films with the NBE peak position at about 350 nm were determined to have the c(+)-polarity.

RESULTS AND DISCUSSION

It was ascertained by an XRD $2\theta-\omega$ scan that all the films prepared in this study had the c-axis orientation. Table I shows the film polarity determined by the PL measurements.

Table I. Polarity of heavily doped films with a 1 mol% Al-doped ZnO buffer layer grown at 700 °C and 400 °C.

Target	Polarity of the buffer layer	Polarity of the film
Al 1mol% doped ZnO	c(+)	c(+)
	c(-)	c(-)
Al 3mol% doped ZnO	c(+)	c(+)
	c(-)	c(+)
Al 4mol% doped ZnO	c(+)	c(+)
	c(-)	c(+)
Ga 2mol% doped ZnO	c(+)	c(+)
	c(-)	c(+)
Ga 4mol% doped ZnO	c(+)	c(+)
	c(-)	c(+)

The films prepared using a 1 mol% Al-doped target with the buffer layer grown at 700 °C had the c(+)-face, whereas the films with the buffer layer grown at 400 °C had the c(-)-face. This means that polarity control can be successfully carried out using the buffer layer. However, the films prepared using targets doped with more than 1 mol% Al- or Ga had the c(+)-face regardless of the polarity of the buffer layer. This result indicates that a change in polarity occurred during the growth of the ZnO films heavily doped with Al or Ga on the c(-)-polar buffer layer. In other words, the heavily doped ZnO films preferentially have the c(+)-face rather than the c(-)-face. The reason why the polarity changes for the heavily doped ZnO films is not clear at present. Further studies will be needed to clarify the polarity change mechanism due to the heavy doping of Al or Ga into ZnO.

To investigate the effect of the polarity on the electrical properties of III-element-doped ZnO films, Hall measurements were carried out for the 1 mol% Al-doped ZnO films on the c(+)- and c(-)-polar buffer layers. Table II summarizes the electrical properties of the films. The growth rate and full width at half maximum (FWHM) of (0002) and (10$\bar{1}$1) reflections were also investigated. The electron concentration of the Al-doped ZnO film with the c(-)-face was about 1.3 times larger than that of the film with the c(+)-face, as can be seen from table II. One reason for the larger electron concentration in the c(-)-polar film is considered to be the larger Al content in the film with the c(-)-face than in the film with the c(+)-face, as in the case of MgZnO films [8]. The vapor pressure of Al is very low at the growth temperature. Therefore, the sticking coefficients of Al, k_{Al}, on the c(+)- and c(-)-faces are probably close to 1. On the other hand, the vapor pressure of Zn is very high at the growth temperature and therefore the sticking coefficient of Zn, k_{Zn}, significantly depends on the surface structure of the ZnO film. It has been reported that the value of k_{Zn} on the c(+)-face is about three times larger than that on the c(-)-face [10]. The mole fraction x of Al is expressed by the following equation:

$$x = k_{Al}J_{Al}/(k_{Al}J_{Al}+k_{Zn}J_{Zn}) , \qquad (1)$$

where J_{Al} and J_{Zn} are the Al and Zn fluxes ablated from a target, respectively. This equation suggests that the difference in the value of k_{Zn} for Zn on the c(+)- and c(-)-faces results in the difference in the Al concentration in the films with the c(+)- and c(-)-faces. The difference in the value of k_{Zn} can be deduced from the growth rates of the films. The growth rate of the film with the c(+)-face was larger than that of the film with the c(-)-face, which indicates that the value of k_{Zn} on the c(+)-face is larger than that on the c(-)-face. Therefore, we conclude that the Al content in the c(-)-polar film is larger than that in the c(+)-polar film.

Table II. Summary of electrical and structural properties and growth rates for the 1 mol% Al-doped ZnO films with the c(+)- and c(-)-faces: carrier concentration, n; Hall mobility, μ; FWHMs of the (0002) and (10$\bar{1}$1) RCs, $\Delta\omega$(0002) and $\Delta\omega$(10$\bar{1}$1), respectively.

Film polarity	n [cm^{-3}]	μ [cm^{2}V^{-1}s^{-1}]	$\Delta\omega$(0002) [deg]	$\Delta\omega$(10$\bar{1}$1) [deg]	Growth rate [nm/pulse]
c(+)	3.3 x 10^{20}	3.7	0.38	0.49	1.8 x 10^{-2}
c(-)	4.2 x 10^{20}	46.7	0.64	0.78	1.1 x 10^{-2}

The Hall mobility of the films with the c(+)-face was smaller than that of the films with the c(-)-face, in spite of the narrower FWHMs of the (0002) and (10$\bar{1}$1) rocking curves (RCs) for the c(+)-polar film than those for the c(-)-polar film. The broadening of the (0002) and (10$\bar{1}$1) RCs is affected by the tilt and twist, respectively, which are considered to be related to electron scattering in the film. However, the Hall mobilities and FWHMs obtained in this study indicate that the tilt and twist do not significantly contribute to the electron scattering. The considerable difference in the Hall mobilities might be due to the incorporation of defects or impurities, such as hydrogen. It is well known that the electrical properties of ZnO films are affected by native defects, such as oxygen vacancies and interstitial Zn. It has been reported that the incorporation of hydrogen atoms into ZnO also affects its electrical properties [11]. The polarity of ZnO is expected to affect the creation of native defects and the incorporation of hydrogen into ZnO films. To determine the polarity dependence of the electrical properties of ZnO, further studies are required.

CONCLUSIONS

In this study, ZnO films heavily doped with Al or Ga were grown on a polarity-controlled buffer layer using PLD. The films prepared using targets doped with more than 1 mol% Al or Ga had the c(+)-face regardless of the polarity of the buffer layer. ZnO films doped with 1mol% Al with the c(+)- and c(-)-faces were successfully grown. The Al content in the film with the c(-)-face was larger than that in the film with the c(+)-face.

ACKNOWLEDGMENTS

This study was partially supported by a Grant-in-Aid for Scientific Research, KAKENHI Grant No. 23626035 from the Japan Society for Promotion of Science, and the Ministry of Education, Culture, Sports, Science and Technology, Japan.

REFERENCES

1. K. Ellmer, *J. Phys. D; Appl. Phys.* **34**(21), 3097 (2001).
2. A. Wander, F. Schedin, P. Steadman, A. Norris, R. McGrath, T.S. Turner, G. Thornton and N.M. Harrison, *Phys. Rev. Lett.* **86**(17), 3811 (2001).
3. N. Ohashi, K. Takahashi, S. Hishita, I. Sakaguchi, H. Funakubo and H. Haneda, *J. Electrochem. Soc.* **154**(2), D82 (2007).
4. N. Ohashi, K. Kataoka, T. Ohgaki, I. Sakaguchi, H. Haneda, K. Kitamura and M. Fujimoto, *Jpn. J. Appl. Phys. Part 2* **46**(41-44), L1042 (2007).
5. Y. Adachi, N. Ohashi, T. Ohnishi, T. Ohgaki, I. Sakaguchi, H. Haneda and M. Lippmaa, *J. Mater. Res.* **23**(12), 3269 (2008).
6. Y. Adachi, N. Ohashi, T. Ohgaki, T. Ohnishi, I. Sakaguchi, S. Ueda, H. Yoshikawa, K. Kobayashi, J.R. Williams, T. Ogino and H. Haneda, *Thin Solid Films* **519**(18), 5875 (2011).
7. H. Maki, I. Sakaguchi, N. Ohashi, S. Sekiguchi, H. Haneda, J. Tanaka and N. Ichinose, *Jpn. J. Appl. Phys. Part 1* **42**(1), 75 (2003).

8.   Y. Adachi, N. Ohashi, I. Sakaguchi and H. Haneda, presented at the IWZnO 2012, Nice, France, 2012, (unpublished).
9.   A. Ohtomo, M. Kawasaki, T. Koida, K. Masubuchi, H. Koinuma, Y. Sakurai, Y. Yoshida, T. Yasuda and Y. Segawa, *Appl. Phys. Lett.* **72**(19), 2466 (1998).
10.   H. Kato, M. Sano, K. Miyamoto and T. Yao, *J Cryst Growth* **265**(3-4), 375 (2004).
11.   D.M. Hofmann, A. Hofstaetter, F. Leiter, H.J. Zhou, F. Henecker, B.K. Meyer, S.B. Orlinskii, J. Schmidt and P.G. Baranov, *Phys. Rev. Lett.* **88**(4), (2002).

Mater. Res. Soc. Symp. Proc. Vol. 1494 © 2013 Materials Research Society
DOI: 10.1557/opl.2013.135

The Structural, Optical and Electrical Properties of Spray Deposited Fluorine Doped ZnO Thin Films

Kondaiah Paruchuri[1], Vanjari Sundara Raja[1], Suda Uthanna[1] and N. Ravi Chandra Raju[2]
[1]Department of Physics, Sri Venkateswara University, Tirupati – 517 502, India
[2]Department of Electrical Engineering, IIT Bombay, Mumbai – 400 085, India

ABSTRACT

Highly transparent and conducting Fluorine doped zinc oxide thin films were deposited using spray pyrolysis method on glass substrates held at 450 ^{0}C. The X-ray diffraction study revealed that as the dopant concentration increases in ZnO films, the intensity of the preferential orientation of (002) reflection decreased and (101) was found to increase up to 5 at. % F. The crystallite size was varied from 40 to 50 nm with dopant concentration. The optical band gap of the un-doped films was 3.30 eV and it increased to 3.34 eV for 3 at. % F. The refractive index of the films was increased from 2.05 to 2.18 with the increase of dopant concentration from 0 to 5 at. %. The scanning electron microscopy results depicted that the microstructure of ZnO: F films highly influenced by the fluorine doping. After annealing the films in hydrogen atmosphere, the resistivity of the films decreased as increase the dopant concentration and it is 4×10^{-3} Ω cm for 3 at. % F beyond which it increased. The mobility of the charge carriers was 14 cm^2/ V sec and the carrier concentration was 7.8×10^{19} cm^3 obtained for the films doped with 3 at. % of fluorine concentration in the starting solution.

INTRODUCTION

Transparent conducting oxides (TCO) thin films have potential applications in various fields such as thin film solar cells, light emitting diodes and gas sensors [1-3]. Several metal oxide semiconductors such as In_2O_3, SnO_2, ZnO and TiO_2 have been employed to fabricate TCO films [4]. Among these, ZnO is most promising n-type semiconductor with wide direct band gap of 3.2 eV and have high exciton binding energy (60 m eV). It was alternative to Indium doped SnO_2 (ITO) due to low cost, abundant, non toxicity and stable under hydrogen plasma [5]. Under normal conditions, ZnO shows n- type conductivity due to the presence of native defects like oxygen vacancies and/or interstitial Zn atoms [6]. The electrical conductivity can be improved by suitable dopants of cations and anions. Extensive work on doped ZnO thin films with cations such as boron, aluminium, gallium and indium has been reported [7-9]. However, there are few reports on fluorine (anion) doped ZnO thin films. ZnO thin films have been deposited by various techniques, such as sputtering [10], chemical vapour deposition [11], sol–gel [12], pulsed laser deposition [13], In addition to these techniques, spray pyrolysis method has received much attention because of its simplicity and cost-effectiveness as it does not require sophisticated vacuum apparatus [14]. In this investigation fluorine doped ZnO thin films deposited by using spray pyrolysis technique and study the influence of fluorine concentration in the starting solution on structural, optical and electrical properties.

EXPERIMENTAL DETAILS

The starting solution used to deposit ZnO films by spray pyrolysis contains zinc acetate (0.2M) dissolved in a mixture of distilled water and methanol in the ratio 3:1. A few drops of acetic acid were added to improve the solubility. For fluorine doping, NH_4F is used as a dopant. The fluorine concentration in the starting solution was varied from 0 to 5 at%. All the films were deposited on Corning glass substrates held at 450^0C. Compressed air was used as carrier gas and a spray rate of 6 ml /min was maintained. The nozzle to substrate distance was kept to be 30 cm for optimum coverage. A chromel - alumel (Type - K) thermocouple was introduced just below the surface of the substrate heater at the centre of the block. The temperature could be maintained to an accuracy of $\pm 10^0C$ during deposition. The thicknesses of the films were measured by means of weight gain method. The optical transmittance at normal incidence of the films, were recorded with a Perkin Elmer double-beam spectrophotometer (LAMBDA – 950) in the UV–Vis–NIR region (300 – 1000 nm). The crystalline structures of the films were analyzed by X-ray powder diffractometer (Model - 3003TT). The surface morphological studies were made using scanning electron microscopy (EVO - MA 15). The electrical conductivity measurements of the films were employing van der Pauw technique using (Aplab Model - 1087) digital multi meter and (Keithley- 617) programmable electrometer.

DISCUSSION

ZnO films deposited on Corning glass substrates were found to be uniform, transparent, and strongly adherent. Figure 1shows the XRD pattern of un-doped and fluorine doped ZnO thin films.

Structural properties and surface morphology

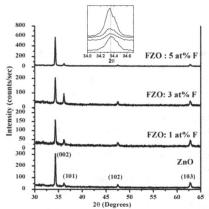

Figure 1. XRD pattern of ZnO and FZO films. inset shows (002) peak shift

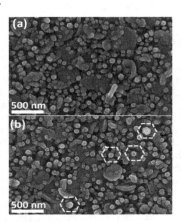

Figure 2. SEM images of (a) ZnO and (b) FZO : 3 at. % F

From the figure it can be seen that all the films shown polycrystalline with a microstructure of hexagonal wurtzite phase of ZnO. No extra phases involving fluorine compounds such as ZnF_2 are found from the XRD patterns, which implies that F substitutes O in the hexagonal lattice and can contribute a free electron from each F atom. The intensity of (002) peak decreases for the films 1at% F doping and then increases with the increase of Fluorine concentration. From this result it can be seen that proper doping level of Fluorine can cause the grain growth and improve the crystallinity of FZO films. These results are also confirmed by the average crystallite size or the full width at half maximum (FWHM) of the films. The average crystallite size can be calculated by Debye - Scherrer's formula.

$$D = 0.9 \lambda / \beta \cos\theta \qquad (1)$$

where λ, θ and β are the Cu Kα X-ray wavelength (0.15406 nm), diffraction angle and FWHM of the ZnO (002) peak, respectively. Table I shows parameters of the (002) diffraction peak of un-doped ZnO and FZO films.

Table I. Parameters of the (002) diffraction peak of ZnO films

(002) Peak	Un doped ZnO	FZO : 1% F	FZO : 3% F	FZO : 5% F
2θ (Degrees)	34.37	34.34	34.32	34.31
FWHM (Degrees)	0.18	0.21	0.19	0.17
Crystallite size (D) (nm)	47	40	45	49
c-axis (nm)	0.5219	0.5225	0.5228	0.5228
Strain (ε)	0.0023	0.0034	0.0040	0.0040

In our experiment the calculated average crystallite size of FZO films is in the range of 40-50nm. It can be found that the FWHM of the (002) diffraction peak decreases firstly and then increases with the increase of F concentration and shifted to lower angle as shown in figure 1 (inset). This may indicates strain developed in the films. It can also be seen from Table I that the diffraction angle of (002) diffraction peaks shifts to small angle compared to the standard XRD spectrum of ZnO powder (34.4310), implying the evolution of the lattice constant c in the FZO films. This change reaches the maximum at the doping concentration of 3.0 at % F, shown in Table I. It implies that the solid solubility of F in ZnO is about 3 at %. The change of the lattice constant c leads to the strain in FZO films. The strain also significantly influences the microstructures and properties of ZnO films [15]. For ZnO films with wurtzite structure, the strain in the c-axis direction can be obtained by following formula,

$$\varepsilon = (C_{film} - C_{bulk}) / C_{bulk} \qquad (2)$$

where C_{film} and C_{bulk} are the c-axes for film and bulk ZnO respectively. The strain in FZO films, as shown in Table I, was calculated according to Eq. (2). All the samples have a positive strain which indicates a compress stress in the films.

The scanning electron microscope images of ZnO and FZO: 3 at. % F were shown in Figure 2. The images show that the films are composed of small hexagonal grains along with flower like structures. It can be seen that the FZO : 3 at. % F films consists of hexagonal small grains and have smooth morphology, which improves electrical transport properties and the average grain size estimated to be in the range 100 - 120 nm.

Optical properties

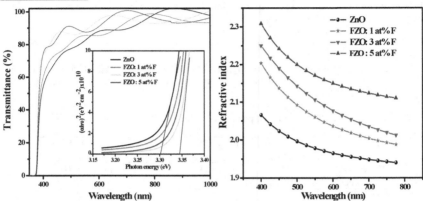

Figure 3. Transmittance spectra of ZnO and FZO films **Figure 4** Variation of refractive index and plot of $(\alpha h\nu)^2$ versus hυ (inset)   with wavelength

Figure 3 depicts the transmission spectra in the wavelength range of 300–1000 nm for ZnO and FZO thin films. All the films exhibit an average optical transmittance higher than 90% and a sharp fundamental absorption edge in the ultraviolet range of 300–400 nm, which shifts to the shorter wavelength as the dopant concentration increases up to 3 at % F. Using the fundamental absorption edge, the optical band gap of the films can be determined from the $(\alpha h\nu)^2$ versus hυ plot by extrapolating the linear portion of the curve to $\alpha = 0$. Variation in the FZO optical band gap with increasing dopant concentration is shown in the inset of Figure 3. It is clear that the optical band gap of the film increases from about 3.30 to 3.34 eV as the dopant concentration increasing up to 3 at % and then decreased to 3.33eV with further increase in dopant concentration. The estimated optical band gap of the FZO thin films is larger than that of un-doped ZnO (3.2eV) due to the Burstein–Moss effect [16]. According to the Burstein–Moss effect, the increase of Fermi level in the conduction band leads to energy band gap broadening by the increase of carrier concentration. According to the Hall data in Figure 5, the films doped with 3 at % of fluorine have highest carrier concentration. The refractive indices of all the films were calculated using swanepoel method and are shown in Figure 4. It was observed that the refractive indices of the F doped ZnO thin films are higher than that of un-doped ZnO film in the measured wavelength range and increased with increase of F doping concentration. The increase of refractive index with the increase of F doping concentration can be attributed to the increase of density in the deposited films.

Electrical properties

For the application of ZnO as a transparent conductor its electrical resistivity must be quite low and transparent must be high. The as deposited films exhibit moderate resistivity about 10^2 Ω cm. It has been reported that the resistivity can be decreased by vacuum or H_2 annealing [17-18]. Figure 5 illustrates the variation in resistivity (ρ), carrier concentration (n), and Hall mobility (μ) as a function of dopant concentration in the starting solution for FZO films

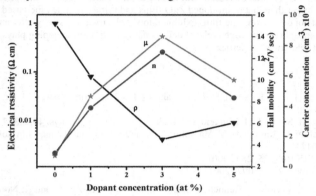

Figure 5. Resistivity, Hall mobility and carrier concentration as function of dopant concentration for ZnO films

It can be seen that resistivity is strongly influenced by the dopant concentration after annealing. As the fluorine concentration increased from 1 to 3 at%, the resistivity of the films decreased rapidly from 7×10^{-2} Ω cm to the minimum value of about 4×10^{-3} Ω cm with a carrier concentration of 7.8×10^{19} cm^3, Hall mobility of 14 cm^2/ V sec there after it increased with further increase of 5 at. % F. As is well known, resistivity is proportional to the reciprocal of the product of carrier concentration and mobility. Therefore, the change in resistivity with dopant concentration is attributed to the change in carrier concentration and/or mobility, which are characteristic parameters reflecting the film structure and/or the impurity contents. When the films annealed in H_2 atmosphere, two types of H related point defects can be identified one is isolated hydrogen atoms located at lattice interstitial sites and second located at anti bonding oxygen sites [19]. These defects along with certain fluorine doping can cause to increase the carrier concentration. As can be seen in Figure 5, the initial decrease in resistivity is ascribed to increase of carrier concentration and Hall mobility. As the dopant increased from 1 to 3 at. %, proper incorporation of fluorine atoms in the oxygen site gives one free electron and the improvement of crystallinity due to lattice structural disorders, will result in increase in the mobility. However, it should be noted that with the further increase of dopant concentration 5 at % F, resistivity, carrier concentration and Hall mobility decreases markedly. These results were well agreed with the previous reported results [20].

CONCLUSIONS

Fluorine doped zinc oxide films were successfully deposited by spray pyrolysis method. The effect of variation of fluorine concentration on the structural, optical and electrical properties of the films was studied. XRD study revealed that fluorine substitute oxygen site in the hexagonal lattice cause compressive stress in the films. The average crystallite size of FZO films decreased for 1 at. % F and then increased with further doping concentration. The optical band gap of FZO films larger than that of un-doped ZnO films and refractive index increased with increase of dopant concentration. The films exhibited lower electrical resistivity after annealing in the hydrogen atmosphere. Hydrogen related defects along with fluorine doping plays major role to improve the electrical properties.

REFERENCES

1. D. W. Kang, S. H. Kuk, K. SunJi, H. M. Lee and M. KooHan, Solar Energy Mater. Solar Cells **95**, 138 (2011).
2. D. K. Hwang, M. S. Oh, J. H. Lim and S. J. Park, J. Phys. D: Appl. Phys. **40**, R387 (2007).
3. K. Vijayalakshmi, C. Ravi dhas, V. Vasanthi Pillay and D. Gopalakrishna, Thin Solid Films **519**, 3378 (2011).
4. T. Minami, MRS Bull. **25**, 28 (2000).
5. K. Ellmer, J. Phys. D **34**, 3097 (2001).
6. D. C. Look, G. C. Farlow, P. Reunchan, S. Limpijumnong, S. B. Zhang and K. Nordlund, Phys. Rev. Lett. **95** 225502-1 (2005).
7. L. Gao, Y. Zhang, J. M. Zhang and K. W. Xu, Appl. Surf. Sci. **257**, 2498 (2011).
8. Y. Kim, W. Lee, D. R. Jung, J. Kim, S. Nam, H. Kim, and B. Park, Appl. Phys. Lett. **96**, 171902 (2010).
9. Z. Yang and J. L. Liu, J. Vac. Sci. Technol. B **28**, No. 3, C3D6 (2010).
10. R. Menon, V. Gupta, H. H. Tan, K. Sreenivas, and C. Jagadish, J. Appl. Phys. **109**, 064905 (2011).
11. W. M. Hlaing Oo, L. V. Saraf, M. H. Engelhard, V. Shutthanandan, L. Bergman, J. Huso, and M. D. Mc Cluskey, J. Appl. Phys. **105**, 013715 (2009).
12. S. Ilican, F. Yakuphanoglu, M. Caglar and Y. Caglar, J. Alloys Comp. **509** 5290 (2011).
13. K. Samanta, P. Bhattacharya and R. S. Katiyar, J. Appl. Phys. **108**, 113501 (2010).
14. M. Krunks, A. Katerski, T. Dedova, I. Oja Acik and A. Mere, Solar Energy Mater. Solar Cells **92**, 1016 (2008).
15. H. X. Chen, J. J. Ding, and S. Y. Ma, Physica E **42**, 1487 (2010).
16. E. Burstein, Phys. Rev. **93**, 632 (1954).
17. W. W. Liu, B. Yao. Y. F. Li, B. H. Li, Z. Z. Zhang, C. X. Shan, D. X. Zhao, J. Y. Zhang, D. Z. Shen and X. W. Fan, Thin Solid Films **518**, 3923 (2010).
18. M. Bouderbala, S. Hamzaoui, M. Adnane, T. Sahraoui and M. Zerdali, Thin Solid Films **517**, 1572 (2009).
19. B. L. Zhu, J. Wang, S. J. Zhu, J. Wu, R. Wu, D. W. Zeng and C. S. Xie Thin Solid Films **519**, 3809 (2011).
20. R. Anandhi, R. Mohan, K. Swaminathan and K. Ravichandran, Superlattices Microstruct. **51**, 680 (2012).

Non-ZnO Oxides

Mater. Res. Soc. Symp. Proc. Vol. 1494 © 2013 Materials Research Society
DOI: 10.1557/opl.2013.5

Crystal Structure of Non-Doped and Sn-Doped α-(GaFe)₂O₃ Thin Films.

Kentaro Kaneko[1,2 *] Kazuaki Akaiwa[1,2], and Shizuo Fujita[2]

[1]Department of Electronic Science and Engineering, Kyoto University, Katsura, Nishikyo-ku, Kyoto 615-8510, Japan
[2]Photonics and Electronics Science and Engineering Center, Kyoto University, Katsura, Nishikyo-ku, Kyoto 615-8520, Japan

ABSTRACT

Corundum structured α-(GaFe)$_2$O$_3$ alloy thin films were obtained on c-plane sapphire substrates by the mist chemical vapor deposition method. Wide range of X-ray diffraction 2θ/θ scanning measurements indicated that these crystals were epitaxially grown on c-plane sapphire substrates and these are no other crystal oriented phase. The cross-sectional and plane-view transmission electron microscope images showed the growth along the c-axis of α-(GaFe)$_2$O$_3$ thin films on sapphire substrates, forming joint of columnar structure. The non-doped α-(GaFe)$_2$O$_3$ thin films showed ferromagnetic properties at 300 K, though the origin of ferromagnetism still remained unresolved. In order to enhance the spin-carrier interaction, Sn doped α-(GaFe)$_2$O$_3$ alloy thin films were fabricated on c-plane sapphire substrates. X-ray diffraction 2θ/θ and ω scanning measurement results indicated that the highly-crystalline films were epitaxially grown on substrates in spite of the Sn-doping.

INTRODUCTION

Gallium oxide (Ga$_2$O$_3$) is a wide band gap semiconductor and has five types of crystal structures as α-, β-, γ-, δ-, and ϵ-[1]. It is also one of the candidate materials which are expected to be applied for conventional and novel electric devices, for example, high voltage power devices[2], solar-blind deep ultraviolet photo-detectors[3-6], gas sensors[7,8], as well as novel applications utilizing spintronic[9-11] and multiferroic[12-14] properties. In terms of making an alloy system, corundum structured α-Ga$_2$O$_3$ has a big advantage over typically-researched conductive oxides such as ZnO, In$_2$O$_3$, or SnO$_2$. Besides α-Ga$_2$O$_3$, there are other various metal oxides with corundum shape as α-Al$_2$O$_3$, α-In$_2$O$_3$, α-Fe$_2$O$_3$, α-Cr$_2$O$_3$, α-V$_2$O$_3$, and α-Ti$_2$O$_3$. These corundum-structured oxides can make a new alloy system realizing a variety of heterostructures and unique multifunctions[15].

Figure 1 shows the relationship between optical band gaps and lattice lengths of a-axis composed of typical corundum-structured metal oxides. Alloys of α-Al$_2$O$_3$, α-Ga$_2$O$_3$, and α-In$_2$O$_3$ will exhibit semiconducting properties[16] with different band gap, and alloys of these semiconductors and other transition-metal oxides will show unique multifunctions. We have paid attention and investigated the alloy of α-Ga$_2$O$_3$ and α-Fe$_2$O$_3$ because the ionic radii of Ga^{3+} and Fe^{3+} are close[17] and the lattice mismatch values are relatively small, while the alloy may exhibit unique properties fusing semiconductor and magnetic properties. From our previous

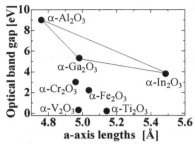

Fig. 1. Relationship between optical band gaps and lattice lengths of a-axis composed of corundum structured typical metal oxides, α-Al_2O_3, α-Ga_2O_3, α-In_2O_3 and transition metal oxides, α-Fe_2O_3, α-Cr_2O_3, α-V_2O_3 and α-Ti_2O_3.

studies, highly-crystalline α-$(GaFe)_2O_3$ alloy thin films were obtained[18] and α-$(GaFe)_2O_3$ (Fe= 58 %) showed ferromagnetic properties at 300K[15]. More recently we showed that Sn-doping, which had been found to be effective for creating donors[19], enhanced the magnetization hysteresis, and this encouraged future application of α-$(GaFe)_2O_3$ for spintronic devices. However, it is necessary to discuss the crystal structures of α-$(GaFe)_2O_3$ in detail in order to clarify the origin of ferromagnetism and to realize enhanced ferromagnetic properties. In this paper we discuss their crystal structure based on X-ray diffraction (XRD) and transmission electron microscope (TEM) observation.

EXPERIMENTS

α-$(GaFe)_2O_3$ alloy thin films were grown by the mist chemical vapor deposition (CVD) technique, which has been originally developed by our laboratory[20,21]. C-plane sapphire (α-Al_2O_3) was used as substrates which were cleaned by acetone, methanol, and deionized water. Gallium acetylacetonate [$(C_5H_8O_2)_3Ga$] and iron acetylacetonate [$(C_5H_8O_2)_3Fe$] were used as sources for Ga and Fe, respectively; they were solved in deionized water. For Sn-doping, tin(II) chloride dehydrate [$SnCl_2 \cdot 2H_2O$] was solved in the source. The concentration of $SnCl_2$ in the solution was fixed at 0.4 %. The growth temperature was 500 °C and the growth time was 50 min. XRD (Rigaku ATX) and TEM (JEOL JEM-2100F) observation were carried out to evaluate the crystallinity and structural properties of thin films.

RESULTS AND DISCUSSIONS

Figure 2 is the XRD 2θ/θ scanning spectra from 20 to 90 degrees of α-$(Ga_{1-x}Fe_x)_2O_3$ alloy thin films whose Fe ratios were $x = 0$, 0.07, 0.26, 0.58, and 1 in the thin films. There are only {0001} diffraction peaks which originated from c-plane. The {0001} peak positions of α-$(GaFe)_2O_3$ were gradually shifted to lower angles as Fe concentrations increased. No other crystal phase was detected. We reported that the full-width at half maximum (FWHM) values of XRD ω scanning profiles for α-$(Ga_{1-x}Fe_x)_2O_3$ alloy thin films were much smaller than 100 arcsec

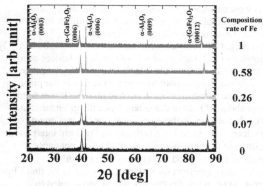

Fig, 2. XRD $2\theta/\theta$ scanning spectra of α-Ga$_2$O$_3$ (bottom) and α-(Ga$_{1-x}$Fe$_x$)$_2$O$_3$ (x= 0.07, 0.26, 0.58, 1) films grown on c-plane sapphire substrates at 500 °C.

in the entire range of Fe concentrations[18]. These results suggest that single-phase α-(GaFe)$_2$O$_3$ thin films were epitaxially grown on c-plane sapphire substrates.

Microstructure analysis were also conducted to reveal the detailed defect structure of α-(GaFe)$_2$O$_3$ thin films on sapphire substrates. Figure 3 is the cross-sectional and plane-view TEM images at the α-(Ga$_{0.56}$Fe$_{0.44}$)$_2$O$_3$/α-Al$_2$O$_3$ interface and the α-(Ga$_{0.56}$Fe$_{0.44}$)$_2$O$_3$ surface. There are clear lines in a vertical direction which is parallel to c-axis. These lines seem to divide α-(Ga$_{0.56}$Fe$_{0.44}$) $_2$O$_3$ crystals and these lines are very different from dislocation lines; they seem to be the edges of columnar crystals of α-(Ga$_{0.56}$Fe$_{0.44}$)$_2$O$_3$, that is, the film is composed with columnar structures. The plain-view image showed the grains whose size was estimated to be

Fig, 3. Cross-sectional TEM images of α-(Ga$_{0.56}$Fe$_{0.44}$)$_2$O$_3$/α-Al$_2$O$_3$ interface viewed along [11$\bar{2}$0] on the left side and plane-view image on the right side.

about 100 nm in diameter. These grains were expected to be the upper parts of the column crystals. These results indicate that the perpendicularly oriented growth is dominant in this crystal growth condition, resulting in collective structure of columnar crystals.

In the previous paper where we reported TEM observation results at the α-Ga$_2$O$_3$/α-Al$_2$O$_3$ interface, we pointed out the domain matching of 20 lattices of α-Ga$_2$O$_3$ with 21 lattices of α-Al$_2$O$_3$, which eliminate the generation of dislocations from the interface originating from the lattice mismatch[22]. However, the domain matching cannot be expected on entire the plane, so it is natural that the grown layers show the grain structure with grain boundaries. It was found that the grain size formed in this way was about 100 nm in diameter. Once the grain structure was formed at the initial stage of the growth, the growth should occur toward c-axis direction, resulting in perpendicular columnar structure. The TEM images shown in Fig.3 are interpreted by the mechanism given above. The growth in the perpendicular direction may not cause fluctuation of lattice planes, and this result in a fairly good XRD ω scanning profile.

We have shown room temperature ferromagnetism for α-(Ga$_{0.56}$Fe$_{0.44}$)$_2$O$_3$ alloy thin films, that is, it showed clear hysteresis curves at 300 K when the magnetic field was applied to perpendicular to c-axis[15]. As an origin of the ferromagnetism we are speculating that s and p orbitals of α-Ga$_2$O$_3$ and d orbital of α-Fe$_2$O$_3$ cause spin-carrier interactions. However the long moment values were much less than the effective moment of Fe^{3+} ($S = 5/2$).
One of the estimated reasons about the small long moment value is that the origin of ferromagnetism is Dzyaloshinskii-Moriya interaction because the long moment values are close to same to α-Fe$_2$O$_3$[23,24]. We are planning to discuss the origin of ferromagnetism in the future study, nevertheless the donor doping may be important as a successive research because the spin-carrier interaction force is expected to be enhanced by the increase of carrier concentration. Since Sn-doping can form donors in α-Ga$_2$O$_3$ thin films and the films exhibited n-type[19,25], attempts were made to dope Sn in α-(GaFe)$_2$O$_3$ expecting the increase of the coercivity or long moment values.

Figure 4(a) is the XRD $2\theta/\theta$ scanning spectra of a Sn-doped α-(GaFe)$_2$O$_3$ thin film on a c-plane sapphire substrate, where the Sn concentration was 0.4 % in the solution source. The XRD profile indicated the epitaxial growth on sapphire substrate. An example of the FWHM values of ω scanning profile is 64.8 arcsec, that is, the obtained Sn-doped α-(GaFe)$_2$O$_3$ thin film

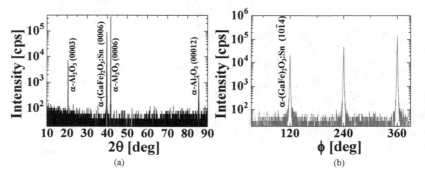

Fig.4. XRD $2\theta/\theta$ scanning spectra of Sn doped α-(Ga$_{0.42}$Fe$_{0.58}$)$_2$O$_3$ thin films. (a) $2\theta/\theta$ and (b) ϕ scanning profile.

is highly-crystalline. The φ scanning profile, listed in Fig. 4(b), showed three-symmetric peaks derived from (10$\bar{1}$4) planes; this is a strong evidence that the obtained crystal is corundum structured.

CONCLUSIONS

We reported the structure analysis of non-doped and Sn-doped α-(GaFe)$_2$O$_3$ thin films on α-Al$_2$O$_3$ substrates. The XRD scanning spectra indicated that the films, both non-doped and Sn-doped, were epitaxially grown on α-Al$_2$O$_3$ substrates. Microstructure analysis by TEM observation revealed the joint of columnar structure. The formation of the columnar structure is speculated to be the domain mismatch at the initial stage of the growth on the α-Al$_2$O$_3$ substrates. The successive growth occurs perpendicularly along the c-axis, without resulting in fluctuation of crystal planes.

ACKNOWLEDGMENTS

The authors acknowledge Professor Minoru Suzuki of Kyoto University for arrangements of magnetization measurements and critical discussions, and Dr. Hitoshi Kawanowa of Ion Technology Center for critical supports in TEM observation and discussions. This work was partly supported by Grant-in-Aid for JSPS Fellows.

REFERENCES

1. R. Roy, V. G. Hill, and E. F. Osborn: J. Am. Chem. Soc. **74** (1952) 719.
2. M. Higashiwaki, K. Sasaki, A. Kuramata, T. Masui, and S. Yamakoshi: Appl. Phys. Lett. **100** (2012) 013504.
3. T. Oshima, T. Okuno, N. Arai, N. Suzuki, H. Hino, and S. Fujita: Jpn. J. Appl. Phys. **48** (2009) 011605
4. Y. Kokubun, K. Miura, F. Endo, and S. Nakagomi: Appl. Phys. Lett. **90** (2007) 031912.
5. T. Oshima, T. Okuno, and S. Fujita: Jpn. J. Appl. Phys. **46** (2007) 7217.
6. T. Oshima, T. Okuno, N. Arai, N. Suzuki,S. Ohira, and S. Fujita: Appl. Phys. Express **1** (2008) 011202.
7. C.-C. Chen and C.-C. Chen: J. Mater. Res. **19** (2004) 1105.
8. M. Ogita, K. Higo, Y. Nakanishi, and Y. Hatanaka: Appl. Surf. Sci. **175** (2001) 721.
9. H. Hayashi, R. Huang, H. Ikeno, F. Oba, S. Yoshioka, I. Tanaka, and S. Sonoda.: Appl. Phys. Lett. **89** (2006) 181903.
10. R. Huang, H. Hayashi, F. Oba, and I. Tanaka: J. Appl. Phys. **101** (2007) 063526.
11. H. Hayashi, R. Huang, F. Oba, T. Hirayama, and I. Tanaka: J. Mater. Sci. **46** (2011) 4169
12. J. P. Remeika: J. Appl. Phys. **31** (1960) 263S.
13. N. Kida, Y. Kaneko, J. P. He, M. Matsubara, H. Sato, T. Arima, H. Akoh, and Y. Tokura: Phys. Rev. Lett. **96** (2006) 167202.
14. T. Arima: J. Phys.: Condens.Matter **20** (2008) 434211
15. K. Kaneko, I. Kakeya, S. Komori, and S. Fujita: J. Appl. Phys. (*submitted*)
16. H. Ito, K. kaneko, and S. Fujita: Jpn. J. Appl. Phys. **51** (2012) 100207
17. R. D. Shannon: Acta Cryst. A**32** (1976) 751
18. K. Kaneko, T. Nomura, I. Kakeya, and S. Fujita: Appl. Phys. Express **2** (2009) 075501

19. K. Akaiwa and S. Fujita: Jpn. J. Appl. Phys. **51** (2012) 070203
20. H. Nishinaka, T. Kawaharamura, and S. Fujita: Jpn. J. Appl. Phys. **46** (2007) 6811
21. T. Kawaharamura, H. Nishinaka, and S. Fujita: Jpn. J. Appl. Phys. **47** (2008) 4669
22. K. Kaneko, H. Kawanowa, H. Ito, and S. Fujita: Jpn. J. Appl. Phys. **51** (2012) 020201
23. I. Dzyaloshinsky: J. Phys. Chem. Solids 4 (1958) 241-255.
24. T. Moriya: Phys. Rev. Lett. 4 (1960) 228.
25. T. Kawaharamura, G. T. Dang, and M. Furuta: Jpn. J. Appl. Phys. **51** (2012) 040207

Mater. Res. Soc. Symp. Proc. Vol. 1494 © 2012 Materials Research Society
DOI: 10.1557/opl.2012.1650

Deposition of tin oxides by Ion-Beam-Sputtering

Martin Becker, Angelika Polity, Davar Feili and Bruno K. Meyer
1st Physics Institute, Justus-Liebig University, Heinrich-Buff-Ring 16, 35392 Giessen, Germany

ABSTRACT

Synthesis of both p-type and n-type oxide semiconductors is required to develop oxide-based electronic devices. Tin monoxide (SnO) recently has received increasing attention as an alternative p-type oxide semiconductor because it is a simple binary compound consisting of abundant elements. Another phase of the tin oxygen system, SnO_2, is of great technological interest as transparent electrodes and as heat-reflecting filters. The preparation of tin oxide thin films has been performed by many different procedures. Radio-frequency (RF) ion-thrusters, as designed for propulsion applications, are also qualified for thin film deposition and surface etching, because different gas mixtures, extraction voltages and RF power can be applied. Tin oxide thin films were grown by ion beam sputtering (IBS) using a 3" metallic tin target. Different aspects of the thin film growth and properties of the tin oxide phases were investigated in relation to flux of oxygen fed into the gas discharge in the ion thruster. Results on thin film growth by IBS will be presented, structural, vibrational and optical properties of the films will be discussed.

INTRODUCTION

Tin dioxide (SnO_2, stannic oxide) is one of the most studied metal oxide semiconductors, which exhibits a wide band gap of 3.8 eV and belongs to the class of transparent conductive oxides. Oxygen vacancies as well as hydrogen and fluorine donors may lead to n-type doping in SnO_2 and, accordingly produce a fairly high electrical conductivity. Such properties lead to several technological applications among them gas sensors [1–3] and solar cells [4,5]. Tin monoxide (SnO, stannous oxide) exhibits p-type conductivity with relatively high hole mobilities [6]. The p-type conductivity of SnO was suggested to originate from tin vacancies [7]. Although SnO may be a promising p-type material, the physical and electrical properties of SnO have not been studied in great detail. In the last years SnO has been used in a variety of applications, anode materials for lithium rechargeable batteries [8] and precursor for the production of SnO_2 [9] among them. Recently, SnO has received particular attention because of the difficulty in obtaining stable and high quality p-type semiconductors based on other oxides like ZnO or Cu_2O. To deposit thin films of tin oxides various techniques have been applied among them sol gel, spray pyrolysis, thermal evaporation, electron beam evaporation, pulsed laser deposition, chemical vapor deposition techniques, molecular beam epitaxy, dc sputtering, RF magnetron sputtering and ion beam sputtering. Among all other deposition techniques, ion beam sputtering [3,10] offers significant advantages over other forms of sputtering [11]. Deposition parameters such as RF power or extraction voltages are not only tunable over a wide range, but can also be varied nearly independently of each other. Furthermore IBS is suitable for the use of a great number of target materials and compositions, since the sputter species in the ion beam can be easily and accurately controlled by adjusting the gas flow into the ion source.

The deposition system generates a high density RF argon plasma in the RIM source ("radio frequency ion thruster for material processing"). In those thrusters the working gas is fed by the flow controller into the discharge chamber made of quartz, which is surrounded by the RF-induction coil of the RF-generator (2 MHz). The inductively coupled eddy electromagnetic RF-field generates a selfsustaining, electrodeless ring discharge. The ion extraction is accomplished by a multihole three-grid extraction system using the accel-decel technique. RF ion sources for material processing have been developed with several ionizer diameters. Therefore these ion sources are excellently suited to be operated with reactive as well as with nonreactive gases. For further details the reader might be referenced to [12].

EXPERIMENT

Deposition of tin oxides is performed using the system described above with intentional heating the substrate up to 400°C from a 3" diameter metallic tin target of 99.999% purity (*Kurt J. Lesker Company*) in an atmosphere of argon and oxygen gases, both of which have a 99.999% purity. The chamber is pumped to a base pressure of 2×10^{-6} mbar. The RF launch power was chosen to be 200 W, whereas the extraction voltages were 2.4 kV applied to the screen grid and 0.2 kV applied to the accel grid. Layers were synthesized on soda lime glass, using different O_2 flux ranging from 0 to 15 sccm. Structural properties of the deposited layers were investigated with X-ray diffraction (XRD) measurements, carried out on a SIEMENS D5000 diffractometer with a Cu K_a X-ray source operating in Bragg-Brentano geometry. Raman measurements were performed on a Renishaw InVia microscope system. The spectra were recorded in backscattering geometry at room temperature. A polarized 525 nm excitation laser was focused onto the sample surface using a 50× objective. The same objective was used to collect the scattered light, which was then dispersed by a spectrometer with a focal length of 250 mm and recorded by a CCD detector. The system's spectral resolution is about 1.5 cm^{-1}. Thickness and optical transmittance were determined using a PerkinElmer Lambda 900. The UV-visible transmission data were then used to extract the optical band-gap (E_g) by following Tauc's method [13].

DISCUSSION

To optimize the properties of tin oxide thin films, a series of tin oxide samples were prepared by varying the reactive gas flow rate. The effect of this deposition parameter on film properties will be discussed here.

X-ray diffraction pattern taken from the layers indicated that it is possible to switch between the growth of different tin oxide phases. Figure 1a shows X-ray diffractograms obtained from thin films grown at 400°C under different oxygen flows, figure 1b the evaluated lattice constants, respectively. The vertical lines indicate the literature position of different lattice planes and were taken from the ICDD PDFs 00-004-0673 for Sn, 01-072-1012 for SnO and 00-041-1445 for SnO_2. It can be observed that between 7 and 9 sccm O_2 flow, polycrystalline SnO layers were synthesized. No additional phases are visible in between this region. At lower oxygen flows additional reflexes of metallic tin can be observed, while at oxygen flux higher than 10 sccm traces of SnO_2 can be found. The lattice constants of both phases determined from these measurements fit extremely well with the values known from literature.

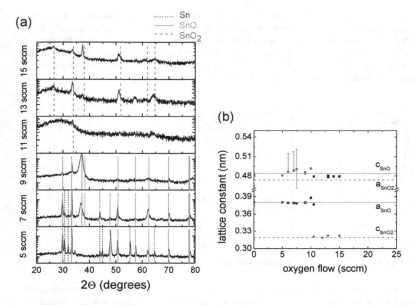

(a)

.... Sn
——— SnO
– – – SnO$_2$

(b)

Figure 1. X-ray diffraction analysis: (a) stack of samples representative for layers deposited at substrate temperature of 400°C under oxygen flux of 5 - 15 sccm and additional inert gas flux of 7.5 sccm; (b) lattice constants of tin oxide thin films calculated from XRD reflex positions in comparison with values from literature indicated by horizontal lines.

The vibrational characteristics of the tin oxide layers were performed via Raman spectroscopy under polarized 525 nm excitation wavelength. One observes modes which according to literature can be assigned to either stannic oxide (SnO$_2$) or stannous oxide (SnO). The process window for SnO around 7 to 9 sccm oxygen flow is slightly smaller than in X-ray diffraction, indicating that Raman spectroscopy might be more sensitive to small inclusions of other phases. Another explanation might be given due to surface oxidation. The optical Γ-point phonon modes of SnO have been discussed in several papers throughout the last decades, where the A$_{1g}$, B$_{1g}$ and E$_g$ modes are Raman active, while A$_{2u}$ and E$_u$ are said to be infrared active. Raman scattering experiments confirmed SnO active Raman modes at 113 and 211 cm^{-1} [14] assigned to B$_{1g}$ and A$_{1g}$, respectively, whereas theory predicts these assignments to be incorrect justified due to calculations in the frozen-phonon approach and frequencies observed in isostructural α-PbO [15]. Figure 2a shows the Raman spectra of different samples analyzed under prospect to be SnO. In addition to volume modes at 113 and 211 cm^{-1} one also detects two modes at around 140 and 165 cm^{-1}, which can neither be attributed to SnO nor SnO$_2$. This phenomenon is well-known from annealing series on SnO and the bands are usually ascribed to an intermediate oxidation state SnO$_x$ [2,14].

According to group theory SnO$_2$ exhibits 18 vibrational modes. Two modes are infrared active (the single A$_{2u}$ and the triply degenerate E$_u$), four are Raman active (the three nondegenerate A_{1g}, B_{1g}, B_{2g} modes and the doubly degenerate E_g one) and two others are silent (the A_{2g} and B_{1u} modes). Figure 2b shows the Raman spectra of samples identified as SnO_2 (at least deduced from XRD). In comparison with the Raman shift of the most important bands observed in SnO_2 [16] one observes a dominant A_{1g} mode and several shoulders indicating bands in the region between 300 and 800 cm^{-1}. According to literature it seems reasonable to assign these modes to IR modes whose Raman activity is induced by disorder [17,18].

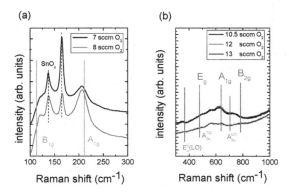

Figure 2. Raman spectra of SnO (a) and SnO_2 (b) samples. Besides volume modes one observes modes, which might be ascribed to intermediate oxidation states and IR modes whose Raman activity is induced by disorder.

These compositional changes are consistent with the modification observed in the X-ray diffraction measurements, which also reflected similar changes in film composition, as represented by variations in the relative intensity of the diffraction peaks corresponding to different phases. Since one might assume from these data (Figure 2), at low oxygen partial pressures the film composition is dominated by a mixture of metallic tin and stannous oxide phase. As the oxygen flux is increased, this phase mixture gives way to stannous oxide.

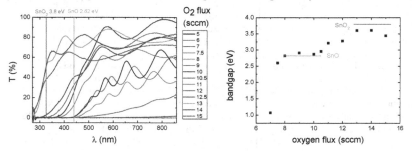

Figure 3. Transmission spectra (left) and calculated band gaps (right) for SnO_x layers prepared at room temperature.

Transmission spectra are given in figure 3a. Owing to the absorption of the glass substrate and the limits of the utilized photomultiplier the transmission is measured only in the range of 0.25 to 0.86 μm. Since it is known from other semiconductors sputtered from metallic targets, that the layer thickness decreases with increasing the oxygen content in the plasma [19], a correction concerning this point has to be performed. Layer thicknesses were calculated from the extrema utilizing the Sellmeier approach and optical constants of SnO_2 taken from [20] and references therein. Optical constants of SnO were calculated from the inverse Sellmeier approach utilizing other methods like X-ray reflectometry and profilometry for determination of thicknesses (not shown here). Once the film thickness is known the Lambert-Beer equation can be used to determine the absorption coefficient and therefore the band gap of a transparent thin film, shown in figure 3b. As can be seen from these spectra the band gap shifts to higher energies with increasing oxygen flux during deposition accompanying the phase transitions from mixed SnO and metallic tin over stannous oxide towards nearly stoichiometric stannic oxide. The slight decrease in band gap observed at high oxygen flux might be due to the uncertainty arising from the decreasing number of maxima / minima accompanying the decrease of thickness. Since evaluated from Tauc plots the absolute values can only be seen as a crude approximation, whereas the plateaus agree with the different oxide phases concluded from X-ray diffraction and Raman spectroscopy.

CONCLUSIONS

This paper reports the ion beam deposition of tin oxide thin films utilizing a RF ion thruster. It is known from the literature survey that the formation of pure SnO film is often difficult, due to the more favored formation or co-formation of SnO_2. Our investigation revealed the importance of precise control of the oxygen flux in order to attain transparent and semiconducting films that are suitable for electronic applications such as TFTs and solar cells. More specifically, at deposition temperature of 400°C the choice of around 10 sccm, which equates to an oxygen partial pressure of 40%, favors the formation of SnO_2. By precise management of oxygen and power settings, transparent SnO_2 films have been synthesized at deposition rates up to 5 nm per minute.

ACKNOWLEDGMENTS

This work was supported in the framework of "LOEWE-Schwerpunkt RITSAT".

REFERENCES

1. T. Oyabu, "Sensing characteristics of SnO2 thin film gas sensor," Journal of Applied Physics **53**, 2785 (1982).
2. L. Sangaletti, L. . Depero, A. Dieguez, G. Marca, J. . Morante, A. Romano-Rodriguez, and G. Sberveglieri, "Microstructure and morphology of tin dioxide multilayer thin film gas sensors," Sensors and Actuators B: Chemical **44**, 268–274 (1997).
3. B. K. Min and S. D. Choi, "SnO2 thin film gas sensor fabricated by ion beam deposition," Sensors and Actuators B: Chemical **98**, 239–246 (2004).

4. T. Feng, A. K. Ghosh, and C. Fishman, "Spray-deposited high-efficiency SnO2/n-Si solar cells," Applied Physics Letters **35**, 266 (1979).

5. N. Nang Dinh, M.-C. Bernard, A. Hugot-Le Goff, T. Stergiopoulos, and P. Falaras, "Photoelectrochemical solar cells based on SnO2 nanocrystalline films," Comptes Rendus Chimie **9**, 676–683 (2006).

6. Y. Ogo, H. Hiramatsu, K. Nomura, H. Yanagi, T. Kamiya, M. Hirano, and H. Hosono, "p-channel thin-film transistor using p-type oxide semiconductor, SnO," Applied Physics Letters **93**, 032113 (2008).

7. A. Togo, F. Oba, I. Tanaka, and K. Tatsumi, "First-principles calculations of native defects in tin monoxide," Phys. Rev. B **74**, 195128 (2006).

8. R. Kohler, H. Besser, M. Hagen, J. Ye, C. Ziebert, S. Ulrich, J. Proell, and W. Pfleging, "Laser micro-structuring of magnetron-sputtered SnO2 thin films as anode material for lithium ion batteries," Microsystem Technologies **17**, 225–232 (2011).

9. W. Choi, H. Sung, K. Kim, J. Cho, S. Choi, H.-J. Jung, S. Koh, C. Lee, and K. Jeong, "Oxidation process from SnO to SnO2," Journal of Materials Science Letters **16**, 1551–1554 (1997).

10. T. Suzuki, T. Yamazaki, H. Yoshioka, and K. Hikichi, "Influence of thickness on H2 gas sensor properties in polycrystalline SnO2 films prepared by ion-beam sputtering," Journal of Materials Science **23**, 1106–1111 (1988).

11. P. Y. Yu, Y. R. Shen, Y. Petroff, and L. M. Falicov, "Resonance Raman Scattering at the Forbidden Yellow Exciton in Cu2O," Physical Review Letters **30**, 283–286 (1973).

12. J. Freisinger, J. Heland, D. Kramer, H. Loeb, and A. Scharmann, "Performance of the rf-ion sources RIM for reactive and nonreactive gases," Review of Scientific Instruments **63**, 2571–2573 (1992).

13. J. Tauc, R. Grigorovici, and A. Vancu, "Optical Properties and Electronic Structure of Amorphous Germanium," physica status solidi (b) **15**, 627–637 (1966).

14. J. Geurts, S. Rau, W. Richter, and F. J. Schmitte, "SnO films and their oxidation to SnO2: Raman scattering, IR reflectivity and X-ray diffraction studies," Thin Solid Films **121**, 217–225 (1984).

15. E. L. Peltzer y Blancá, A. Svane, N. E. Christensen, C. O. Rodríguez, O. M. Cappannini, and M. S. Moreno, "Calculated static and dynamic properties of β-Sn and Sn-O compounds," Phys. Rev. B **48**, 15712–15718 (1993).

16. R. S. Katiyar, P. Dawson, M. M. Hargreave, and G. R. Wilkinson, "Dynamics of the rutile structure. III. Lattice dynamics, infrared and Raman spectra of SnO2," Journal of Physics C: Solid State Physics **4**, 2421–2431 (1971).

17. L. Abello, B. Bochu, A. Gaskov, S. Koudryavtseva, G. Lucazeau, and M. Roumyantseva, "Structural Characterization of Nanocrystalline SnO2 by X-Ray and Raman Spectroscopy," Journal of Solid State Chemistry **135**, 78–85 (1998).

18. A. Diéguez, A. Romano-Rodríguez, A. Vilà, and J. R. Morante, "The complete Raman spectrum of nanometric SnO2 particles," Journal of Applied Physics **90**, 1550–1557 (2001).

19. F. M. Li, R. Waddingham, W. I. Milne, A. J. Flewitt, S. Speakman, J. Dutson, S. Wakeham, and M. Thwaites, "Low temperature (<100°C) deposited P-type cuprous oxide thin films: Importance of controlled oxygen and deposition energy," Thin Solid Films **520**, 1278–1284 (2011).

20. R. D. Shannon, "Refractive Index and Dispersion of Fluorides and Oxides," Journal of Physical and Chemical Reference Data **31**, 931 (2002).

Mater. Res. Soc. Symp. Proc. Vol. 1494 © 2012 Materials Research Society
DOI: 10.1557/opl.2012.1580

Growth of Ultra-thin Titanium Dioxide Films by Complete Anodic Oxidation of Titanium Layers on Conductive Substrates

Karsten Wolff[1], Petri Heljo[1] and Donald Lupo[1]

[1]Tampere University of Technology, Department of Electronics, Printed and Organic Electronics Group, P.O. Box 692, 33101 Tampere, Finland

ABSTRACT

The growth of thin and ultra-thin titanium dioxide layers was investigated. Oxide films were grown by galvanostatic and potentiodynamic anodisation of evaporated titanium layers on conductive substrates. It is shown that thin-film oxidation differs significantly from anodic oxidation of solid foils or plates, due to the sudden stop of anodisation process before complete oxidation of the thick films. Depending on the pH value and the potential sweep rate, the effective defect density and the dielectric constant of the anodized layers vary from $3 \cdot 10^{19}$ cm^{-3} to 10^{20} cm^{-3} and from 16 to 27, respectively, whereas the electrolyte temperature plays only a minor role.

INTRODUCTION

Titanium dioxide (TiO$_2$) is a widely studied material and it is used in many applications including gate insulators, dye-sensitized solar cells, photo-catalysis, anti-corrosion coatings as well as sensors. In contrast to oxides of other valve metals (e.g. Al or Hf), TiO$_2$ exhibits rather semiconducting than insulating properties. It has a relatively small bandgap ($E_g \approx 3.2$ eV) and oxygen deficiencies cause n-type conductivity. Very thin layers of TiO$_2$ are needed for some applications like gate dielectrics or organic tunnelling diodes. Anodisation is known to be a cost-effective technique to grow the thin oxide layers [1, 2]. In some applications the full oxidation of the Ti layer is required. Since we do not know any report on the complete anodisation of Ti on conductive substrates, we have focused on the growth behavior and the influence of the process parameters on the effective defect density and the dielectric constant.

EXPERIMENT

Titanium electrodes were fabricated on glass substrates (2.5cm x 2.5cm) coated with a conductive layer of either gold, platinum or indium tin oxide. For Au and Pt as conductive film, microscope slides were degreased in isopropanol, rinsed with water and dried using an ionized air gun. Prior to deposition of the conductive films, UV/ozone treatment was performed. Then, 5 nm adhesion layer of Ti and 50 nm of Au or 100 nm of Pt were e-beam evaporated. The platinum layer was thicker in order to compensate the lower conductivity compared to gold. Polished and ITO-coated glass substrates were purchased from Delta Technologies, Ltd. (USA) and cleaned following the previously mentioned procedure. On top of the respective conductive layer, Ti was e-beam evaporated at different layer thicknesses. Measurement of the sheet resistances revealed values of $R_{sh} \approx 0.3$ Ω/sq and $R_{sh} \approx 21$ Ω/sq for the Au/Pt and ITO samples, respectively.

Anodisation was performed in citric acid solutions (1 mM and 0.1 M) using a Zennium electrochemical workstation (Zahner Elektrik GmbH, Germany). The samples were mounted in a sample holder, which provides electrical contacts and front side exposure to the electrolyte. The exposed area was fixed at 1.0 cm^2. A platinum mesh counter electrode and an Ag/AgCl (3M KCl) reference electrode completed the experimental setup. It has to be noted that a reference electrode was only available for the potentiodynamic anodisation and related measurements, so the potential shift ($\Delta V = 0.68$ V $- 0.71$ V) was measured and added subsequently. The correction limited the accuracy, but did not change the actual qualitative results. All potentials are presented as potentials versus normal hydrogen electrode, although measured versus Ag/AgCl or Pt counter electrode. Due to good control of the oxidation process, anodisation was done in galvanostatic (constant current density) and potentiodynamic mode (linear potential ramp).

The effective density of defect states was determined by electrochemical impedance spectroscopy in 0.1M NaCl and subsequent analysis of the Mott-Schottky plots. The measurement frequency was set between 2 kHz and 5 kHz, because the double layer capacitance can be neglected and the frequency is still low enough to ionize most of the defect and trap states.

RESULTS AND DISCUSSION

Galvanostatic anodisation

Anodisation in galvanostatic mode should lead to a linearly increasing cell potential in order to maintain a sufficiently high electric field, which drives a constant ion flux through the oxide layer. The applied potential is therefore proportional to the oxide layer thickness. Figure 1(a) shows the anodisation characteristics of 6 nm Ti and 200 nm Ti on Au-coated glass substrates as well as of solid titanium foil for comparison. The oxidation of the foil causes an increase of the cell potential. A deviation from an ideal linear increase was reported before [3]. After Faraday's law, the thickness of the oxide layer was approximately 110 nm. In contrast to Ti foil, thin layers of titanium do not show such a steep potential increase. Quite the contrary, the anodisation processes seemed to stop and the cell potentials approach limit values of 4.1 V and 2.4 V for 200 nm Ti and 6 nm Ti, respectively. This behavior is most surprising. Assuming a growth constant of 1.5 nm/V for a 200 nm film in citric acid [1], less than 3% of the maximum possible layer thickness are oxidized at this point and a reasonable explanation is not known. A significant potential rise for the 6 nm sample was only observed during the first 10 seconds of the oxidation, where the potential increased from 1.0 V to 2.3 V. It can be assumed that the 6 nm layer was entirely oxidized, if the native oxide is taken into account. The starting cell potential is due to the native oxide, too.

For some applications, smooth oxide layers are essential. However, during the anodisation porous layers of titanium dioxide can be formed [4]. Figure 1 (b) and (c) show the AFM topographies of the pristine Ti surface (200 nm) and 6 nm TiO$_2$ (grown out of 200 nm Ti), respectively. The difference between both topographies is not significant, but the anodized layer seems to have larger grain sizes. This is also mirrored by the RMS roughness, which increased from 3.1 nm to 3.6 nm during anodisation. Nevertheless, the change is only of little account and no porosity is visible by neither AFM nor optical inspection.

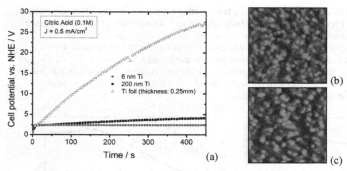

Figure 1. (a) Voltage-time characteristics of galvanostatic anodisation of Ti on Au-coated glass substrates in citric acid, (b) AFM topographies (non-contact mode) of non-anodized Ti and (c) 6 nm TiO_2. Scan width is 1 μm, full z-scale is 30 nm.

Anodisation was also tested for lower acid concentrations. Decreasing the ion concentration did not affect the anodisation but caused an additional voltage drop across the electrolyte. The electrolyte resistance was 2 kΩ – 4.5 kΩ higher for 1 mM citric acid compared to 0.1 M citric acid. Variation of the current density led primarily to more oxygen evolution at the working electrode. The anodisation characteristics are displayed in Figure 2. In order to make the experimental results comparable, potentials are plotted versus charge instead of time. The 6 nm samples show no dependency on the current density except of the electrolyte voltage drop. The measured potential differences between the samples are due to additional voltage drop across the electrolyte, which underlines the assumptions of full oxidation. However, certain dependence occurs for the 200 nm samples. The potential differences are in the range of 0.7 V to 1.4 V, which is too high to be caused by electrolyte resistance only. Nevertheless, we do not assign this potential increase to the oxide growth, because the expected change of layer color could not be observed.

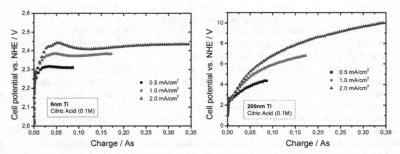

Figure 2. Anodisation characteristics for different current densities (anodisation duration: 600 s). Left: 6 nm Ti on Au, right: 200 nm on Au.

161

The galvanostatic anodisation was also tested with Pt- and ITO-coated glass substrates. As can be seen from Figure 3, different conductive substrates do not show essential differences of the anodisation behavior. It is remarkable that a peak occurs during anodisation on ITO. This peak is caused by partial peeling off of Ti/TiO$_2$. The inset in the right plot displays an optical micrograph after anodisation with point-shaped layer defects, which are only visible on samples anodized in excess of the peak or close-by. Adhesion of Ti on ITO and volumetric expansion stress seems to be a severe problem, in particular for a layer thickness > 5 nm, whereas thin layers are more robust against mechanical stress.

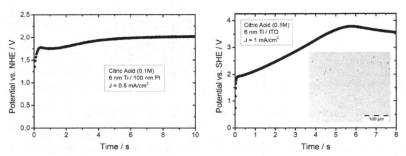

Figure 3. Anodisation characteristics for Pt-coated (left) and ITO-coated glass substrates (right). The inset shows an optical micrograph after anodisation.

Potentiodynamic anodisation

The potentiodynamic mode, where the voltage is increased with constant slope, is a more straightforward technique to analyze the anodisation process, because the charge transport can directly be read from the characteristics. A typical JV-curve is shown in Figure 4 (left). When starting the sweep from 0 V, the electric field strength is too low in order to cause the ion current through the native oxide. Then, anodisation begins slowly and current stabilizes (in the ideal case, the current would be constant). Right after the beginning of the process, a current peak was observed for all the samples, the position of the peak was dependent on the native oxide thickness. When all of the Ti was oxidized and the potential is further increased, the applied electric field exceeds the breakdown field strength and a linear increase of the current sets on, i.e. electron transport through the layer and electrolyte electrolysis.

The right plot in Figure 4 displays a typical Mott-Schottky plot for different measurement frequencies. While the upper plateau (> 0.2 V) represents the full depletion of the TiO$_2$ layer, the depletion condition is not fulfilled on the lower plateau (< 0 V). Between both levels, there is a narrow linear regime, whose slope is used for calculation of the trap density N$_t$ [5]. Under the assumption of a plate capacitor, the dielectric constant is derived from capacitance values in the full depletion portion.

The effective trap density and the dielectric constant were determined in dependence on various process parameters, which can potentially affect the physical properties [6]. Since adhesion is a severe problem and first pin-holes become visible at breakdown, the samples were only anodized up to a maximum potential just below breakdown.

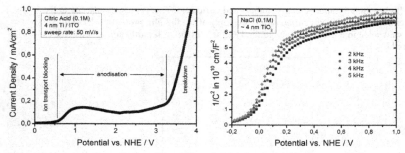

Figure 4. Potentiodynamic anodisation of Ti on ITO: typical JV-characteristics (left), Mott-Schottky plot (right)

Different sweep rates and therefore oxidation rates should result in different degrees of defect density. As can be seen from Figure 5 (left), the dependence on the sweep rate is only slightly pronounced, but indeed the effective trap density increases with faster sweeping, while the dielectric constant, which is an indicator for crystallinity, drops from 27 to 20.5. Also the

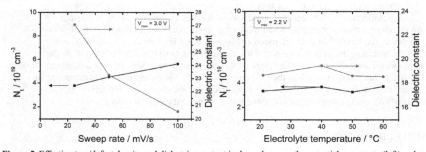

Figure 5. Effective trap/defect density and dielectric constant in dependence on the potential sweep rate (left) and on the anodisation electrolyte (0.1M citric acid) temperature (right). The initial Ti layer thicknesses in the left and right figure were 4 nm and 3 nm, respectively. Capacitance-voltage measurements at 2 kHz were performed at room temperature.

electrolyte temperature should have a certain impact on the electronic properties, because it will affect the oxidation rates, too. There is however no significant influence observable in Figure 5 (right). Either the effective defect density or the dielectric constant do not show a clear tendency for the variation of the temperature.

Beside the sweep rate and the temperature, it is known that the pH value affects the anodic growth of oxides and their electronic properties. Instead of pure citric acid, buffered solutions were prepared using sodium citrate as buffering agent in order to maintain a constant pH value during anodisation. Sodium hydroxide was used to adjust the pH value. In the range from pH 2 to pH 5, the effective defect density varies between $5 \cdot 10^{19}$ cm^{-3} and 10^{20} cm^{-3} with a decreasing tendency (Figure 6). Again, the dielectric constant goes up for less defective layers. The influence of the pH value is thus stronger than the influence of the sweep rate. Note that N_t

is generally higher than for anodisation in unbuffered citric acid. This is presumably due to the higher concentrations of Na^+ and citrate ions which may affect the film properties. Nevertheless, all reported values are in a reasonable range for TiO_2, in particular, if partly amorphous phases are considered [4].

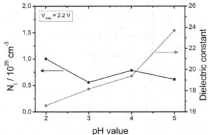

pH value

Figure 6. Effect of the electrolyte pH value on the defect density and dielectric constant. Anodisation in buffered citric acid; initial Ti layer thickness was 4 nm.

CONCLUSIONS

The anodisation of thin and ultra-thin Ti layers on conductive substrates behaves differently in comparison to conventional anodisation of solid Ti foils or plates. While the oxidation process stops automatically when oxidizing layers thicker than 10 nm, ultra-thin layers were completely oxidized. The detailed mechanism for the sudden process disruption is not known yet. In addition, the impact of sweep rate, electrolyte temperature and pH value on the defect density and the dielectric constant were investigated for potentiodynamic anodisation. While the temperature does not affect the properties distinctly, the defect density is increased for higher sweep rates and lower pH values. The dielectric constant shows the respective opposite tendency confirming the measured effect of the parameters on the defectiveness.

ACKNOWLEDGMENTS

The authors thank P. Berger (Ohio State University, USA) and M. Grell (Sheffield University, UK) for the valuable discussions. K. Wolff acknowledges the financial support from the German Research Foundation DFG (project number WO 1530/2-1). P. Heljo and D. Lupo acknowledge the financial support from the Academy of Finland (project number 251983).

REFERENCES

1. L. A. Majewski, R. Schroeder, M. Grell, Adv. Funct. Mat. **15**, 1017 – 1022 (2005)
2. W.-J. Yoon,S.-Y. Chung, P. R. Berger, S. M. Asar, Appl. Phys. Lett. **87**, 203506 (2005)
3. J.-L. Delplancke, R. Winand, Electrochimica Acta **33**, 1539 – 1549 (1988)
4. P. Roy, S. Berger, P. Schmuki, Angew. Chem. Int. Ed., **50**, 2904 – 2939 (2011)
5. S. M. Sze, K. K. Ng, Physics of Semiconductor Devices, Wiley & Sons, New York, (1981)
6. A. Goswami, Thin Film Fundamentals, New Age International, New Delhi (1996)

Mater. Res. Soc. Symp. Proc. Vol. 1494 © 2012 Materials Research Society
DOI: 10.1557/opl.2012.1581

Optical and electrical properties of Cu_2O, Cu_4O_3 and CuO

Daniel Reppin[1], Angelika Polity[1], Bruno K. Meyer[1], and Sviatoslav Shokhovets[2]

[1] Institute of Experimental Physics I, Justus-Liebig-University, Heinrich-Buff-Ring 16, 35392 Giessen, Germany

[2] Institute of Physics, Ilmenau University of Technology, Weimarer Strasse 32, 98693 Ilmenau, Germany

ABSTRACT

We deposited copper oxides by rf magnetron sputtering from a 4N Cu-target at room temperature, varying the oxygen flux and keeping the argon flow constant.
Dependent on the oxygen flux Cu_2O, Cu_4O_3 or CuO were synthesized. The different compounds were characterized by XRD. The dielectric functions of the oxides were determined by spectroscopic ellipsometry and show significant differences between the compounds.
The electrical properties, like the carrier concentration, of each compound can be tuned by adjusting the oxygen flux. We discuss the structural, optical and electrical properties of the copper oxides in terms of phase purity and stoichiometry deviations.

INTRODUCTION

With the band gaps ranging from 1.4 eV (CuO) to 2.2 eV (Cu_2O) and with the intrinsic p-type conductivity the copper oxides are interesting materials for a wide range of applications like thin film transistors or photovoltaic devices. Thus in the last years copper oxides gained a renewed interest of the research community due to their semiconducting properties and sustainability. The system of copper oxides comprises the phases cuprite (Cu_2O), paramelaconite (Cu_4O_3) and tenorite (CuO) [1–5]. Whereas the optical properties of Cu_2O are well known this does not apply for the other two compounds. Especially for Cu_4O_3, the published band gap energies range from 1.3 to 2.5 eV [3,6–8]. The non-stoichiometry of the copper oxides may also be useful in other applications like oxygen storage [9,10]. We investigated the properties of the three different compounds Cu_2O, Cu_4O_3 and CuO by XRD, Hall and spectroscopic ellipsometry (SE) measurements.

EXPERIMENT

The copper oxides were deposited on soda lime glass by rf magnetron sputtering technique from a copper target of 4N purity. The base pressure of the chamber was $5 \cdot 10^{-6}$ Pa for all depositions; the sputtering pressure was around $4.8 \cdot 10^{-1}$ Pa, depending on the oxygen flux. Sputtering power was kept constant at 75 W, the argon flux was 35 sccm and the oxygen flux was varied in 0.1 sccm steps. The deposition time was 10 minutes and the expected substrate temperature is below 100 °C. The crystallographic properties were measured using a Siemens D5000 diffractometer. Electrical properties were determined by Hall effect measurements. A Perkin Elmer Lambda 900 spectrometer was used for measuring the transmission and reflection spectra. The complex dielectric function was measured with a rotating analyzer Woollam VASE ellipsometer with an autoretarder.

EXPERIMENTAL RESULTS AND DISCUSSION

By varying the oxygen flux in the sputter process in the range of 3.0 to 8.0 sccm, it is possible to synthesize all three copper oxides. Figure 1 shows the diffractograms of representative samples for the three oxides and for samples in the intermediate regions.

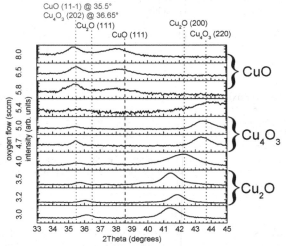

Figure 1. Diffraction patterns of samples which are representative for the different oxides. The samples made at 3.0, 4.0 and 5.4 sccm oxygen flow are intermediates and do not exhibit the optical and electrical properties of the pure phases.

The dashed and dotted lines indicate the literature position of different lattice planes and were taken from the ICDD PDFs 00-005-0667 for Cu_2O, 01-083-1665 for Cu_4O_3 and 00-048-1548 for CuO. Compared to the ICDD values all reflexes are slightly shifted to lower diffraction angles which may be a consequence of stoichiometric deviations. The ranges of the oxygen flow for growth of Cu_2O (3.2-3.5 sccm) and Cu_4O_3 (4.7-5.0 sccm) are comparatively narrow whereas the CuO phase shows up for an oxygen flux of 5.8 to 8.0 sccm.

Although the reflexes of the samples grown at 4.0 and 5.4 sccm are at the literature positions the samples are not single phase. This was concluded from the optical and electrical measurements. Also the reflexes are broad in comparison with the other samples, which is a further indication that the samples are not of a pure phase.

Figure 2 shows the band gap energies versus the oxygen flow, as they were obtained from Tauc plots [11,12]. Three plateaus are observed, where the values of the band gap energies for the different oxides is constant over a range of oxygen flows. These plateaus agree with growth ranges of the oxides in figure 1. The approximate band gap values for the three oxides at room temperature are 2.2 eV for Cu_2O, 1.75 eV for Cu_4O_3 and 1.55 eV for CuO.

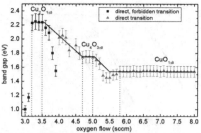

Figure 2. Band gap energies of the different oxides as obtained from Tauc plots. The graph shows three plateaus which can be correlated to the three copper oxide phases observed in the XRD measurements.

The values of the band gap energies as estimated from the Tauc plots are slightly higher than in literature which might be caused by the small crystallites < 50 nm in diameter as measured by atomic force microscopy. Another source of error is the transition type used in the Tauc plot, for Cu_2O it is well known that the lowest transition is direct and forbidden, for Cu_4O_3 and CuO a direct allowed transition was assumed.

Spectroscopic ellipsometry measurements were performed to obtain the absolute values of the absorption coefficients and the dielectric functions. The absorption coefficient reaches a value of $1 \cdot 10^5$ cm^{-1} at 2.7 eV for Cu_2O, 2.4 eV for Cu_4O_3 and 2.1 eV for CuO. Figure 3 shows the real and imaginary part of the complex dielectric function for the three copper oxides. One can clearly see that the spectral dependencies differ for all phases. The results of the dielectric function for Cu_2O and CuO are in good agreement with the results from Ito [5,13]. For Cu_4O_3 the dielectric function was measured for the first time. A more detailed discussion of the dielectric functions can be found in [2,3].

Figure 4 shows the conductivity as observed by Hall measurements. The areas of phase pure growth are marked in grey hatchings. With increasing oxygen flow the conductivity increases and reaches maxima outside the phase purity ranges. Then it decreases before a phase change appears and the oxide is phase pure again. The carrier concentration (not shown) increases with increasing oxygen flow and reaches a saturation before the phase changes. The electrical properties are summarized in table I.

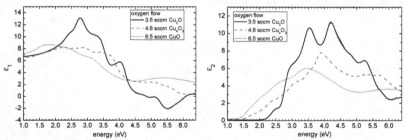

Figure 3. Real and imaginary part of the dielectric function as obtained from SE measurements.

167

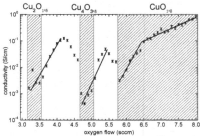

Figure 4. Conductivity of the samples as obtained by Hall measurements. The grey boxes mark the regions of nearly phase pure growth of the different oxides as obtained by the XRD and optical measurements. One sees a rise of the conductivity with increasing oxygen flux within the different phases which may be correlated to an increase of copper vacencies.

This behavior may also evolve from the generation of copper vacancies with increasing oxygen flow; to prove this, positron annihilation spectroscopy measurements are suitable. The carrier mobility for the Cu_2O samples is in the range of 1 to 20 cm²/Vs, for the other oxides the measured mobilities were below 1 cm²/Vs. All samples showed p-type conductivity behavior.

CONCLUSIONS AND OUTLOOK

It is possible to synthesize all three copper oxides (Cu_2O, Cu_4O_3 and CuO) by sputtering at different oxygen flows. The electrical properties can be adjusted without affecting the optical properties. All copper oxides are good candidates for use in photovoltaic and optoelectronic materials if the characteristics of the absorption coefficient are properly accounted.

Due to their non-stoichiometry and excess of oxygen, the use of the studied compounds as an oxygen reservoir may be possible. This would open a new field of possible applications, one may think of using the oxygen richest copper oxide (CuO) as solid state reservoir for oxygen in all solid state ionics devices. To verify these ideas studies of the ion conduction behavior of the copper oxides are under way.

Investigations of the vacancy mechanisms may also give a better understanding of the source of the p-type conduction. If this work is successful one has nontoxic, cheap and easy to synthesize materials for a wide range of applications.

Table I. Electrical properties of the copper oxides. The labeling "saturation" refers to the carrier concentration, at these oxygen flows Cu_2O and Cu_4O_3 are not phase pure.

	Cu_2O	Cu_2O Saturation	Cu_4O_3	Cu_4O_3 Saturation	CuO
O_2-flux (sccm)	$3.2 - 3.5$	4.0	$4.7 - 5.0$	5.4	$5.8 - 8.0$
Conductivity (S/cm)	$1.4 \cdot 10^{-3} - 10 \cdot 10^{-3}$	$1 \cdot 10^{-1}$	$5 \cdot 10^{-4} - 5 \cdot 10^{-3}$	$5 \cdot 10^{-2}$	$2 \cdot 10^{-3} - 1$
Carrier conc. (cm^{-3})	$7 \cdot 10^{14} - 2 \cdot 10^{16}$	$1 \cdot 10^{20}$	$1 \cdot 10^{16} - 3 \cdot 10^{17}$	10^{19}	$2 \cdot 10^{17} - 1 \cdot 10^{21}$
Carrier mobility (cm^2/Vs)	$17.5 - 0.3$	0.01	$0.5 - 0.05$	0.02	$0.1 - 0.01$

ACKNOWLEDGMENTS

This work was supported in the framework of LOEWE-Schwerpunkt RITSAT.

REFERENCES
[1]   S. Brahms, S. Nikitine, and J. P. Dahl, Physics Letters **22**, 31 (1966).
[2]   C. Malerba, F. Biccari, C. Leonor Azanza Ricardo, M. D'Incau, P. Scardi, and A. Mittiga, Solar Energy Materials and Solar Cells **95**, 2848 (2011).
[3]   B. K. Meyer, A. Polity, D. Reppin, M. Becker, P. Hering, P. J. Klar, T. Sander, C. Reindl, J. Benz, M. Eickhoff, C. Heiliger, M. Heinemann, J. Bläsing, A. Krost, S. Shokovets, C. Müller, and C. Ronning, Physica Status Solidi (b) **249**, 1487–1509 (2012).
[4]   W. Ching, Y.-N. Xu, and K. Wong, Physical Review B **40**, 7684 (1989).
[5]   T. Ito, H. Yamaguchi, T. Masumi, and S. Adachi, Journal of the Physical Society of Japan **67**, 3304 (1998).
[6]   J. F. Pierson, A. Thobor-Keck, and A. Billard, Applied Surface Science **210**, 359 (2003).
[7]   J. F. Pierson, E. Duverger, and O. Banakh, Journal of Solid State Chemistry **180**, 968 (2007).
[8]   A. Thobor and J. F. Pierson, Materials Letters **57**, 3676 (2003).
[9]   O. Porat and I. Riess, Solid State Ionics **74**, 229 (1994).
[10] O. Porat and I. Riess, Solid State Ionics **81**, 29 (1995).
[11] J. Tauc, R. Grigorovici, and A. Vancu, Physica Status Solidi (b) **15**, 627–637 (1966).
[12] J. Tauc, Materials Research Bulletin **3**, 37 (1968).
[13] T. Ito, T. Kawashima, H. Yamaguchi, T. Masumi, and S. Adachi, Journal of the Physical Society of Japan **67**, 2125 (1998).

Mater. Res. Soc. Symp. Proc. Vol. 1494 © 2013 Materials Research Society
DOI: 10.1557/opl.2013.158

Expanded Thermochromic Color Changes in VO$_2$ Thin Film Devices Using Structured Plasmonic Metal Layers

Yan Wang[1, 2] and John F. Muth[1]
[1]Department of Electrical and Computer Engineering, North Carolina State University, Raleigh, North Carolina 27695, USA
[2]Department of Physics, North Carolina State University, Raleigh, North Carolina 27695, USA

ABSTRACT

We investigate metallic thin films on VO$_2$ and show that the magnitude of the reflected color change in that visible portion of the spectrum as VO$_2$ undergoes the insulating to metallic phase transition can be controlled by changing the type of metal, the thickness of the metal and by patterning the metal at the nano scale. We consider the role of surface plasmas in the metal film and show that in the near infrared, the magnitude of the reflectivity increase for metal coated VO$_2$ films, but decrease for uncoated VO$_2$ thin films. This is explained in the context of Fresnel equations and considering the large change in the imaginary part of the dielectric constant as the VO$_2$ changes state from the insulating to metallic phase.

INTRODUCTION

Thermochromic material which is well known for the optical property of being able to change reversibly upon temperature has attracted much attention. Thermochromic thin films whose reflectance depends on temperature can be used as coating materials for smart windows or chromic glass [1]. Vanadium dioxide (VO$_2$) has been one of the most interesting thermochromic materials because of its high infrared transmittance in insulating phase and high infrared reflectance in metallic phase. Thermochromics in infrared region have been observed from single VO$_2$ films or VO$_2$ films coating by oxides [2, 3]. VO$_2$-based nanothermochromics has also been reported before [4, 5].

In this paper, we report an optical observation of the reflectivity change at the interface between VO$_2$ and gold film. Three device structures were fabricated with Pulsed Laser Deposition (PLD) and e-beam deposition. The reflection and transmission spectra of these devices are compared at the phase transition region of the VO$_2$ film. Reflectance of VO$_2$ film decreases as we increase the temperature. However, we have seen opposite reflectance change after coating the VO$_2$ film with a layer of gold as we change the temperature from low to high. This effect can be explained by dielectric constant change of the VO$_2$ film during phase transition and the increasing in electron damping at the interface.

EXPERIMENT

Pulsed Laser Deposition (PLD) was used to deposit the VO$_2$ films on an ITO covered glass substrate. A KrF excimer laser at a wavelength of 248 nm was focused on a homemade V$_2$O$_5$ target. The repetition rate and energy level of laser pulses were kept at 10 Hz and 200 mJ during deposition. The base pressure of the deposition chamber was pumped to 10^{-7} Torr. The depositions were performed in an ambient of 20 mTorr oxygen and the substrate temperature was maintained at 800 °C. After the deposition, the samples were cooled down at the deposition oxygen pressure to room temperature. The resistivity of 120nm thick VO_2 films on sapphire and ITO glass has been measured. (Figure 1a) For films on sapphire, the resistivity has changed up to 3 orders through the insulating to metallic transition. The thermal hysteresis is about 8 °C. The derivative of the resistivity was plotted in Figure 1b. The transition temperature of heating and cooling are 70 °C and 62 °C respectively. But for VO_2 thin films on ITO glass, the hysteresis is broader, the transition temperature is higher and has smaller amplitude. The different transition properties were induced by the effect of strain on VO_2. The lattice mismatch between C-plane sapphire and VO_2 is much smaller than that between ITO and VO_2. The stronger strain effect change the transition properties of the VO_2 films on ITO glass.

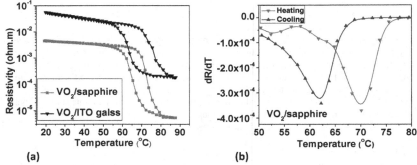

Figure 1. (a) Resistivity of VO_2 thin films on sapphire and ITO glass as temperature change; (b): Derivative of resistivity over temperature as temperature change for VO_2/sapphire

Three devices were fabricated for comparison, as shown in Figure 2. The first device is simply the VO_2 /ITO/Glass structure. The second device has a layer of e-beam deposited gold (180 nm) on the VO_2 layer to form a Gold/VO_2 /ITO/Glass structure. The third device was fabricated by patterning the gold layer with hexagonal array nanobump patterns for surface plasmon excitation. The periodic nanobump arrays were fabricated by nanosphere lithography method. In this process, polystyrene nanospheres were self-assembled on top of VO_2 as a template for subsequent gold deposition. The detail procedure was described in our previous paper [6].

The optical properties of these devices were measured at the insulating-metallic phase transition region. A tungsten light source with 430-900 nm wavelength range was used to generate a collimated light beam and normally incident onto the VO_2 film. The transmission and reflection light were collected by an Ocean Optics USB 4000 spectrometer, while the temperature of the VO_2 film was changed between 30 °C and 80 °C. The device active area was around 1 mm² for device response collection, and the spectra show little variation for different spots on the same film. All the reflective light response was collected from the glass substrate side, as shown in Figure 2. For the devices with gold film on top of VO_2, although the VO_2 /ITO interface also reflected the incoming light, its effect was weaker comparing to the reflection at

Gold/VO$_2$ interface. Thus we consider the observed optical effects are mainly caused by the changes that happen at the Gold/VO$_2$ interface.

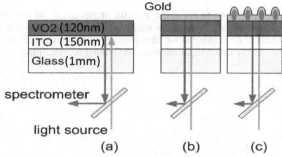

Figure 2. Schematic of the three devices under test: (a) a VO$_2$ /ITO/Glass structure, (b) a gold/VO$_2$ /ITO/glass structure, and (c) a nanobump array patterned gold/VO$_2$ /ITO/glass structure. The reflected light was majorly from the VO$_2$ /ITO interface for device (a) and from the gold/VO$_2$ interface for device (b) and (c).

DISCUSSION

With only VO$_2$ on top of ITO glass, the optical reflectivity is higher at low temperature insulating state than high temperature metallic state at wavelength longer than 630 nm. The reflectivity at wavelength around 630 nm does not change after the temperature increase from 30 °C to 80 °C. The point where the reflectivity curves cross for the different temperature has also been observed after coating the gold layer. Figure.3 (b) shows the optical transmission at insulating state and metallic state. As expected, the transmission at longer wavelength beyond 700nm decreases as the temperature increases. The changes in transmission in the visible region form 400nm to 700nm are small, and when examined by eyes there is almost no change in color.

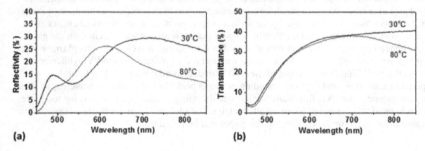

Figure 3. Spectral reflectivity (a) and transmission (b) at 30 °C and 80°C for VO$_2$ (120nm)/ITO

(130nm) glass.

After depositing a layer of gold on top of the VO$_2$, the magnitude of the reflectivity at high temperature increase to as much as 2 times of the reflectivity at low temperature at wavelength longer than 630nm instead of decreasing as we have seen in Figure.3a, with The point where the reflectivity curves cross for the different temperature still around 630nm. (Figure 4)

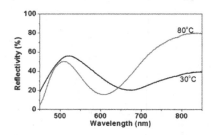

Figure 4. Spectral reflectivity at 30 °C and 80°C for flat gold (180nm) /VO$_2$ (120nm)/ITO

(130nm) glass.

By coating the VO$_2$ with nano bump gold layer, we observed that, as the temperature increases from 30 ˚C to 80 ˚C, the magnitude of the optical reflectivity at high temperature metallic state increases to nearly 4 times of the reflectivity at low temperature insulating state at wavelength around 675nm.This is in visible range which means this change is more sensitive to human eyes. The steady point moves to 610nm due to the nano-gold surface plasmon tunability. Figure 5 (b) shows the optical transmission at 30 ˚C and 80˚C without significant temperature dependence observed. We have much lower transmission in the visible range because of the gold layer. However, due to the patterned gold surface with nano bumps, we observed enhanced transmission at wavelength around 800nm. Surface plasmon modes were excited on the gold film. Enhanced transmission effect was observed in the infrared wavelength. The plasmon peak was pretty wide because the two lowest order modes on the gold/air and gold/VO$_2$ interfaces were overlapped. Dielectric constant of the VO$_2$ film decreased during the low to high temperature transition, and the corresponding plasmon peak shifted to shorter wavelength. This effect compensates the VO$_2$ film transmission rate dropping around 700 nm, thus the total rate kept unchanged at this region. More detailed analysis of surface plasmon effects and plasmonic crystals similar to the ones deposited in this paper can be found in [7].

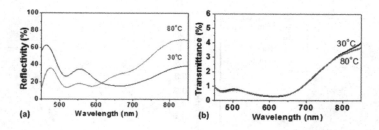

Figure 5. Spectral reflectivity (a) and transmission (b) at 30 °C and 80°C for gold (90nm) with

nano bumps /VO_2 (120nm)/ITO (130nm) glass.

The color change traces for reflection at the Gold/VO_2 and Aluminum/VO_2 interfaces during the VO_2 phase transition were plotted in the CIE 1931 color space, as shown in Figure 6a. Several metal films, including 50nm and 180nm flat gold, 90nm gold with nanobump and 90nm flat aluminum, were deposited onto flat VO_2 films, respectively. The reflected light spectra were collected with a collimated tungsten light source. Comparing to the reflected color of VO_2 without any metal, the reflected color of 50nm gold and 90nm aluminum on top of VO_2 has larger tunability, which are located at the right to the edge of the color space. The trace for 90nm gold with nanobumps on top of VO_2 is at the middle of the color space with the same tunability of the flat gold/VO_2. All of the films show a yellow color at low temperature when the VO_2 is in the insulating phase, the reflection color moves to blue-green with an increasing in temperature. The traces for heating up and cooling down processes are overlapped in the CIE plot, but the correspondent color points for each temperature are not at the same location. These devices show a hysteresis effect that is commonly seen for VO_2 film during the phase transition. The reflected light intensities for the nanobump patterned gold/VO_2 interface at 675 nm are plotted for the two phase change directions, as shown in Figure 6b. For the insulator to metallic transition, the transition temperature (medium) is at 75 °C and the intensity change slop is 7.9% per degree. The transition temperature for metallic to insulator transition is at 60 °C with a slop of 6.3% per degree. The average transition temperature is 67.5 °C with a 15 °C hysteresis width.

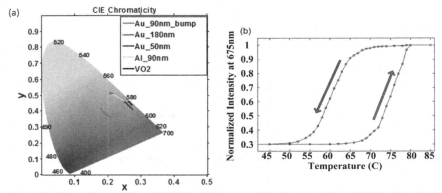

Figure 6. Color change during the VO$_2$ insulating-metallic phase transition. (a) Color changing trace on CIE 1931 color space. Upper end of the trace is at metallic phase and lower end is at insulating phase. (b) Light intensity changes for 675 nm wavelength during the phase transition.

The observed reflection spectra change can be explained by Fresnel equations, which describe the reflection and refraction of light moves from one medium to another medium with different optical index. For our devices, the optical constants of metallic state VO$_2$ and insulating state VO$_2$ is different, especially in the near infrared region. On the other hand the dielectric function of gold is dramatically different in the ranges of shorter wavelength and longer wavelength. The different optical constants of gold and VO$_2$ at different wavelength cause the opposite shifting behavior in the near infrared region.

CONCLUSIONS

In summary, the reflection spectrum of a Metal/VO$_2$ thin film structure was studied. The device spectrum shows modulation effect around the VO$_2$ insulating-metallic phase transition temperature and the direction of change is opposite with VO$_2$ /ITO film spectrum. The presence of the metal films enhances the thermochromic modulation of VO$_2$ films. Reflectivity from the interface between nano-bump array patterned gold film and VO$_2$ shows a larger modulation effect at red and NIR wavelength due to surface plasmon enhanced field at the gold surface. This effect can be utilized to design temperature sensor with optically remote data collection and also electrically controlled light modulators.

REFERENCES

1. C. Granqvist, Thin Solid Films **193**, 730–741 (1990).
2. N.R. Mlyuka, G. a. Niklasson, and C.G. Granqvist, Physica Status Solidi (a) **206**, 2155 (2009).

3. M. Tazawa, K. Yoshimura, P. Jin, and G. Xu, Applied Physics A: Materials Science & Processing **77**, 455 (2003).
4. S.-Y. Li, G. a. Niklasson, and C.G. Granqvist, Journal of Applied Physics **108**, 063525 (2010).
5. M. Maaza, O. Nemraoui, C. Sella, A.C. Beye, P.O. Box, S. Africa, U. Pierre-, M. Curie, P. Vi, and P. Cedex, Gold Bulletin 100 (2005).
6. Y. Lou, L.M. Lunardi, and J.F. Muth, Sensors Journal, IEEE **10**, 617–620 (2010).
7. Y. Lou, Doctor of Philosophy Dissertation, North Carolina State University (2011)

Mater. Res. Soc. Symp. Proc. Vol. 1494 © 2012 Materials Research Society
DOI: 10.1557/opl.2012.1698

Atomic Layer Deposition of SrO: Substrate and Temperature Effects

Han Wang, Xiaoqiang Jiang, and Brian G. Willis

Department of Chemical, Materials & Biomolecular Engineering, University of Connecticut, Storrs, Connecticut 06269.

ABSTRACT

The atomic layer deposition (ALD) of SrO was conducted on various oxide surfaces by using strontium bis(tri-isopropylcyclopentadienyl) and water at deposition temperatures of 200 and 250°C. The initial and steady growth behaviors were studied by *in-situ* spectroscopic ellipsometry and *ex-situ* X-ray photoelectron spectroscopy. For initial growth, the growth per cycle (GPC) of SrO not only depends on the concentration of hydroxyl groups but also the formation of interfacial Sr-O-Si bonds. For the steady growth, *in-situ* annealing was used to enhance the growth rate and multiple growth regions were identified.

INTRODUCTION

Atomic layer deposition (ALD) has attracted much attention due to its capability for accurate thickness control and superior conformal growth [1]. Challenges for ALD include non-ideal nucleation and substrate effects often encountered at the interfaces between dissimilar materials [2-4]. These substrate effects are particularly problematic for the growth of more complicated materials including ternary systems because of the difficulty encountered in achieving steady, predictable growth [5]. The growth of metal oxides is one of the most extensively studied and promising areas of ALD. SrO ALD is critical for the ALD of the ternary strontium titanate (STO), which is of major interest for use with high-density metal-insulator-metal (MIM) capacitors [6-8]. SrO is also of interest for the growth of epitaxial perovskite oxides on semiconductors where it acts as a buffer layer between the reactive semiconductor and the metal oxide layers [9,10].

In this work, we investigate strontium oxide (SrO) as a model system to better understand the fundamental origins of substrate and temperature effects on ALD. Especially, we study both initial and steady growth of SrO on two insulating oxide substrates at 200 and 250°C using *in-situ* real-time spectroscopic ellipsometry (RTSE) [11]. We observe complex ALD growth characteristics with several different ALD operating regimes.

EXPERIMENT

SrO thin films were deposited in a warm wall stainless steel reactor on approximately 1 cm^2 1.6 nm native SiO_2, or 28 nm ALD-grown $Al_2O_3/Si(100)$ substrates [11]. The Sr-metal precursor was the THF adduct of $Sr(C_5^iPr_3H_2)_2$ (Air Liquide). H_2O vapor (Millipore 18 MΩ) was used as the oxidizer. The solid Sr source compound was heated to 140–145°C in a glass container where it becomes a light yellow liquid. The temperature of the H_2O container was 25°C. Both inert gas purging and vacuum pumping were used between reactant exposures. Argon was regulated to maintain 1 Torr reactor pressure using leak valves; vacuum pumping was to below 10 mTorr. The whole ALD system was heated to a temperature higher than 150°C to prevent precursor condensation on the reactor and tubing walls. Typical purging and pumping times were 40 and 30 s due to the relatively large reactor volume [11]. The whole ALD process was monitored via an M-2000 V™ spectroscopic ellipsometer with the COMPLETE EASE 4.32 data analysis software from J. A. Woollam Co., Inc. Ellipsometric Ψ and Δ data were acquired at a fixed angle of incidence (68°) over the spectral range 370–1000 nm (1.20–3.34 eV). Crystalline strontium oxide has an indirect bandgap of approximately 5.7 eV and is transparent (k=0) over the whole photon energy range of our ellipsometer.

XPS studies were carried out using a Kratos Axis-165 XPS instrument equipped with an Al $K\alpha$ x-ray source monochromated to 1486.6 eV. The base pressures of the transfer and analysis chambers are $\sim 10^{-9}$ and $\sim 10^{-10}$ Torr, respectively. Pass energies of 160 and 20 eV were used for survey and core-level spectra, respectively, and all spectra were referenced to Si $2p_{3/2}$ at 99.4 eV. The sensitivity factors used for quantitative analysis were obtained directly from the XPS instrument manufacturer.

DISCUSSION

Hydroxyl groups have been widely identified as the main reactive sites for the adsorption of metal precursors in metal oxide ALD processes [1]. Thus, the GPC is partially dependent on the number of OH groups on the surface and for a specific ALD process any non-linear growth behavior may result from the variation in the number of reactive sites.

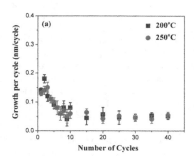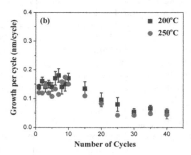

Figure 1. Growth per cycle (GPC) as a function of number of ALD cycles on **(a)** ALD-grown $Al_2O_3/Si(100)$ and **(b)** native $SiO_2/Si(100)$ substrates at 200 and 250°C.

As shown in Figure 1(a), SrO growth is promoted during the first few cycles compared to later ALD cycles on ALD-grown Al_2O_3 surface. The substrate-enhanced growth on Al_2O_3 surface probably indicates that the concentration of OH groups on SrO is less than on Al_2O_3 at both 200 and 250°C. Also, the slowly decline of GPC during the first few cycles means a full coverage of SrO film is not achieved after the very first cycle probably due to the steric hindrance from the relatively large cyclopentadienyl complex. Steady growth characteristic of SrO homo-ALD is observed after 5-7 cycles.

By contrast, as shown in Figure 1(b), a more persistent substrate effect is observed for growth on SiO_2. On SiO_2, the initial GPC also shows enhanced growth similar to Al_2O_3, but the effect is steady for a full 10 cycles. After 10 cycles, the GPC steadily decreases until reaches its final value after 30 and 25 cycles at 200 and 250°C, respectively. When combined with the steady GPC value in Figure 1(a), it is clear that once the substrate is fully covered by the Sr ALD films, the steady GPC is the same on both Al_2O_3 and $SiO_2/Si(100)$ substrates. Based on the ellipsometric analysis, the optical properties (reflective index here) are the same within the ellipsometric error for Sr films in the steady growth region on both Al_2O_3 and $SiO_2/Si(100)$ substrates [11]. However, the difference in the number of OH groups cannot fully explain such substrate-enhanced growth behavior on the native SiO_2 surface. Methaapanon et al. studied TiO_2 ALD growth on oxide-terminated silicon and reported that the initial growth was accelerated by the formation of Ti-O-Si bonds at the interface between SiO_2 and TiO_2 [4]. Therefore, XPS was used to examine the layers to determine if a similar effect is occurring for SrO ALD. Because SrO is very sensitive to air (CO_2 and H_2O), a protective layer (Al_2O_3 here) is used to encapsulate the SrO films to preserve the chemical states of the interface between SiO_2 and SrO.

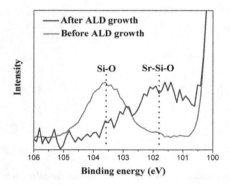

Figure 2. Si 2p XPS core level spectra before and after ALD growth on $SiO_2/Si(100)$ substrate at 250°C.

In the XPS fine scan of the Si 2p core level peaks, Figure 2, the single peak at 103.6 eV is attributed to native SiO_2. With 1.5 nm ALD oxide, a new peak with binding energy (~101.8 eV) shows up, which is the signature of the formation of Sr-O-Si structure widely observed in molecular beam epitaxy studies [7,8]. According to the relative areas of Sr 3d, Si 2p, and O 1s silicate components, the silicate composition is close to Sr_2SiO_4. Different from TiO_2 ALD on SiO_2, several monolayers of Sr silicate form during the SrO ALD process. Calculations based on the change in Si 2p and O 1s XPS core level spectra of SiO_2 components indicate that more than 70% and 90% of SiO_2 has converted into Sr silicate after 20 and 30 ALD cycles, respectively. The continuous formation of Sr silicate is enabled by facile diffusion of Sr atoms through the top silicate layers into the underlying SiO₂ network, leaving the surface terminated by silicate. Likely, Sr silicate itself has OH density comparable to SiO_2 surface, which leads to the transient enhanced GPC observed in Figure 1 (b). Spontaneous silicate formation is observed even at the lowest possible deposition temperatures of 150°C, limited by precursor vaporization.

In addition to sensitivity to the substrate material, we have found that SrO ALD growth is sensitive to substrate process treatments. We observed an unusual effect of annealing on the GPC of SrO films grown at 250 and 200°C. We firstly deposited a 3.5 nm film (about 40 cycles) on native SiO₂ substrates for both 250 and 200°C, followed by *in-situ* annealing of each sample at 350°C under 1 torr Ar flow for 30 min. Prior to annealing, the samples reached the normal homo-SrO GPC of 0.06 nm/cycle. After annealing, additional SrO was deposited at the original deposition temperatures of 250 or 200°C. As seen in Figure 3, the annealing process has a large effect on the growth rate. At both temperatures, the first post-anneal GPC is more than a factor of 2 higher than normal growth. The effect is short lived for films deposited at 200°C, but persists for more than 120 cycles for growth at 250°C. Eventually, the effect decays back to normal growth for both temperatures. The observations show that the GPC is sensitive to the changes of the substrate induced by the *in-situ* annealing treatment.

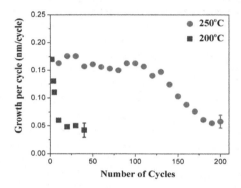

Figure 3. Annealing effect on GPC of SrO films grown at 200 and 250°C. *In-situ* annealing was carried out at 350°C under Ar condition for 30 min.

Other studies have reported a related effect where the GPC varies with the film thickness [12]. For example, in TiO₂ ALD, film crystallization is thickness and temperature-dependent, and at higher deposition temperature above a critical thickness the GPC is enhanced by nearly 140% due to an increase in OH density after crystallization [12]. In this work, no GPC enhancement is observed with film thickness up to 20 nm for 250 and 200°C growth without annealing. SrO films have been reported to crystallize above 300°C with preferred (111) orientation [13]. Thus, the thin Sr film layers may crystallize during annealing at 350 °C and serve as a seed layer for modified growth at 200 and 250°C [13]. The crystallization may modify the density and/or reactivity of the hydroxylated surface to enhance growth. At 200°C, the decay is rapid, but at 250°C the added thermal energy may enhance surface processes that allow the crystallization effect to decay more slowly. The persistence of the effect for > 120 cycles strongly suggests that a change in the film structure is involved. More experimental work on structure and crystallinity is under way for understanding the effect of *in-situ* annealing on growth behaviors in SrO ALD process.

CONCLUSIONS

Initial and steady growth behaviors of SrO ALD process were investigated in detail on various oxide surfaces. It was demonstrated that in addition to hydroxyl groups solid-sate reactions play a key role in substrate effects for growth on SiO_2. A Sr silicate forms at the interface with a stoichiometry close to Sr_2SiO_4 and enhances the adsorption of Sr precursors, possibly due to a dense OH concentration at the silicate surface. The underlying SiO_2 layers are continuously transformed into silicate by diffusion of Sr ions into the SiO_2 network. We also find that *in-situ* annealing leads to another unusual effect with significant enhancement of the growth rate. Films grown at 250°C after annealing to 350°C show a transient GPC enhancement by over 160% up to 120 cycles. The transient enhancement is likely due to a crystallization effect that decays with film thickness at lower deposition temperatures.

ACKNOWLEDGMENTS

This work was supported by the National Science Foundation, Award # CBET-0932834.

REFERENCES

1. R. Puurunen, *J. App. Phys.* **97**, 121301 (2005)
2. M. Green, M. Ho, B. Busch, G. Wilk, T. Sorsch, T. Conard, B. Brijs, W. Vandervorst, P. Raisanen, D. Muller, M. Bude, and J. Grazul, *J. Appl. Phys.* **92**, 7168 (2002).
3. A. Rahtu, T. Hanninen, and M. Ritala, *J. Phys. IV* **11**, 923 (2001).
4. R. Methaapanon and S. Bent, *J. Phys. Chem. C* **114**, 10498 (2010).
5   S. Riedel, J. Neidhardt, S.Jansen, L. Wilde, J. Sundqvist, E. Erben, S. Teichert, and A. Michaelis, *J. Appl. Phys.* **109**, 094101 (2011).
6   R. McKee, F. Walker, and M. Chisholm, *Phys. Rev. Lett.* **81**, 3014 (1998).
7. X. Hu, H. Li, Y. Liang, Y. Wei, Z. Yu, D. Marshall, J. Edwards, R. Droopad, X. Zhang, A. Demkov, and K. Moore, *App. Phys. Lett.* **82**, 203 (2003).
8. M. Kazzi, G. Delhaye, C. Merckling, E. Bergignat, Y. Robach, G. Grenet, and G. Hollinger, *J. Vac. Sci. Technol. A* **25**, 1505 (2007).
9. C. Zhang, L. Wielunskib, and B. Willis, *Appl. Surf. Sci.* **257**, 4826 (2011).
10. C. Marchiori, M. Frank, J. Bruley, V. Narayanan, and J. Fompeyrine, *Appl. Phys. Lett.* **98**, 052908 (2011).
11. H. Wang, X. Qiang, and B. Willis, *J. Vac. Sci. Technol. A* **30**, 01A133 (2012).
12. S. Kim, S. Hoffmann-Eifert, M. Reiners, and R. Waser, *J. Electrochem. Soc.* **158**, D6 (2011).
13. M. Vehkamaki, Ph.D. thesis (University of Helsinki, 2007).

Mater. Res. Soc. Symp. Proc. Vol. 1494 © 2013 Materials Research Society
DOI: 10.1557/opl.2013.6

Role of Oxidation State of Vanadium in Vanadium Oxide-Hematite Nanoparticles

Monica Sorescu, Tianhong Xu and Collin Wade

Duquesne University, Department of Physics, Fisher Hall, Pittsburgh, PA 15282-0321, USA

ABSTRACT

Single phase, a $FeVO_4$ triclinic crystalline structure was successfully synthesized by annealing the mechanochemically milled $xV_2O_5 \cdot (1-x)\alpha\text{-}Fe_2O_3$ composites (x = 0.5) at 550 °C for 1 h. X-ray powder diffraction (XRD) and Mössbauer spectroscopy were combined for a detailed study of the assisting role of the mechanochemical milling process. Mechanochemical milling homogeneously mixed the starting materials of $\alpha\text{-}Fe_2O_3$ and V_2O_5 and substantially decreased their average grain sizes. The partially V^{5+}-substituted $\alpha\text{-}Fe_2O_3$ phase and Fe^{3+}-substituted V_2O_5 could be the important intermediate phases in the production of $FeVO_4$ single phase. In addition, $xV_2O_3 \cdot (1-x)\alpha\text{-}Fe_2O_3$ (x = 0.1, 0.3, 0.5, and 0.7) solid solutions were successfully synthesized by mechanochemical activation of V_2O_3 and $\alpha\text{-}Fe_2O_3$ mixtures. Complete solid solutions exist after 12 h ball-milling time for all studied x values. The synthesized $xV_2O_3 \cdot (1-x)\alpha\text{-}Fe_2O_3$ solid solutions with x = 0.5 and 0.7 were mainly paramagnetic at room temperature. The study demonstrates that the transformation pathway is related to the valence state of the metallic specie of the oxide used in connection with hematite.

INTRODUCTION

Hematite is one of the most used oxides, with various applications in scientific and industrial fields. It can be used as semiconductor compound [1], magnetic material [2], catalyst [3], and gas sensor [4].

The vanadium oxides form a fascinating class of materials that exhibit a large variety of structures with different physical and chemical properties. This diversity makes vanadium oxides technologically relevant and leads to various applications, such as light detectors, electrical and optical switching devices, as well as heterogeneous catalysis [5, 6]. From fundamental research interest and wide application points of view, vanadium oxides possess various vanadium oxidation states ranging from +2 to +5 as well as different oxygen geometries in the crystalline structures. These make vanadium oxides attracting materials due to their rich electronic and magnetic structures as well as phase transitions.

Iron (III) vanadate, $FeVO_4$, appears in four polymorphs and has attracted attention because of its interesting physical and chemical properties with wide applications [7-13]. The $FeVO_4$ materials have been synthesized via a surfactant-assisted sol-gel method, hydrothermal method, and a conventional solid state reaction. However, these methods either require expensive chemical reagents or sintering processes at high temperature. The high temperature synthesis leads to aggregation and sintering of the reaction products, which is expected to lower the catalytic activities of $FeVO_4$ due to the reduction of surface area.

High energy ball-milling is a well established method for preparing extended solid solutions, composite and nanostructure systems using commercially obtained oxides as starting materials. This preparation method is promising for production scale use due to its relatively low cost and simple operation. In this work, we report the successful synthesis of single-phase $FeVO_4$ by a mechanochemical milling assisted method through the ball-milling of V_2O_5-α-Fe_2O_3 mixtures. The mixture consists of a stoichiometric ratio of starting materials, with the reaction being carried out at room temperature, and then followed by a heat treatment at 550 °C for 1 hour. X-ray powder diffraction and Mössbauer spectroscopy have been employed to investigate the phase evolution, structural and magnetic properties of the as-synthesized $FeVO_4$ as well as the ball-milled oxides at various ball-milling times.

In addition, $xV_2O_3(1-x)\alpha$-Fe_2O_3 ($x = 0.1$, 0.3, 0.5, and 0.7) solid solutions were successfully synthesized by mechanochemical activation of V_2O_3 and α-Fe_2O_3 mixtures. The phase sequence and properties of the as-milled materials evidenced the crucial role played by the oxidation state of vanadium in determining the final products of the vanadium oxide-hematite ball milling process.

EXPERIMENTAL

The sources of vanadium (V), vanadium (III) and iron (III) oxides were commercially purchased from Alfa Aesar: Vanadium (V) pentoxide (99.2% metals basis, average particle size is about 93.6 nm), Vanadium (III) trioxide (95% metal basis, average particle size 67.8 nm) and hematite (α-Fe_2O_3, 99% metal basis, average particle size is about 49.2 nm). Powders of hematite and vanadium oxides were milled at various molar ratios in a hardened steel vial with 12 stainless-steel balls (type 440; eight of 0.25 in diameter and four of 0.5 in diameter) in the SPEX 8000 mixer mill for time periods ranging from 2 to 12 h. The ball/powder mass ratio was 5:1. Prior to their introduction in the ball milling device, the powders were manually ground in air to obtain a homogeneous mixture.

The X-ray powder diffraction patterns of samples were obtained using a Panalytical X'pert Pro MPD powder diffractometer with CuK_α radiation (45 kV/40 mA, $\lambda = 1.54187$ Å) with a nickel filter on the diffracted side. A silicon-strip detector called X'Celerator was used. The scanning range was 10 - 80° (2θ) with a step size of 0.02°. The average grain size was determined by the Scherrer method. The Scherrer formula was used on the most intense reflection only. The lattice parameters were extracted from Rietveld structural refinement of the XRD patterns using GSAS software to perform least-square fitting.

Room temperature transmission Mössbauer spectra were recorded using an MS-1200 constant acceleration spectrometer with a 10 mCi ^{57}Co source diffused in Rh matrix. Least-squares fittings of the Mössbauer spectra were performed with the NORMOS program.

RESULTS AND DISCUSSION

V_2O_5-doped hematite, $xV_2O_5(1-x)\alpha$-Fe_2O_3 in which $x = 0.5$, was milled from 2 to 12 h (Figure 1). The starting materials were pure α-Fe_2O_3 and V_2O_5, no diffraction peaks from other phases were detected after the physical mixing process (Figure 1a). For the ball-milled composites (Figure 1b-e), the patterns show progressive peak broadening with increased milling time. This peak broadening is associated with the decrease in the grain size of both hematite (from 49.2 to 30.1 nm) and V_2O_5 samples (from 93.6 to 14.9 nm). It can also be seen that the diffraction peak intensities of α-Fe_2O_3 and V_2O_5 decrease with the increase of ball-milling time, indicating the possible ion substitutions between V^{5+} and Fe^{3+} in the corresponding hematite and

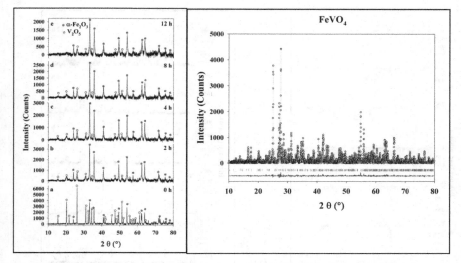

Fig. 1: XRD patterns of vanadium pentoxide
hematite series

Fig. 2: Rietveld refinement of the iron
vanadate XRD pattern

V_2O_5 lattices. Only α-Fe_2O_3 and V_2O_5 phases are present in the XRD patterns up to 12 h of milling time. This indicates that there is no solid state reaction between α-Fe_2O_3 and V_2O_5 under mechanochemical milling up to 12 h.

For the 12 h ball-milled sample after being annealed in air at 550 °C for 1 hour, none of the peaks corresponding to the starting materials of V_2O_5 or α-Fe_2O_3 were observed, indicating the completion of the solid state reaction between V_2O_5 and α-Fe_2O_3. X-ray analysis, alongside the Rietveld refinement of the mechanochemically milled composites after annealing process are shown in Figure 2, indicating that $FeVO_4$ was successfully obtained after a very short time annealing process for the 12 h ball-milled composites. A model with a single phase of $FeVO_4$ (triclinic P-1 space group) was employed to perform the refinement. The refined lattice parameters of the synthesized $FeVO_4$ have values of $a = 6.7114$ Å, $b = 8.0561$ Å, $c = 9.3545$ Å, $\alpha = 96.730°$, $\beta = 106.672°$, $\gamma = 101.565°$, and unit cell $V = 466.837$ Å^3, respectively, which are similar to the previously reported results [1]. The average grain size of the obtained $FeVO_4$ is estimated from Scherrer formula to be ~ 84.0 nm. In contradistinction to our method, conventional solid state reaction requires temperatures of 800-1000 °C and times greater than 10 hours.

The room temperature transmission Mössbauer spectra of the $xV_2O_5 \cdot (1-x)\alpha$-$Fe_2O_3$ composites (x = 0.5) after ball milling for 0, 2, 4, 8 and 12 hours, respectively, are represented in Figure 3(a)-(e). At 0 h of milling time, the spectrum was fitted with 1 sextet (Figure 3a), corresponding to α-Fe_2O_3. After 2 h of milling time, the Mössbauer spectrum was fitted with 3

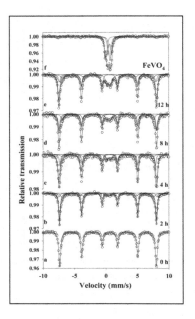

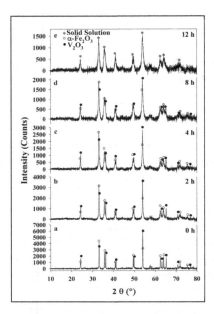

Fig. 3: Mössbauer spectra of vanadium pentoxide-hematite series

Fig. 4: XRD patterns of vanadium trioxide hematite series for x=0.5.

sextets and 1 doublet. The three sextets can be attributed to the α-Fe_2O_3 phase and the vanadium-substituted α-Fe_2O_3 phase with V^{5+} substitution of Fe^{3+} in the α-Fe_2O_3 lattice, respectively. The doublet can be assigned to Fe^{3+} ions which substitute V^{5+} in the V_2O_5 lattice. When part of V^{5+} substitutes Fe^{3+} in α-Fe_2O_3 lattice, antiferromagnetic ordering among Fe^{3+} can still be induced by an exchange interaction, which gives sextets in Mössbauer spectrum with different strengths in hyperfine magnetic fields. However, when a small amount of Fe^{3+} substitution of V^{5+} in the V_2O_5 lattice occurs, no antiferromagnetic ordering among Fe^{3+} in the V_2O_5 lattice can be induced by an exchange interaction, therefore, only a quadrupole doublet appears. The isomer shift of Fe^{3+} extracted from the doublet is higher compared to the value of Fe^{3+} extracted from the sextet pattern of Fe_2O_3; this may be attributed to the change of neighbors in these two different cases. The relative abundance of a magnetic phase with highest hyperfine field (48.23 T) is 67.97%, and the relative population of a magnetic phase with lower hyperfine fields of 47.79 and 45.17 T is 10.63% and 17.27%, respectively.

Three doublets are observed in the Mössbauer spectrum of as-synthesized $FeVO_4$ (Figure 3f). The absence of a sextet corresponding to α-Fe_2O_3 indicates the complete consumption of α-Fe_2O_3 components, which is in good agreement with the XRD results, as no diffraction peaks corresponding to the starting materials are observed. The existence of three doublets for the as-synthesized $FeVO_4$ sample is in great agreement with the three inequivalent iron positions in the triclinic crystalline structure of $FeVO_4$. The three independent iron atoms create a doubly bent

188

chain of six edge-sharing polyhedral, the chains are joined by VO^{3-}_4 tetrahedra. The populations of these three different sites are 37.45%, 32.10%, and 30.43%, respectively, which is not in an exact 1:1:1 ratio, indicating the slight distortion of the iron sites. The difference in the isomer shift, quadrupole splitting values, and relative populations of Fe^{3+} sites depends significantly on the preparation methods which produce the $FeVO_4$, with considerable difference in the degree of crystallinity and structural imperfections [2].

From the variations in lattice parameters and average grain sizes of α-Fe_2O_3 and V_2O_3, it can be inferred that the mechanochemical activation of the α-Fe_2O_3-V_2O_3 mixtures reduces the average grain sizes and introduces the ion substitutions between V^{3+} and Fe^{3+} in α-Fe_2O_3/V_2O_3 lattices and the continuous formation of α-Fe_2O_3-V_2O_3 solid solution (Figure 4). When the ball milling time is up to 12 hours, a complete solid solution forms under mechanochemical activation. All of the formed solid solutions have the average grain size of ~ 12 nm, as evidenced from the similar intensities and widths in corresponding diffraction peaks. Compared to the XRD patterns of the original Fe_2O_3 corundum material (Figure 5), it was found that the diffraction peaks of the formed solid solutions shift to higher 2θ angles systematically with the increase in the molar fraction x of V_2O_3, owing to the difference in radius between Fe^{3+} and V^{3+}.

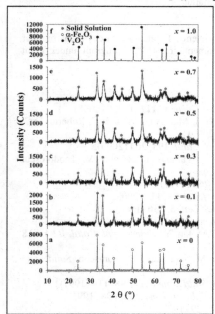

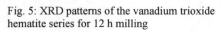

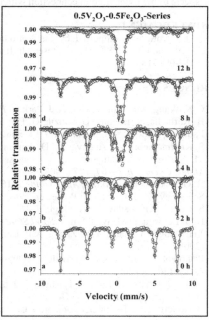

Fig. 5: XRD patterns of the vanadium trioxide hematite series for 12 h milling

Fig. 6: Mossbauer spectra of the vanadium trioxide-hematite series at x=0.5.

At 0 h of milling time, the Mössbauer spectrum was fitted with 1 sextet (Figure 6), corresponding to α-Fe_2O_3. After 2 h of milling time, the Mössbauer spectrum was fitted with 3 sextets and 1 doublet. The three sextets can be attributed to the α-Fe_2O_3 phase and the vanadium-substituted α-Fe_2O_3 phase with V^{3+} substitution of Fe^{3+} in the α-Fe_2O_3 lattice, respectively, and the doublet can be assigned to Fe^{3+} ions which substitute V^{3+} in the V_2O_3 lattice. All Mössbauer spectra of the formed solid solutions (for 12 hours of milling and various x) showed both sextets and doublet, indicating the co-existence of both ferromagnetic and paramagnetic properties. The percentage of paramagnetic components increases with the increase in V_2O_3 molar fraction x, owing to the decrease in exchange coupling between Fe^{3+} ions with the increase in V^{3+} concentration.

CONCLUSIONS

The study demonstrates the formation of a solid solution for vanadium trioxide and that of an iron vanadate ($FeVO_4$) for vanadium pentoxide ball milled with hematite. It has been demonstrated that the transformation pathway is directly related to the oxidation state of the metallic specie of the oxide used in connection with hematite.

ACKNOWLEDGMENT

This work was supported by the National Science Foundation under grant number DMR-0854794 and the NSF Major Research Instrumentation (MRI) program under grant number CHE-0923183.

REFERENCES

1   G.X. Wang, X.L. Gou, J. Horvat, J. Park, *J. Phys. Chem. C* **112**, 15220 (2008).
2   B. Raffaella, S. Etienne, G. Cinzia, G. Fabia, G.H. Mar, A.G. Miguel, C. Roberto, D.C. Pantaleo, *Phys. Chem. Chem. Phys.* **11**, 3680 (2009).
3   S. Krishnamoorthy, J.A. Rivas, M.D. Amiridis, *J. Catal.* **193**, 264 (2000).
4   M. Sorescu, L. Diamandescu, A. Tomescu, D. Tarabasanu-Mihaila, V. Teodorescu, *Mater. Chem. Phys.* **107**, 127 (2008).
5   B. Grzybowska-Swierkosz, *Appl. Catal. A* **157**, 263 (1997).
6   S. Suenev, M.G. Ramsey, F.P. Netzer, *Prog. Surf. Sci.* **73**, 117 (2003).
7   S. Gupta, Y.P. Yadava, R.A. Singh, *J. Mater. Sci. Lett.* **5**, 736 (1986).
8   L.M. Levison, B.M. Wanklyn, *J. Solid State Chem.* **3**, 131 (1971).
9   J.H. Deng, J.Y. Jiang, Y.Y. Zhang, X.P. Lin, C.M. Du, Y. Xiong, *Appl. Catal. B* **84**, 468 (2008).
10   P. Poizot, E. Baudrin, S. Laruelle, L. Dupont, M. Touboul, J.M. Tarascon, *Solid State Ionics* **138**, 31 (2000).
11   M. Hayashibara, M. Eguchi, T. Miura, T. Kishi, *Solid State Ionics* **98**, 119 (1997).
12   G. Mangamma, E. Prabhu, T. Gnanasekaran, *Bull. Electrochem.* **12**, 696 (1996).
13   B. Robertson, E. Kostiner, *J. Sol. State Chem.* **4** , 29 (1972).

Mater. Res. Soc. Symp. Proc. Vol. 1494 © 2012 Materials Research Society
DOI: 10.1557/opl.2012.1651

Metal Oxide Nanowire Growth via Intermediate Hydroxide Formation: A Thermochemical Assessment

Avi Shalav and Robert G. Elliman

Department of Electronic Materials Engineering, Research School of Physics and Engineering, Australian National University, Canberra, ACT0200, Australia

ABSTRACT

In this study we apply reaction thermodynamics to show that a significant volatile hydroxide vapor partial pressure forms at a metal-oxide interface and is a likely precursor source for nanowire growth. The growth of WO_3 and CuO nanowires are used as examples for reactions dependent on only H_2O and O_2+H_2O, respectively. Optimal temperatures, H_2O (and O_2) partial pressures for volatile hydroxide formation are calculated and experimentally investigated. We conclude that metal oxide nanowires can be readily grown at relatively low temperatures (close to or less than 500°C) over short anneal times (tens of minutes). The growth of these metal oxide nanowires, with many oxidation states, by this simple thermal technique is readily suited for a range of emergent large surface area nanostructured optical and electrical applications, including sensing, photocatalysis and ultracapacitors.

INTRODUCTION

Many transition metals, for example Fe, Cu, V, Zn, Al, W and Mo, have been shown to readily form oxide nanowires upon annealing at elevated temperatures under atmospheric conditions [1]. The growth mechanism is widely accepted to be dependent on metallic-oxygen reactions with atmospheric O_2, where metal cations diffuse along the grain boundaries from the metal-oxide interface to the surface. The role of H_2O vapor is often neglected, even though it is well known within corrosion science that metals readily oxidize in the presence of H_2O vapor and that one or more volatile hydroxide species can be produced [2].

The high temperature oxidation of metals under O_2 and/or H_2O vapor conditions, although well researched, is particularly complex due to the formation of a number of stable and unstable oxide scales. For nanowires to grow via a vapor-liquid-solid (VLS) or a vapor-solid-solid (VSS) mechanism, a vapor transport precursor is required. Evidence suggests that if an underlying substrate is to provide this precursor, a metastable surface oxide is required to separate catalytic metal particles from the substrate [3-5] and/or to provide nucleation sites for diffusion mediated growth [6, 7].

The aim of this study is to investigate the annealing conditions required to grow metal oxide nanowires directly from a metal substrate, where the substrate itself provides the required precursor for nanowire growth. Two different metals have been selected and thermochemically investigated to illustrate slightly different growth conditions, namely H_2O and O_2+H_2O dependencies. Firstly, W and its oxides are both unstable at higher temperatures in the presence of H_2O vapor producing the volatile hydroxide $WO_2(OH)_2$ [6, 8]. Although oxide formation depends on the amount of O_2 present, the production of $WO_2(OH)_2$ is independent of the O_2 partial pressure, that is, the reaction only depends on the W metal/oxides and the H_2O partial pressure with an additional H_2 vapor by-product. Secondly, Cu is often used to typify the oxidation mechanism for many of the transition metals. Its hydroxide, CuOH, and therefore associated nanowire growth, is dependent on both H_2O and O_2 partial pressures, consistent with literature observations [9].

THEORY

The W-H₂O system

The metastable stoichiometric oxide sequence on W typically includes $WO_3|WO_{2.9}|WO_{2.7}|WO_2$ with the O_2 deficient WO_2 phase next to the W metal. The reactions of each layer with H_2O vapor, resulting in the formation of volatile $WO_2(OH)_2$ can be described using the following equations [6]:

$$WO_3(s) + H_2O(g) = WO_2(OH)_2(g) \tag{1}$$
$$10WO_{2.9}(s) + 11H_2O(g) = 10WO_2(OH)_2(g) + H_2(g) \tag{2}$$
$$3.571WO_{2.72}(s) + 4.571H_2O(g) = 3.571WO_2(OH)_2(g) + H_2(g) \tag{3}$$
$$WO_2(s) + 2H_2O(g) = WO_2(OH)_2(g) + H_2(g) \tag{4}$$
$$W(s) + 4H_2O(g) = WO_2(OH)_2(g) + 3H_2(g) \tag{5}$$

where (s) and (g) denote the solid and gaseous phases respectively. Using standard thermochemical data, the partial vapor pressure for each volatile species under equilibrium conditions can be calculated. Figure 1(a,b) shows the calculated $WO_2(OH)_2$ (g) partial pressure at 500 and 700°C versus humidity (assuming that the sum of the partial pressures of H_2O (g) and H_2 (g) is unity).

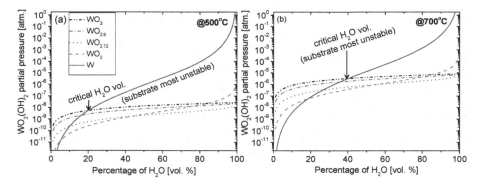

Figure 1. Calculated $WO_2(OH)_2$ (g) partial pressures at (a) 500°C and (b) 700°C. At a given humidity the substance with the highest partial pressure is the most unstable.

Figure 2 shows the the critical $WO_2(OH)_2$ (g) and H_2O (g) partial pressures as a function of temperature for the W substrate becoming the most unstable layer, that is, having the highest $WO_2(OH)_2$ (g) partial pressure of all the layers. These results suggest that higher temperatures produce more $WO_2(OH)_2$ (g) from all layers and at high H_2O (g) partial pressures, the W substrate provides most of the volatile species.

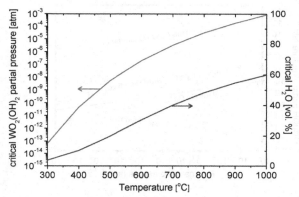

Figure 2. Critical $WO_2(OH)_2$ (g) and H_2O (g) partial pressures as a function of temperature for tungsten metal, values that have also been indicated in Figure 1.

The Cu-H₂O-O₂ system

Similarly to the W-H_2O system, the stoichiometric oxide sequence on Cu can be described by $CuO|Cu_2O$ with the O_2 deficient Cu_2O phase next to the Cu metal. The reactions of each layer with H_2O and O_2 vapor, resulting in the formation of volatile CuOH(g) can be described using the following equations:

$$4CuO(s) + 2H_2O(g) = 4CuOH(g) + O_2(g) \tag{6}$$
$$Cu_2O(s) + H_2O\ (g) = 2CuOH(g) \tag{7}$$
$$4Cu(s) + 2H_2O(g) + O_2(g) = 4CuOH(g) \tag{8}$$

Assuming that O_2(g) is rapidly consumed at the metal-oxide or oxide-oxide interface, then the partial pressures of O_2(g) at each interface can be calculated at a given temperature using the following equations:

$$2Cu_2O(s) + O_2(g) = 4CuO(s) \tag{9}$$
$$4Cu(s) + O_2(g) = 2Cu_2O(s) \tag{10}$$

At $500°C$, O_2 partial pressures (in atm.) are calculated to be of the order of 10^{-7} and 10^{-14} respectively from equations (9) and (10) respectively. Substituting these values into equations (6) and (8), corresponding to the $Cu_2O|CuO$ and $Cu|Cu_2O$ interfaces respectively, it is possible to estimate the CuOH partial pressures at each interface as a function of H_2O, and shown in Figure 3. Similarly to the W-H_2O results, these results suggest that a reaction directly with the metal substrate provides most of the volatile hydroxide species.

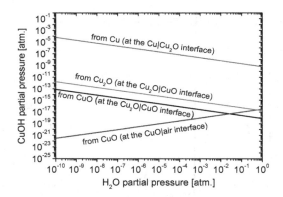

Figure 3. CuOH partial pressures as calculated from reactions (6-8) showing that the vapor pressure is greatest for equation (8), the reaction between Cu(s) and H_2O(g) at the $Cu|Cu_2O$ interface.

EXPERIMENTAL RESULTS AND DISCUSSION

The W-H_2O system

W (99.95% pure) pieces were annealed in a quartz tube furnace for 30 minutes in a N_2 ambient (60-90 sccm) at 500°C and 700°C in a 'wet' and 'dry' atmosphere respectively. The 'wet' atmosphere was obtained by pre-bubbling the N_2 through heated (80-100°C) water while the 'dry' atmosphere was obtained by pre-bubbling the N_2 through unheated (room temperature) water. Figure 4 shows typical scanning electron microscope (SEM) images of the surfaces.

The Cu-H_2O-O_2 system

A Cu foil (with a thickness of 0.9mm) was annealed at 500°C in an open ended furnace tube (ambient conditions) for 30 minutes. As a comparison, 450nm Cu films deposited via thermal evaporation onto untreated quartz and Si pieces were also annealed under the same conditions. A film thickness of 450nm is close to the lower limit where CuO nanowires have been observed to grow [10]. Figure 5 shows typical SEM images of the Cu surfaces after annealing.

Figure 4. SEM images for tungsten oxide nanowires grown at (a) 500°C in wet N_2 and (b) 700°C in dry N_2.

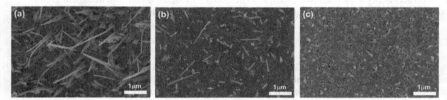

Figure 5. SEM images for copper oxide nanowires grown on (a) Cu foil and thermally evaporated Cu (450nm) on (b) quartz and (c) Si.

A significant amount of hydroxide vapor is required for nanowires to be observed, most likely close to the order of 10^{-10} to 10^{-6} atm [11, 12]. As shown in Figure 1(a), in a wet atmosphere at 500°C, the least stable layer is that of the W substrate, with $WO_2(OH)_2$ partial pressures likely to be orders of magnitude greater than for the oxides ($>>10^{-8}$ atm). Although the stoichiometry of these nanowires are yet to be determined, it is likely that the nanowires are of the WO_2 or $WO_{2.72}$ phase. Similarly, with reference to Figure 1(b), in a dry atmosphere at 700°C, W and WO_2 are the most stable phases, with the partial pressures of the hydroxide of the order of 10^{-6} atm. for WO_3 and $WO_{2.9}$, suggesting that the nanowires will be more W rich than compared to the wires grown in a wet atmosphere at 500°C. The nanowires most likely grow via a condensed phase diffusion transport process, where a concentration gradient of the hydroxide determines (at a given temperature) the final stoichiometry. The W oxide nanowires grown at 500 and 700°C have different lateral dimensions, most likely due to different nucleation crystal sizes of oxides with a different stoichiometry [7]. Interestingly, when annealed in a dry O_2 ambient at 700°C, no nanowires were observed, only a thick yellow film corresponding to WO_3. In this case a film is created over the surface preventing water vapor and hydroxide diffusion from the metal and WO_x (x<3) regions. For the Cu-H_2O-O_2 system, volatile partial pressures of CuOH (g) is orders of magnitude greater for the reaction with the metal rather than the oxides (refer to Figure 3), even under standard atmospheric conditions (2-3% H_2O at room temperature). Sporadic nanowire growth of Cu coated Si and quartz suggests that thicker (> 450nm) metal coatings are required if nanowires are to be grown on secondary substrates, since the metal is completely oxidized and mostly delaminates from the substrate.

CONCLUSIONS

The oxidation kinetics of metal surfaces is complex but can be analyzed thermochemically to ascertain oxide nanowire growth conditions. More specifically, determination for the formation of a metal transport precursor via a volatile hydroxide species needs to be considered. The role of H_2O and/or H_2O and O_2 is important for not only volatile hydroxide formation, but also oxide scale growth. Tungsten and copper oxide nanowires can be readily grown on metal substrates at relatively low temperatures (500°C), and can be described using reactions with H_2O and H_2O+O_2, respectively. A thermochemical assessment suggests that the dominant metal transport species, resulting in nanowire growth, is that of a volatile hydroxide formed at a metal-metastable oxide interface, a mechanism that could potentially be applied to many other metal systems. It is likely that controlling the H_2O and/or H_2O and O_2 it is possible to grow nanowires at even lower temperatures (<500°C) for short anneal times, though their stoichiometry may differ than those grown at higher temperatures. These nanowires can be grown on secondary substrates (for example Si or quartz), though a metal film thickness greater than about 450nm is likely to be required to provide a sufficient hydroxide source.

ACKNOWLEDGMENTS

Financial support from the Australian Research Council (ARC) and access to the facilities of the Centre for Advanced Microscopy (CAM) with funding through the Australian Microscopy and Microanalysis Research facility (AMMRF) is gratefully acknowledged.

REFERENCES

1. S. Rackauskas, A. G. Nasibulin, H. Jiang, Y. Tian, V. I. Kleshch, J. Sainio, E. D. Obraztsova, S. N. Bokova, A. N. Obraztsov and E. I. Kauppinen, *Nanotech.* **20** (16), 165603 (2009).
2. D. J. Young, *High Temperature Oxidation and Corrosion of Metals.* (Elsevier, 2008).
3. T. H. Kim, A. Shalav and R. G. Elliman, *J. Appl. Phys.* **108** (7), 076172 (2010).
4. A. Shalav, G. H. Collin, Y. Yang, T.-H. Kim and R. G. Ellimam, *J. Mat. Res.* **17**(4), 2240-2246 (2011).
5. A. Shalav, T. Kim and R. Elliman, *IEEE Sel. Top. Quant. Elect.,* **17** (4), 785-793 (2011).
6. E. Lassner and W.-D. Schubert, *Tungsten. Properties, Chemistry, Technology of the Element, Alloys, and Chemical Components.* (Plenum Publishers, 1999).
7. L. Yuan and G. W. Zhou, *J. Electrochem. Soc.* **159** (4), C205-C209 (2012).
8. W. D. Schubert, B. Lux and B. Zeiler, *Int. J. Refract. Met. & Hard Mater.* **13** (1-3), 119-135 (1995).
9. J. P. Wang and W. D. Cho, *ISIJ International* **49** (12), 1926-1931 (2009).
10. K. L. Zhang, C. Rossi, C. Tenailleau, P. Alphonse and J. Y. Chane-Ching, *Nanotechnology* **18** (27), 275607 (2007).
11. R. G. Elliman, T. H. Kim, A. Shalav and N. H. Fletcher, *J. Phys. Chem. C* **116** (5), 3329-3333 (2012).
12. A. Shalav and R. G. Elliman, *in preparation.*

Mater. Res. Soc. Symp. Proc. Vol. 1494 © 2012 Materials Research Society
DOI: 10.1557/opl.2012.1656

Semiconductive Properties of Alternating Mg/C Multi-layer Films

with Hydroxylation Treatment

Masafumi Chiba[1], Daisuke Endo[1], Kenichi Haruta[2], Hideki Kimura[2], and Hideo Kiyota[3]
[1] Department of Materials Chemistry, Tokai University,
317 Nishino, Numazu, Shizuoka, 410-0395, Japan
[2] Department of Electrical and Electronic Engineering, Tokai University,
4-1-1 Kitakaname, Hiratsuka, Kanagawa, 259-1292, Japan
[3] Department of Mechanical Systems Engineering, Tokai University,
9-1-1 Toroku, Kumamoto, Kumamoto, 862-8652, Japan

ABSTRACT

A $Mg(OH)_2$–C transparent conductive film was prepared using the sputtering method by the initial formation of a Mg-C film generated by the alternate layering of Mg and C on a rotating substrate and subsequent exposure of the film to atmospheric water vapor. To examine the influence exerted by the Mg/C layers of the starting film sample on semiconductivity, evaluations of the electrical conductivity properties of the film during the hydroxylation process and the optical properties after the hydroxylation process were carried out. As a result, although no effects on the characteristics of the electrical conductivity properties associated with the composition or number of layers in the films could be confirmed, it was determined that the films possessed the characteristics of semiconductors. On the other hand, the optical properties were found to be affected by the composition and number of layers of the Mg/C films.

INTRODUCTION

Currently, among the various transparent electrically conductive materials used in flat-panel displays (FPD) and the like, indium tin oxide (ITO) has become the most standard material [1]. However, In, which is the main raw material for these applications, is a rare metal, and its Clarke number is 1×10^{-5} [2]. Accordingly, owing to resource depletion and the consequential soaring prices, research into alternative materials is being actively pursued [3-5]. Our research group has succeeded in the development of $Mg(OH)_2$ doped with C as a new transparent electrically conductive material to replace ITO [6-8]. This material is prepared by the incorporation of C into the $Mg(OH)_2$ lattice by a process involving the alternate layering of Mg and C on a rotating substrate by the sputtering method followed by hydroxylation, which affords electrical conductivity. The film preparation method by the rotating-substrate used here facilitates the variation of the amount of Mg and C deposited on the substrate. At the same time, the approach makes it possible to adjust the composition, the layer thickness, and the number of layers of each atomic species. As a result, even if the film thickness is consistent, it is possible to prepare materials in which the number of layers of each atom and the thickness of each layer differ. For this reason, adjustments to the rotation speed of the substrate and the C concentration at the time of layer formation are expected to affect the material properties.

Thus, in the present research, because differences in the layer thickness and the composition of Mg and C can be expected to affect the transparency and electrical conductivity of the material, an evaluation of the electrical conductivity properties during the hydroxylation step of the film-forming process and the optical properties after hydroxylation was carried out. Based on the results, we studied the influence of the composition and number of layers on semiconductivity.

EXPERIMENTAL DETAILS

For the preparation of the starting samples, a radio-frequency magnetron-sputtering system SPC-350 (Nichiden ANELVA) was used. The film samples were elaborated on square glass substrates 25 mm on a slide (Corning, #7059). The multi-layered film formation was carried out using Mg and C (both of high purity: 3N) disk targets. The starting sample was set so that the film thickness was 1 μm. To study the effects of the number of layers on the sample, the setting was adjusted so that the composition would be $Mg_{63}C_{37}$, and the experiment was conducted under various substrate rotation rates. The film formation was conducted over the rotation range from 15 rpm, in which a single atom layer of C was deposited, to 60 rpm, in which C was unevenly distributed in the layer. Additionally, to study the effects of the film composition, the rotation rate of the substrate was fixed at 60 rpm, and the reaction time was extended. Also, film formation was conducted over a carbon concentration range from 40 at% C concentration to the reduced concentration of $Mg_{65}C_{35}$. After the starting film formation, hydroxylation was carried out by exposure to atmospheric water vapor. The detail of these film preparations and hydroxylation conditions are written in previous work [9]. An ultraviolet absorption spectrophotometer (V-570; Nihon Bunko, Inc.) was used to measure the transparency of the final sample, and the Hall effect measurement equipment (HL5500PC; Nanometrics Japan, Inc.) was used to measure and calculate the conductivity and mobility.

RESULTS AND DISCUSSIONS

Electrical properties

The results of the electrical conductivity of the Mg/C films measured by using the Hall effect measurement equipment are shown below. The results of carrier density and conductivity measurements are shown in figure 1. In all the samples in which the C concentration and rotational speed during the film formation were varied, the conductivity tended to increase with the carrier density. Moreover, the C concentration in the film and the rotation rate during film formation exert almost no influence on these properties.

The measurements of the carrier density and mobility are shown in figure 2. These results show a tendency for the mobility to decrease as the carrier density increases in the samples generated under the above conditions. However, as seen previously for the carrier density and conductivity, the C concentration in the film and the rotation rate during film formation exert almost no influence on these properties.

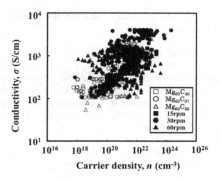

Figure 1. Relationship between conductivity and carrier density of the Mg/C film in which the composition and rate of rotation of the substrate have been adjusted.

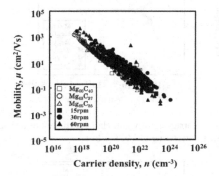

Figure 2. Relationship between mobility and carrier density of the Mg/C film in which the composition and rate of rotation of the substrate have been adjusted.

The results from the measurements of carrier density and Hall coefficient are shown in figure 3. These results show that for the studied samples, when the carrier density increases, the Hall coefficient tends to converge from either a negative or positive value to 0 cm^2/C. It is suggested that the Hall coefficient may be positive or negative because during film formation, the sample is in a metallic state and also because there is no difference between the number of electrons and holes. Furthermore, it is expected that when Mg hydroxylation occurs, because the mobility decreases as a result of the increasing carrier density, as shown in figure 2, the Hall coefficient value converges to 0 cm^2/C. Again, similar to the results described above, the C concentration in the film and the rotation rate during film formation exert almost no influence on these properties.

In figure 4, a representation of the Hall coefficients from the vertical axis of figure 3 is shown using absolute values. By disregarding the sign of the Hall coefficient, a linear relationship is revealed and can be used to calculate the correlation between the carrier density and the Hall coefficient. See equation 1.

$$R_H = 0.96/e \cdot n \approx 1/e \cdot n \quad (1)$$

When using the equation to compare the general carrier density with the Hall coefficient, the error is 4%. From the above mentioned results, it can be said that the Mg/C thin film possesses semiconductor characteristics afforded by the hydroxylation process. On the other hand, it is not clear whether the electrical conductivity of the sample is dependent upon the atomic composition and the number of layers.

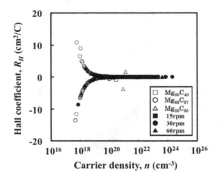

Figure 3. Relationship between carrier density and Hall coefficient as a function of the composition and the substrate rotation rate of Mg/C films.

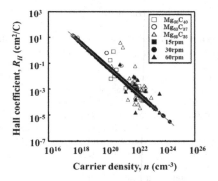

Figure 4. Relationship between carrier density and Hall coefficient as a function of the composition and the substrate rotation rate of Mg/C films.

Optical properties

Since it is thought that the modifications to the C-atom distribution in the film afforded in the substrate rotation rate described in the previous section, it is interesting to consider the effects exerted on the sample's optical properties.

Thus, to measure the influence on the transparency of the thin-film material afforded by hydroxylation, the average transparency in the transmission region of 380–750 nm wavelength (λ) is calculated. The relationship between the average transparency and the rotation speed of the substrate is shown in figure 5. The results show that the average transmission rate tends to increase as the rotation rate of the substrate increases. It is expected that this occurs because the quantity of light absorbed by C atoms increases owing to an increase in both the number of layers in the film and the thickness of these layers caused by the reduction in the rotation rate of the substrate.

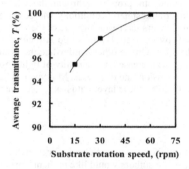

Figure 5. Dependence of the average transmittance of Mg/C films within the visible wavelength on the substrate rotation rate.

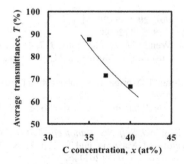

Figure 6. Dependence of the transmittance of Mg/C films within the visible wavelength on the C concentration.

The relationship between the C concentration and the average transparency of the sample is shown in figure 6. The results show that the average transmittance decreases with increasing C concentration. Because the transmittance of the Mg/C thin film decreases with a higher carbon concentration and an increase in the number of layers, it is clear that the C concentration and the rotation rate of the substrate affect the optical properties of the sample.

CONCLUSIONS

In this study, a Mg/C film is prepared by the sputtering method. In addition, an evaluation of the electrical conductivity during the subsequent hydroxylation process and the optical properties after the hydroxylation process is performed as well as an experiment to determine the dependence of semiconductivity on the composition and layer number. The results are summarized below.

Mg/C films acquire semiconductor properties through the conversion process from Mg to $Mg(OH)_2$ by hydroxylation under atmospheric conditions. Moreover, even though the C concentration and the number of C layers increases, the size of the C atoms is small, and because they are positioned in the lattice between $Mg(OH)_2$ units, no influence on the electrical conductivity of the sample is observed. On the other hand, when the C atom concentration in the sample that has become transparent is increased, the quantity of light absorbed by the C atoms increases, and as a result, the transmission rate decreases. Moreover, when multiple C-atom layers form, because light is absorbed by these layers, it is expected that this further reduces the transmittance.

ACKNOWLEDGMENTS

The part of this work was supported on "A Subsidy for Activating Educational Institutions" by Tokai University. The authors would like to thanks professor Dr. Toshiro Kuji of Tokai University for useful technical advice.

REFERENCES

1. T. Minami, *Semicond. Sci. Techn.*, **20**, S35 (2005).
2. Y. Baba, *J. Surf. Sci. Soc. Jpn.*, **29**, 578 (2008).
3. K. Ellmer, R. Cebulla, and R. Wendt, *Thin Solid Films*, **317**, 413 (1998).
4. K. Tominaga, T. Takao, A. Fukushima, T. Moriga, and I. Nakabayashi, *Vacuum*, **66**, 505 (2002).
5. K. Hayashi, K. Kondo, K. Murai, T. Moriga, I. Nakabayashi, H. Fukumoto, and K. Tominaga, *Vacuum*, **74**, 607 (2004).
6. T. Kuji, T. Honjo, M. Chiba, T. Nobuki, and J. -C. Crivello, *e-J. Suff. Sci. Nanotech.*, **6**, 15 (2008).
7. T. Honjo, M. Chiba, and T. Kuji, *e-J. Suff. Sci. Nanotech.*, **7**, 791 (2009).
8. M. Chiba, M. Higashi, H. Kiyota, M. Maizono, and T. Kuji, *TMS (The Minerals, Matals and Materials Society), 2011 Supp. Proc.*, **3**, 605 (2011).
9. M. Chiba, D. Endo, M. Maizono, M. Higashi, and H. Kiyota, *Mater. Res. Soc. Symp. Proc.*, **1406**, 119 (2012).

Mater. Res. Soc. Symp. Proc. Vol. 1494 © 2012 Materials Research Society
DOI: 10.1557/opl.2012.1699

Intense Ultraviolet Photoluminescence Observed at Room Temperature from NiO Nano porous Thin Films Grown by the Hydrothermal Technique

Sachindra Nath Sarangi[1], Dongyuan Zhang, Pratap Kumar Sahoo[2], Kazuo Uchida, Surendra Nath Sahu[3] and Shinji Nozaki

Graduate School of Informatics and Engineering, The University of Electro-Communications, 1-5-1 Chofugaoka, Chofu-shi, Tokyo 182-8585, Japan

[1]On leave from Institute of Physics, Bhubaneswar, 751005, India

[2]National Institute of Science Education and Research (NISER), Institute of Physics Campus, Bhubaneswar-751005, Orissa, India

[3] The National Institute of Science and Technology, Palur Hills, Berhampur 761008, India

ABSTRACT

We have successfully formed high-quality nanoporous NiO films by the hydrothermal technique and observed intense ultraviolet (UV) luminescence at room temperature. The SEM image reveals nanoporous NiO films with pore diameters from 70 to 500 nm. The results of XRD, Micro Raman and FTIR characterizations confirm the cubic structure of NiO. The optical band gaps estimated from the absorption spectrum are found to be 3.86 and 4.51 eV. The former is similar to that of bulk NiO, while the latter is much higher than that of bulk NiO. The increased band gap was attributed to the quantum confinement in the NiO nanocrystals, which may be present in the nanoporous NiO film. The room-temperature photoluminescence (PL) spectrum shows a peak of intense luminescence at 3.70 eV and several other peaks in the UV and near-UVwavelength regions. The intense UV luminescence at 3.70 eV was associated with the near band-edge emission and the others with defect-related emission. The high-quality wall of nanoporous NiO with a large surface-to-volume ratio provided the intense UV emission.

INTRODUCTION

Some of metal-oxide semiconductors have attracted increasing attention of engineers and scientists developing ultraviolet (UV) - LEDs because of their wide band gaps and may become materials of UV-LEDs in future. Among such metal-oxide semiconductors, Nickel oxide (NiO) is one of few oxide semiconductors which can have the p-type conductivity. However, there are only few reports on intense UV luminescence observed in NiO at room temperature. It is well known that p-type conductivity of NiO can be realized by nickel vacancies and/ or oxygen interstitials. Among the p-type oxide-semiconductors, the greatest attention has been paid to NiO thin films because of their excellent optical, electrical and magnetic properties. A number of studies have been made on high dielectric permittivity [1], thermoelectric behavior [2], and ferromagnetism [3], whereas very few reports are found on its luminescence property.

The technique mostly commonly employed to deposit NiO is sputter deposition. However, the plasma damage caused by sputter-deposition may form defects and oxygen deficiency in the deposited NiO films. Although other techniques, such as ammonia precipitation [3,4], hydrothermal method [5], sol–gel method [2,6], electrochemistry [7], microemulsions [8] methods. Among them, the hydrothermal method is the simplest, most cost-effective and able to obtain high-quality NiO films.

In this study, we employed a hydrothermal method to synthesize NiO nanoporous thin films, from which intense UV luminescence was observed and associated with the quantum confinement in the thin wall.

EXPERIMENTAL DETAILS

NiO thin films have been deposited on sapphire and Si substrates at 100 °C by the hydrothermal technique. In a typical deposition, 0.2 M of Nickel (II) Nitrate Hexahydrate (Ni $(NO_3)_2$. 6 H_2O) and Hexamethylenetetramine ($C_6H_{12}N_4$) reactants were dissolved in deionized water in a stainless steel bottle. The hydrothermal synthesis was carried out by heating the bottom of the bottle at 100 °C for 4 h. The substrates of size 5 mm x 10 mm were suspended in the solution with the face down during the growth after being cleaned in acetone, ethyl alcohol and deionized water in an ultrasonic bath for 5 min each. The Si substrates were used for the micro Raman scattering measurements to calibrate the monochromator using the optical phonon peak of Si, and the sapphire substrates were used for the rest of characterizations. The morphology and crystal structure were studied by the field emission scanning electron microscopy (FE-SEM) and x-ray diffraction (XRD), respectively. The SEM images were taken with a JEOL FE-SEM operated at 5 kV. The XRD patterns were collected with a Philip X-ray diffractometer (X'Pert) using the Cu-K_α x ray. The optical properties were characterized by micro Raman scattering, photoluminescence (PL), and absorption and Fourier Transform Infrared Absorption (FTIR) measurements. Micro Raman spectra were measured at room temperature through 10× microscope objective lens using a Jasco micro-Raman spectrometer equipped with laser (532 nm, power = 5 mW). The spectral signals were collected by a Peltier-cooled (−50°C) CCD detector with the 1800 grooves/mm grating. The PL spectra were obtained at room temperature using the 250 nm line of a UV lamp.

RESULTS AND DISCUSSION

The X-ray diffraction (XRD) pattern is shown in Fig. 1. The XRD pattern clearly shows the diffraction peaks at 2θ of 37.80, 44.03 and 61.09, corresponding to the reflection from the (111), (200) and (220) planes, respectively, of the face-centered-cubic (fcc) structured NiO. The Bragg angles and relative intensities of the peaks are in good agreement with that reported for fcc-structured NiO [9]. The result suggests that the film is not well oriented.

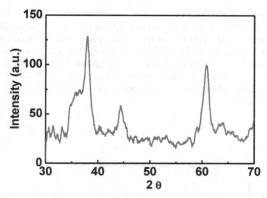

Figure 1. X-ray diffraction pattern of the NiO nanoporous film deposited on a sapphire substrate.

Figure 2 shows the SEM image of the growth morphology of the deposited NiO film. The inset shows the higher-magnification image of the same sample. The image reveals a sponge-like nanoporous film formed by interconnected NiO nanoflakes with a thickness of about 50 to 300 nm. These nanoflakes grow on the substrate, forming a network with 70 – 500 nm pores.

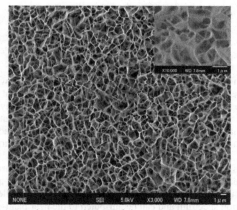

Figure 2 Scanning electron microscopy image of the as-grown NiO nanoporous film. The inset shows the higher-magnification image.

The Raman spectrum of the NiO nanoporous film is shown in Fig. 3. Four peaks are observed at 572, 736, 950 and 1150 cm^{-1} in the spectra. The former two peaks could be assigned to the first-order transverse optical (TO) and longitudinal optical (LO) phonon modes of NiO, respectively. The latter two peaks could be attributed to TO+LO and 2LO modes of NiO [10]. The optical phonon mode of the Si substrate is also seen at 520 cm^{-1}. The FTIR spectrum of the NiO nanoporous film (the figure is not shown here) shows three peaks at 403, 415 and 430 cm^{-1}, corresponding to the stretching vibrations of Ni–O bond. The micro Raman and FTIR spectra confirm NiO with good-crystallinity.

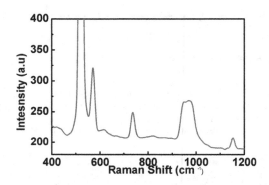

Figure 3.   Micro Raman spectrum of the NiO nanoporous film on Si.

The optical transmission spectrum of the NiO nanoporous film is shown in Fig. 4 (a). The optical band edge estimated from the figure is 3.86 eV which is similar to that of bulk NiO (3.6 – 4.0 eV) [11]. It should be noted that the spectrum has a tail in the high-energy side and the band gap related to this tail is estimated to be 4.51 eV. The increased band gap is attributed to the quantum confinement in nanocrystals which may be present on the pore surface, as often observed in porous silicon. A similar optical band gap was found in NiO nanoparticles with a size of 20 nm [12].

Figure 4 (b) shows a photoluminescence spectrum of the NiO nanoporus film. The sample shows an intense main peak at 3.7 eV (335 nm) and two shoulder peaks at about 4.27 eV (290 nm) and 3.31 eV (374 nm). The peak energy of 3.7 eV is close to the smaller optical band gap obtained from the absorption spectrum and can be attributed to the near band-edge luminescence. The PL peak at 4.27 eV may be associated with the surface defects. Such above-band-gap luminescence was often observed in porous Si and was associated with the surface defects of nanocrystals [13]. The other peak with a longer wavelength is also associated with defect-related luminescence.   It should be noted that the luminescence intensity of near band-edge luminescence is usually very low at room temperature in bulk NiO and becomes higher in nanostructured NiO [14]. Nevertheless, such UV emission with a higher energy and intensity we obtained has been rarely reported and was attributed to the high- quality wall of porous NiO

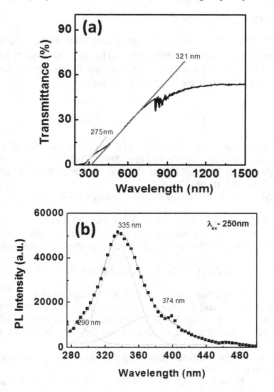

Figure 4 (a) Optical transmission spectrum of the NiO nanoporous film and (b) Photoluminescence spectrum of the NiO nanoporous film.

CONCLUSIONS

Nanoporous NiO films have been successfully synthesized by the hydrothermal technique using Nickel (II) Nitrate Hexahydrate (Ni(NO$_3$)$_2$. 6 H$_2$O) and Hexamethylenetetramine (C$_6$H$_{12}$N$_4$) as the precursors. The XRD pattern confirmed the fcc crystal structure of NiO. The optical absorption spectrum showed two optical band gaps, estimated to be 3.86 and 4.51 eV. The smaller band gap was similar to the band gap of bulk NiO. The larger one was associated with the quantum confinement in NiO nanocrystals, which may be present in the nanoporous NiO. The nanoporous NiO film exhibited an intense near band-edge luminescence at 3.7 eV. A high-quality wall of nanoporous NiO with a high surface-to-volume ratio provided the intense UV emission.

ACKNOWLEDGMENTS

The authors would like to thank Mr. H. Ono for his assistance with the experiments. We acknowledge that the research project was in part supported by the Japan Society for the Promotion of Science (JSPS). One of the authors (S. N. Sarangi) also greatly acknowledges the JSPS Postdoctoral Research Fellowship for Foreign Researchers.

REFERENCES

1. P. Thongbai, T. Yamwong and S. Maensiri, *Appl. Phys. Lett.* **94,** 152905(2009).
2. S. D. Tiwari and K. P. Rajeev, *Thin Solid Films,* **505,** 113 (2006).
3. J. H. He, S. L. Yuan, Y. S. Yin, Z. M. Tian, P. Li, Y. Q. Wang , K. L. Liu and C. H. Wang, *J. Appl. Phys.* **103,** 02390 (2008).
4. M. M. Kashani Motlagh, A. A. Youzbashi and L. Sabaghzadeh, *Int. J. Phys. Sci.* **6(6),** 1471 (2011).
5. S. K. Meher, P. Justin and G. R. Rao, *Nanoscale* **3,** 683 (2011).
6. M.A. Ahmed , *J. Photochemistry and Photobiology A: Chemistry* **238,** 63 (2012).
7. S. Xing, Q. Wang, Z. Ma, Y. Wu and Y. Gao, *Materials Research Bulletin* **47,** 2120 (2012).
8. Y. Du, W. Wang, X. Li, J. Zhao, J. Ma, Y. Liu, G. Lu and Y. Du, *Materials Letters* **68,** 168 (2012).
9. Z. Zhang, Y. Zhao, and M. Zhu, *Appl. Phys. Lett.* **88,** 033101 (2006).
10. M. Lu, T. Y. Lin, T. Weng, and Y. Chen, *OPTICS EXPRESS* **19,** 16266 (2011).
11. Y. Lin, T. Xie, B. Cheng, B. Geng, L. Zhang, *Chem. Phys. Lett.* **380,** 521(2003).
12. Soumen Das, Tandra Ghoshal and M. G. Nambissan, *Phys. Status Solidi C* **6,** 2569 (2009).
13. P. M. Fauchet, *Semiconductor and Semimetals,* Chapter 6, edited by D. J. Lockwood, **49** (Academic Press, 206 (1998).
14. Z. Y. Wu, C. M. Liu, L. Guo, R. Hu, M. I. Abbas, T. D. Hu, and H. B. Xu, *J. Phys. Chem. B* **109,** 2512 (2005).

Mater. Res. Soc. Symp. Proc. Vol. 1494 © 2013 Materials Research Society
DOI: 10.1557/opl.2013.582

Advantages of Hydrogen Peroxide as an Oxidant for Atomic Layer Deposition and Related Novel Delivery System

Jeff Spiegelman[1], Dan Alvarez,Jr[1], Russell J. Holmes[1], Ed Heinlein[1], and Zohreh Shamsi[1]
[1] RASIRC, 7815 Silverton Ave, San Diego, CA, 92126, U.S.A.

ABSTRACT

Proposed is the use of Hydrogen Peroxide (H_2O_2) as the ideal oxidant for atomic layer deposition of metal oxide films. H_2O_2 has similar oxidation properties to Ozone while simultaneously having slightly stronger proton transfer properties than water. Vital to the success of any vapor phase chemistry is delivery of stable compositions, temperature and pressure. This study demonstrates the viability of a new membrane technology for the precise delivery of $H_2O_2/$ H_2O mixtures starting from a liquid range of 30-70%. An in-situ gas phase cleaning process to remove carbon contamination from Ge(100) surfaces using gas phase H_2O_2 has been characterized.

INTRODUCTION

Atomic Layer Deposition (ALD) processing is currently moving toward high volume production for several applications[1] including: High-k gate dielectrics (hafnium oxide, hafnium silicate): High-k capping layers for metal gate work function tuning; High-speed aluminum oxide for magnetic heads; Conformal passivation layers; and High-k for MEMS. A major advantage of ALD is that the self-limiting nature of the process enables fully conformal films. Another advantage is that a variety of materials can be formed at relatively low temperatures. ALD can be applied to produce various oxides, nitrides or other compounds. The surface control achieved with ALD allows single or tailored multiple layer deposition of thin, uniform and pinhole-free films over large areas.

High mobility channel materials and new device structures will be needed to meet the power and performance specifications in future technology nodes.[2] In these new material systems and devices, various electrically active defects are present at or close to the interface between the high-K dielectric and the alternative channel material that are a major concern for both the performance and the reliability of these new devices. For these devices, the use of ALD appears to be the only viable solution, as CVD and PVD lead to a disproportionate number of defects. Several challenges exist regarding manufacturability of ALD. These include precursor and oxidant optimization in order to achieve the highest quality films at the lowest possible temperatures. Moreover, new surface preparation methods are needed for these films, where near perfect surfaces are needed for defect-free initiation of ALD.

The deposition of metal oxide films requires an oxidant precursor, typically water or ozone. Ozone is a strong oxidant and can provide high quality films. However, ozone must be generated in situ, is equipment intensive, hazardous and suffers from high cost. Water vapor delivered under controlled conditions can be used as an alternative with generally similar results for most film thicknesses.

This is especially true with metal organic precursors which incorporate nitrogen containing ligands. The current precursors of choice for the ALD of hafnium oxide based layers are hafnium amides, specifically, tetrakis(dimethylamino)hafnium, $[Hf(NMe_2)_4]$, and tetrakis(ethylmethylamino)hafnium $[Hf(NEtMe)_4]$.

These hafnium amides give "true" self-limiting ALD at growth temperatures below ~275°C. Above this temperature they thermally break down on the wafer surface, destroying the self-limiting growth mechanism. At or below ~275°C, though water can yield high quality films, it is problematic in that reaction is too slow for high throughput manufacturing.

From a thermodynamic standpoint, ozone is a much stronger oxidant than water and should yield a higher sticking ratio than water. However this pathway entails initial oxidation of the carbon containing ligand followed by several convoluted steps to eliminate hydrocarbon by-products.

Proposed is the use of hydrogen peroxide (H_2O_2) as the ideal oxidant. H_2O_2 has similar oxidation properties to ozone (oxidation potential $O_3 = 2.1V$ versus $1.8V$ for H_2O_2) while simultaneously having slightly stronger proton transfer properties than water (H_2O_2 pKa = 6.5 versus pKa = 7.0 for water). However, extreme difficulty lies in the delivery of H_2O_2. Early studies that reported low quality films obtained with the use of H_2O_2 have been limited by unsophisticated H_2O_2 vapor delivery systems.[3] Bubblers, and other initial delivery device designs are highly variable with respect to mass transfer rates and are only able to deliver very low concentrations of H_2O_2. Concentrated liquid H_2O_2 is highly explosive therefore H_2O_2 is commonly used as 30% H_2O_2/H_2O solutions. The inherent low vapor pressure of H_2O_2 versus H_2O in these mixtures leads to delivery of minute concentrations at room temperature of 294ppm H_2O_2 for bubbler systems and 150ppm H_2O_2 for systems under vacuum at 30 torr.[4]

The design and delivery of new vapor phase chemistries has been the initial focus of this work. Vital to the success of any vapor phase chemistry is delivery of stable compositions, temperature and pressure. Limited success obtained by early workers in this area can be attributed to a lack of process control or the inability to deliver high concentrations of low vapor phase chemistries in multi-component systems. This study demonstrates the viability of a new membrane technology for the precise delivery of H_2O_2/H_2O mixtures.

MEMBRANE TECHNOLOGY

In this study, a novel membrane delivery system that incorporates temperature, pressure, and flow rate control as well as chemical precursor purification has been employed. The system adds controlled amounts of vapor chemistry to any carrier gas. The unit may operate in atmospheric and vacuum pressures, and uses liquid solution sources. The device utilizes a hydrophilic membrane that excludes organic and metal contaminants from entering the vapor stream. Liquid sources from the membrane unit do not come into direct contact with the vacuum system or reaction environment.

Chemical Delivery System

A carrier gas flows into the membrane assembly where a liquid chemical diffuses across a membrane into the carrier gas. The temperature of the carrier gas is then measured and fed back to a temperature controller to adjust the concentration level. Internal pressure control maintains independence from variations in downstream process pressures, allowing operation into atmospheric and vacuum pressure environments. The membrane is highly selective, preventing most carrier gases from crossing over into the source (Figure 1). Selectivity of up to a million to one water molecules over nitrogen has been found. Organic and metal contaminants in the liquid source cannot permeate across the membrane or enter the carrier gas stream, resulting in a chemical vapor saturated product that is consistent and pure.

The carrier gas is saturated based on the temperature at the gas/liquid interface, providing accurate delivery of chemical vapor. With the addition of a back pressure regulation device, low vapor pressure gases can be delivered into sub-atmospheric processes.

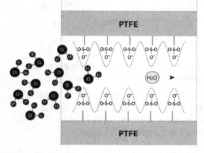

Figure 1. Membrane cross section showing water flow across the membrane. Idealized membrane pore showing the hydronium ion, water molecules, and radially symmetric axially periodic distribution of sulfonate (-SO_3-) fixed sites.

H_2O_2 VAPOR DELIVERY

Extreme difficulty lies in the delivery of vapor phase H_2O_2. Liquid H_2O_2 is commonly used as 30% H_2O_2/ H_2O solutions. The inherent low vapor pressure of H_2O_2 versus H_2O in these mixtures leads to delivery of 8 parts H_2O_2 per 1000 parts water at 20°C. Many published reports exist on the use of H_2O_2 vapor in semiconductor processes, however, in actuality, very little H_2O_2 is delivered to the process.[1,3]

Bubblers, and other initial delivery device designs are highly unstable with respect to mass transfer rates. As the vapor is drawn off by vacuum or carrier gas, water is preferentially removed and the H_2O_2 concentration increases. In addition, our preliminary work showed that evaporative cooling takes place, which when combined with changes in concentration, leads to highly variable mass transfer rates. In the case where temperature is controlled, variable liquid concentration levels still lead to unstable delivery concentrations.

A new technology that is capable of generating and delivering stable concentrations of H_2O_2 vapor has been developed (Figure 2). Pervaporation devices are used in series, where pre-loading a carrier gas first with water and then flowing through a pervaporater (HPDA) filled with H_2O_2/ H_2O mix enables a controlled H_2O_2/ H_2O ratio to be delivered to a process. Here, pre-humidification with water vapor ensures maintenance of the liquid source concentration, which not only ensures process control but also has safety ramifications.

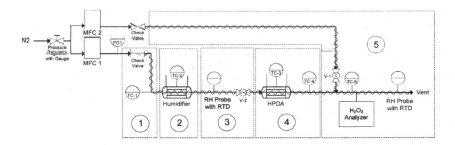

Figure 2. Process and instrumentation diagram of the H_2O_2 delivery system. MFC=mass flow controller, RH=relative humidity, RTD-rapid thermal detector, HPDA=hydrogen peroxide delivery assembly.

Initial studies show that water vapor may be added to the upstream carrier gas in order to keep the liquid H_2O_2 solution concentration constant. However, if too little humidification is applied to the upstream carrier gas, the H_2O_2 liquid concentration increases. If too much humidification is applied upstream, the H_2O_2 liquid concentration decreases. Figure 3 demonstrates this crossover effect.

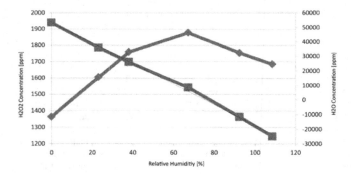

Figure 3. Crossover effect of H_2O_2 liquid concentration with respect to upstream humification levels. Blue line represents H_2O_2. Red line represents change in H_2O across the HPDA.

The preload on the carrier gas was varied from 0% to 110% rH. Initially, as the preload was increased, the amount of H_2O_2 picked up by the carrier gas increased to peak around 70%. This point was slightly below the transition point for water vapor reversing flow from out of the solution to back into the solution.

Figure 4 shows stable delivery of a ~30% H_2O_2 solution at 40°C. The average H_2O_2 concentration was 700ppm with a standard deviation of 16ppm. For the solution, the beginning concentration average was 33.4% and the final concentration average was 33.2%. This demonstrates the ability to control the H_2O_2 mass transfer rate while maintaining a relatively stable liquid solution concentration.

212

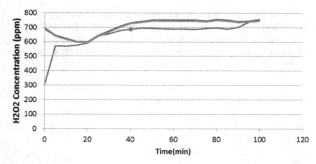

Figure 4. H₂O₂ output at 40°C with 30% solution.Blue line=465L Analyzer Reading. Red line=Raoults Law @ 33.4%. Green line=Raoults Law @ 33.2%.

The output of a ~70% H_2O_2 solution held at 40°C is shown in Figure 5. The average H_2O_2 concentration was 2583ppm with a standard deviation of 23ppm. For the solution, the beginning concentration average was 69.1% and the final concentration average was 69.9%.

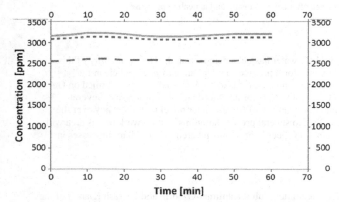

Figure 5. H_2O_2 output at 40°C with 70% solution. Blue large dash line=H2O2. Red short dash line=69.1% with Raoult's Law. Blue solid line=69.9% with Raoult's Law.

GERMANIUM PASSIVATION

Germanium is promising candidate for potential channel materials,[2] however air exposed Ge surfaces have a high density of defects and contaminants. New "dry" cleaning and surface preparation methods are now being sought, where the term "dry" refers to gas phase or vapor chemistries.

Recently (Figure 6), a completely in-situ gas phase cleaning process to remove carbon contamination from Ge(100) surfaces using gas phase H_2O_2 has been characterized. The H_2O_2

gas converts surface carbon to a gas phase product, removing it from the surface while leaving a surface oxide, GeO_x. In addition, to minimize the oxide-semiconductor interfacial defect density, a proper passivation layer must be used before the metal oxide layer is deposited. By using H_2O_2 dosing, the density of Ge–OH sites can be doubled versus using H_2O, thereby increasing the potential nucleation density for metal precursors.

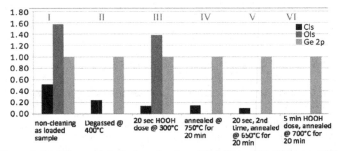

Figure 6. Multidose and single dose cleaning of Ge(100) by 30% $H_2O_2(g)$ at 300°C surface temperature. XPS analysis of the Ge(100) surface after several dosing and annealing processes.

CONCLUSION

Controlled delivery of H_2O_2/water has been demonstrated. The availability of this chemical in stable, controlled vapor form provides a wide range of potential cleaning formulations for advanced structures and metal surfaces. An initial study performed on Ge passivation shows the viability of H_2O_2/ H_2O for advanced cleaning processes. Several challenges exist regarding manufacturability of Ge and other novel materials, however the availability of H_2O_2 vapor can lead to several process solutions. Future work in this area will involve studies with collaborators on specific problems inherent to sub-20nm processes in development.

ACKNOWLEDGMENT

Thanks to Prof. Andrew Kummel, Tobin Kaufman-Osborn, and Kiarash Kiantaj of the Department of Chemistry at the University of California San Diego for their contribution related to Ge surface preparation.

REFERENCES

1. George, S.M.; *Chem. Rev.*, **2110**, *110*, 111-131.
2. S.H. Hsu, C.L. Chu, W. H. Tu, Y. C. Fu, P. J. Sung, H. C. Chang, Y. T. Chen, L. Y. Cho, W. Hsu, G. L. Luo, C. W. Liu, C. Hu, and F. L. Yang, *(IEDM), 2011 IEEE International*, **2011**, *1*, 35.2.1
3. Kumagai, H.; Toyoda, K. *Appl. Surf. Sci.* 82/83 **1994**, 481-486
4. "Hydrogen Peroxide" Ch. 4, p. 189, Schumb, W. C., Satterfield, C.N., Wentworth, R.L. New York : Reinhold Pub. Corp., **1955**.

Mater. Res. Soc. Symp. Proc. Vol. 1494 © 2013 Materials Research Society
DOI: 10.1557/opl.2013.180

Investigations on electrical conduction properties and crystallization conditions of V2O5-P2O5 glass based semiconductors

Akifumi Matsuda[1], Takuya Aoyagi[2], Takashi Naito[1,2], Tadashi Fujieda[2], Kenjiro Ikejiri[3], Koji Koyama[3], Ryosuke Yamauchi[1], Geng Tan[1], Satoru Kaneko[1,4], and Mamoru Yoshimoto[1]

[1]Department of Innovative and Engineered Materials, Tokyo Institute of Technology, 4259-J3-16 Nagatsuta, Midori, Yokohama 226-8502, Japan.
[2]Hitachi Research Laboratory, Hitachi Ltd., 7-1-1 Omika, Hitachi 319-1292, Japan.
[3]Namiki Precision Jewel Co., Ltd., 3-8-22 Shinden, Adachi, Tokyo 123-8511, Japan.
[4]Kanagawa Industrial Technology Center, 705-1 Shimoimaizumi, Ebina 243-0435, Japan.

ABSTRACT

We studied the electrical properties of thermally treated V_2O_5-CuO-Fe_2O_3-P_2O_5 (vanadate) glasses under reducing high-vacuum conditions. The glasses were prepared by using a melt-quenching method and then applied on Al_2O_3 substrates as ~40µm-thick films. The glass films were then heat treated at 375–550°C under a vacuum of 10^{-6} Pa. Powder X-ray diffraction showed the formation of complex oxides of both $M_xV_2O_5$ (M = Cu, Fe; x = 0.12–1.3) and vanadium oxides (VO_x; x = 1.5–2.5). The resistivity of the glass film crystallized at 550°C measured at 50°C and 300°C were 1.8×10^0 Ωcm and 2.8×10^{-1} Ωcm, respectively, which was 10 times lower than that of the film crystallized in air. The Seebeck coefficient was −132 µV/K at 50°C and −130 µV/K at 300°C. These results show that the vanadate glasses crystallized under the appropriate condition become potential candidate materials for semiconductor and thermoelectric application.

INTRODUCTION

Thermoelectric (TE) generation has been recognized as a promising energy-harvesting technology because it provides a direct conversion of thermal energy [1]. Chalcogenide compounds such as Bi_2Te_3- and PbTe-related materials with relatively high figure-of-merit (ZT > 1) have been used at temperatures up to 800°C [2]. However, the major drawbacks of TE devices based on these substances are the consistently low thermal-to-electric conversion efficiency of ~8% (when compared to that of photovoltaic generation) and the toxic and rare elements in their composition. Thus, the search for alternative nontoxic and environmentally friendly TE materials as well as improved conversion characteristics is important. Nonetheless, there have been reports of alternative TE materials such as layered cobalt oxides and heavily doped $SrTiO_3$ [3], and some organic materials [4], e.g., poly(3,4-ethylenedioxythiophene);polystyrene sulfonate (PEDOT-PSS) [5].

On the other hand, vanadate glasses have semiconductor properties such as electric conductivity of about 10^5 $\Omega^{-1}cm^{-1}$ due to the hopping conduction between V^{4+} and V^{5+} ions [6-8]. V_2O_5-P_2O_5 glasses are considered to have a layered structure comprising VO_4, VO_5, and PO_4 polyhedra; moreover, they have a potential for applications in energy devices such as Li-ion batteries as well as $Li_{1+x}V_{1-x}O_2$ layered compounds [9]. Few studies have been conducted regarding the thermoelectric applicability of these oxide-based glass systems; there is a report on the crystallization of V_2O_5-P_2O_5 glasses by microwave heating and the TE properties of the crystallized glass [10,11].

In this paper, we investigated the effect of ambient atmosphere and temperature on the precipitated crystalline phases and crystallinity of V_2O_5-P_2O_5 glasses as well as their electrical properties including the Seebeck coefficient.

EXPERIMENT

Vanadate glasses composed of $65V_2O_5$-$15CuO$-$10Fe_2O_3$-$10P_2O_5$ and $65V_2O_5$-$15CuO$-$5Fe_2O_3$-$15P_2O_5$ were prepared by using a conventional melt-quenching method. V_2O_5 powder (99.9% purity, Kojundo Chemical Laboratory), P_2O_5 powder (>98% purity, Junsei Chemical), Fe_2O_3 powder (99.9% purity, Kojundo Chemical Laboratory), and CuO powder (99.9% purity, Kojundo Chemical Laboratory) were used for the glass synthesis. The source materials were placed in a platinum crucible and heated for 1 hour at 1000°C in an electric furnace. The product melts were quenched by pouring them onto a stainless steel plate at 150°C. Subsequently the glasses were milled into ~4-μm powders.

The vanadate glass powders were applied onto α-Al_2O_3 substrates as pastes by adding diethylene-glycol-monobutyl-ether-acetate (BCA; $C_{10}H_{20}O_4$) and ethylcellulose (EC) as solvent and sintering agents, respectively. The film samples of the aggregated glass powders were then sintered at 365°C in air, and BCA and EC were thus removed. Glass crystallization took place under high-vacuum of 10^{-6} Pa at 375°C, 480°C, and 550°C. Figure 1 shows the electric furnace used for the crystallization, which was equipped with a vacuum system composed of a rotary and an oil diffusion pump, and an Ar gas line used during cooling and sample changing.

The crystallization tendency of the milled glass powders was measured using differential thermal analysis (DTA) at a heat rate of 5 K/min. The precipitated crystalline phases were identified by powder X-ray diffraction (XRD) using a Rigaku RINT-2000 diffractometer and CuKα radiation. The Seebeck coefficient (S) and the electrical conductivity (σ) of the crystallized glass films were measured under helium gas using a commercially available device (ULVAC-RIKO, model ZEM-3). The Seebeck coefficient was determined by measuring ΔV, which is the Seebeck voltage between the two ends of the sample, as a function of the temperature difference ΔT between the two ends. Since $\Delta V = S\Delta T$, the Seebeck coefficient could be determined from the slope. We measured the electrical resistance by using the four-probe DC method.

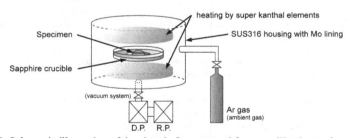

Figure 1. Schematic illustration of the electric furnace used for crystallization under vacuum.

RESULTS and DISCUSSION

The DTA curves of the V_2O_5-CuO-Fe_2O_3-P_2O_5 glasses are shown in Figure 2. The DTA curve for the $65V_2O_5$-$15CuO$-$10Fe_2O_3$-$10P_2O_5$ glass (Fig.2(a)) exhibits large exothermic peaks around 405°C (T_{p1}) and 510°C (T_{p2}) above the glass transition temperature T_g of ~300°C, which are attributed to the first and second crystallization. On the other hand, the curve for the $65V_2O_5$-$15CuO$-$5Fe_2O_3$-$15P_2O_5$ glass (Fig.2(b)) suggests crystallization peak at T_p ~ 420°C above the T_g of ~294°C and a weak endothermic peak around 360°C due to softening. The differences in the thermal characteristics are probably caused by changes in the vitrification content rate (P_2O_5).

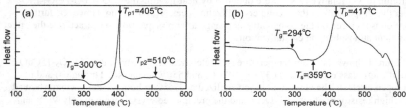

Figure 2. DTA curves of (a) the $65V_2O_5$-$15CuO$-$10Fe_2O_3$-$10P_2O_5$ glass and (b) the $65V_2O_5$-$15CuO$-$5Fe_2O_3$-$15P_2O_5$ glass.

In order to explore the relation between the precipitated crystalline phases and characteristics, the $5V_2O_5$-$15CuO$-$10Fe_2O_3$-$10P_2O_5$ glass was used in the X-ray and electrical properties investigations. Figure 3 shows the XRD spectra for the $65V_2O_5$-$15CuO$-$10Fe_2O_3$-$10V_2O_5$ glass crystallized at (a) 375°C, (b) 480°C, and (c) 550°C under high vacuum. The crystalline phases precipitated at each temperature are summarized in Table I. The double circles represent the main crystalline phases, other major phases by the circles, and the minor phases by the triangles. The data for the glass crystallized at 375°C (Fig.3(a)) suggest formation of crystalline V_2O_5 and VO_2 although the temperature was lower than the T_{p1} and the diffraction peaks are rather broad, and suggest relatively poor crystallinity. On the other hand, peaks assigned to $Cu_xV_2O_5$ ($x = 0.55-1.3$) and $Fe_xV_2O_5$ ($x = 0.12-0.33$), which have conductivities about 10^6 times higher than those of V_2O_5 [12], appeared at 480°C and 550°C as shown in Figs.3(b) and 3(c), respectively.

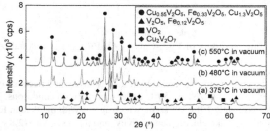

Figure 3. X-ray diffraction profiles for $65V_2O_5$-$15CuO$-$10Fe_2O_3$-$10P_2O_5$ glass samples crystallized at (a) 375°C, (b) 480°C, and (c) 550°C.

Table I. Crystalline phases detected by powder XRD.

Crystallizing temperature	V_2O_5 $Fe_{0.12}V_2O_5$	VO_2	CuV_2O_7	$Cu_{0.55}V_2O_5$	$Cu_{1.3}V_2O_5$	$Fe(PO_3)_3$
375°C	O	◎	Δ	-	-	-
480°C	O	-	-	◎	O	Δ
550°C	O	-	-	◎	O	Δ

Thus, crystalline phases with comparably high electrical conductivity because of the larger concentration of V^{4+} relative to V^{5+} were obtained by heat-treating the V_2O_5-CuO-Fe_2O_3-P_2O_5 glass near the temperature at the second crystallization peak (T_{p2}) in vacuo. However, further investigation is required on the compositional change after precipitation to characterize the effect of the ambient atmosphere on crystallization.

Figure 4 shows the room-temperature (RT) electrical resistivity of the 65V_2O_5-15CuO-10Fe_2O_3-10P_2O_5 glass crystallized at 375°C, 480°C, and 550°C in air (solid triangles) and in vacuum (open circles). The glass samples crystallized at 375°C in both air and vacuum showed relatively high resistivity of 10^3–10^4 Ωcm, and the precipitated crystals were mainly V_2O_5 and VO_2. On the other hand, the resistivity decreased by 10^3 times when complex oxides of $M_xV_2O_5$ (M = Cu, Fe; x = 0.12–1.3) crystallized, with values of 5.2 Ωcm and 2.3 Ωcm for glasses crystallized at 480°C and 550°C in vacuum, respectively. Additionally, there were minimal changes in the latter cases, in which minimal differences were observed in the crystalline phases, whereas there would be variances in the density of grain boundaries, degree of crystallinity, and so on. The decrease in resistivity was also observed when the glass was crystallized in higher temperature in air, however, which is considered due to the grain growth according to the crystallographic analyses.

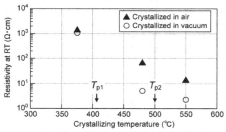

Figure 4. Room-temperature resistivity of V_2O_5-CuO-Fe_2O_3-P_2O_5 glasses crystallized at various temperatures.

Figure 5 shows the temperature dependency of the resistivity and Seebeck coefficient for V_2O_5-CuO-Fe_2O_3-P_2O_5 glass crystallized at 550°C under high vacuum conditions. The resistivity measured at RT was approximately 1.8×10^0 Ωcm, which agrees with the data in Fig. 4. The material showed semiconductor-like electric conductivity, and the resistivity (2.8×10^{-1} Ωcm at 300°C) decreased with rising temperature. The Seebeck coefficient was approximately -130 µV/K and was relatively stable in the temperature range of 50–300°C.

218

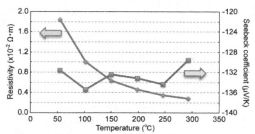

Figure 5. Temperature dependency of the resistivity and Seebeck coefficient of V_2O_5-CuO-Fe_2O_3-P_2O_5 glass crystallized at 550°C in vacuum.

CONCLUSIONS

This study reveals that the crystallization of V_2O_5-CuO-Fe_2O_3-P_2O_5 glasses in reducing high-vacuum atmosphere results in preferential precipitation of $M_xV_2O_5$ (M = Cu, Fe; x = 0.12–1.3) crystals and VO_x (x = 1.5–2.5) crystals. The formation of $M_xV_2O_5$ was found to decrease the resistivity of the crystallized glass by 10^3 times in comparison with the air-crystallized case in which the crystalline phases mostly consisted of VO_2 and V_2O_5. Consequently, these glass/crystal composites were found to have promising resistivity and Seebeck coefficients, and may substitute conventional thermoelectric materials. Therefore, we propose that crystallized V_2O_5-P_2O_5 glasses are potential candidate materials for semiconductor and thermoelectric applications. Further investigations on the physical properties of these materials and their relation with crystallization, ambient atmosphere, and heat treatment is required to assess the role of each precipitated phase.

REFERENCES

1. Francis J. DiSalvo, *Science* **285**, 703 (1999).
2. Terry M. Tritt and M. A. Subramanian, *MRS Bulletin* **31**, 188 (2006).
3. Kunihito Koumoto, Ichiro Terasaki and Ryoji Funahashi, *MRS Bulletin* **31**, 206 (2006).
4. Nidhi Dubey and Mario Leclerc, *J. Polym. Phys. B: Polym. Phys.* **49**, 467 (2011).
5. Naoki Toshima, Nattha Jiravanichanun and Hiromasa Marutani, *J. Electron. Mater.* **41**, 1735 (2012).
6. E. P. Denton, H. Rawson and J. E. Stanworth, *Nature* **173**, 1030 (1954).
7. G. S. Linsley, A.E. Owen and F. M. Hayatee, *J. Non-Crystalline Solids.* **4**, 208 (1970).
8. T. Naito, T. Aoyagi, Y. Sawai, S. Tachizono, K. Yoshimura, Y. Hashiba and M. Yoshimoto, *Jpn. J. Appl. Phys.* **50**, 088002 (2011).
9. A. R. Armstrong, C. Lyness, P. M. Panchmatia and M. S. Islam, *Nature Mater.* **10**, 223 (2011).
10. T. Fujieda, T. Aoyagi and T.Naito, *AIP Adv.* **2**, 022164 (2012).
11. T. Aoyagi, T. Fujieda, Y. Sawai, M. Miyata, T. Naito and H. Yamamoto, *Mater. Res. Soc. Symp. Proc.* **1454**, 15 (2012).
12. T. Toda, K. Kosuge and Y. Kachi, *Nippon Kagakushi* **87**, 1311 (1966).

Mater. Res. Soc. Symp. Proc. Vol. 1494 © 2013 Materials Research Society
DOI: 10.1557/opl.2013.125

Fabrication of Corundum-Structured α-(InFe)$_2$O$_3$ Alloy Films on Sapphire Substrates by Inserting α-Fe$_2$O$_3$ Buffer Layer

Kazuaki Akaiwa[1], Norihiro Suzuki[1], Kentaro Kaneko[1], and Shizuo Fujita[2]

[1] Department of Electronic Science and Engineering, Kyoto University, Kyoto 615-8510, Japan
[2] Photonics and Electronics Science and Engineering Center, Kyoto University, Kyoto 615-8520, Japan

ABSTRACT

We successfully fabricated corundum-structured α-(InFe)$_2$O$_3$ alloy films on sapphire substrates by inserting α-Fe$_2$O$_3$ buffer layers. The ion compositions in the α-(In$_{1-x}$Fe$_x$)$_2$O$_3$ films, x, were artificially tuned for the entire range from 0 to 1 by changing the ion precursor composition in source solution. Magnetic measurements revealed that the α-(In$_{1-x}$Fe$_x$)$_2$O$_3$ (x = 0.13) alloy film showed ferromagnetism at 5 K.

INTRODUCTION

Since ferromagnetism in InMnAs[1] and GaMnAs[2] was reported, semiconductors doped with magnetic elements have attracted much interests owing to their unique property. Our group has proposed a new alloy system composed of corundum-structured oxides, that is, the alloys of oxides semiconductors such as α-Ga$_2$O$_3$ and α-In$_2$O$_3$, as well as and transition metal oxides such as α-Fe$_2$O$_3$, α-V$_2$O$_3$, α-Cr$_2$O$_3$. We expect that this alloy system enables to synthesize a variety of functional materials based on oxide semiconductors. We reported the fabrication of α-(GaFe)$_2$O$_3$ alloy films, which exhibited ferromagnetic property at 110 K[3]. On the other hands, recently, we succeeded in the fabrication of α-In$_2$O$_3$ films on sapphire substrates by inserting α-Fe$_2$O$_3$ buffer layers. We consider that it is possible to make alloys of α-In$_2$O$_3$ and α-Fe$_2$O$_3$ by inserting an α-Fe$_2$O$_3$ buffer layer and such alloys also exhibit ferromagnetism like α-(GaFe)$_2$O$_3$ alloy films. Owing to the low resistivity of α-In$_2$O$_3$, compared to α-Ga$_2$O$_3$, we can expect enhanced spin-electron interaction. In the present paper, we show the successful fabrication of α-(InFe)$_2$O$_3$ alloy films on sapphire substrates with α-(InFe)$_2$O$_3$ buffer layers, together with the result of magnetization measurements.

EXPERIMENTS

We fabricated α-(InFe)$_2$O$_3$ thin films on sapphire substrates, on which α-Fe$_2$O$_3$ was grown as a buffer layer, by using the mist chemical vapor deposition (mist-CVD) method. Fig.1 is a schematic illustration of the mist CVD system. Indium(III) acetylacetonate and iron(III) acetylacetonate were adopted as indium and iron precursors, respectively. As a reaction source, we used water solution of these precursors with addition of a small amount of hydrochloric acid, which helped to solve the precursors completely. The sum of concentrations for indium and iron precursors was fixed at 0.05 mol/L. The iron precursor composition ratios to the sum of iron and indium precursor concentrations in the solution, γ, were changed as 0%, 20%, 40%, 60%, 80%,

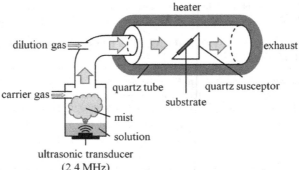

Fig. 1 Schematic illustration of mist CVD system.

and 100%. Nitrogen gas was used as carrier and dilution gases. The flow rates of carrier and dilution gases were set at 3.0 and 0.5 L/min, respectively.

Substrates were c-plane sapphire, on which an α-Fe$_2$O$_3$ buffer layer was fabricated. The substrate temperature was set at 500 $^\circ$C. The growth time was kept at 30 min. As given above, we only changed the value of γ and unchanged the other growth conditions. Buffer layers were fabricated also by the mist-CVD method. Thickness of the buffer layers was about 10 nm.

RESULTS AND DISCUSSIONS

Fig.2 shows the X-ray diffraction (XRD) 2θ/θ-scan profiles of the films. For all samples, the XRD peaks derived from α-(InFe)$_2$O$_3$ (0006) and sapphire (0006) diffractions were observed,

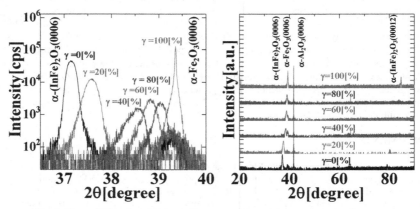

Fig. 2 X-ray diffraction (XRD) 2θ/θ-scan profiles of the α-(InFe)$_2$O$_3$ films as a function of γ, iron precursor concentrations in the solution.

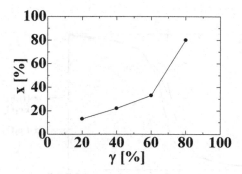

Fig. 3 Iron composition ratios x in α-$(In_{1-x}Fe_x)_2O_3$ films
as a function of γ, iron precursor concentrations in the solution.

without any other peaks derived from other crystalline orientations and phases, and the peak
position for α-$(InFe)_2O_3$ (0006) diffraction monotonically shifted to higher angles with the
increase of γ. On the supposition of the Vegard's low, the iron composition ratios x in the α-$(In_{1-x}Fe_x)_2O_3$ films are calculated from the increase of peak angle for α-$(InFe)_2O_3$ (0006) diffraction
in the XRD $2\theta/\theta$-scan profiles.

Fig.3 shows the calculated iron composition ratios x in α-$(In_{1-x}Fe_x)_2O_3$ films as a
function of γ, iron precursor concentrations in the solution. With the increase of γ, x increases

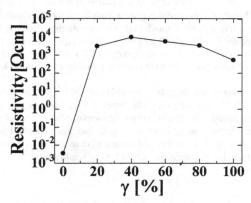

Fig. 4 Resistivity of the films as a function of the iron precursor
concentrations in the source solution, that is, γ

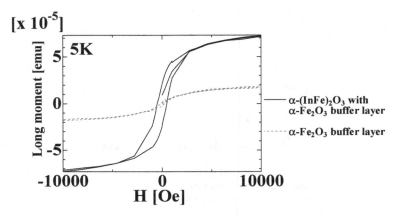

Fig. 5 Magnetization properties (M-H curves) for the sample grown from
the solution of $\gamma = 20\%$ (x = 0.13) and α-Fe_2O_3 buffer layer at 5 K.

monotonically. This suggests that the iron composition ratios x in α-$(In_{1-x}Fe_x)_2O_3$ films can well
be controlled by the iron precursor concentrations in the solution. At the lower iron precursor
concentrations in the solution, the solidus iron composition x does not increase effectively, that is,
the inclusion of iron is lower than the iron precursor concentration. This is probably because of
the higher formation energy of α-Fe_2O_3 compared to α-In_2O_3.

Fig.4 shows the resistivity of the films as a function of the iron precursor concentrations
in the source solution, that is, γ. The samples grown from the solution containing iron precursor,
all samples except for $\gamma = 0\%$, exhibited much higher resistivity than the sample of γ =0%. Hall
measurement was succeeded for the sample of $\gamma = 0\%$. The sample showed n-type conductivity
with Hall mobility of 58.8 cm^2/V s and carrier density of 3.0 x 10^{19} cm^{-3}. With the inclusion of
iron, generation of deep traps seem to degrade electron mobility by forming defects originating
from Fe.

Fig.5 shows the magnetization properties (M-H curves) for the sample grown from the
solution of $\gamma = 20\%$ (x = 0.13) and α-Fe_2O_3 buffer layer at 5 K. The magnetic field was applied
parallel to the film. Diamagnetic contribution from the sapphire substrate was subtracted. Note
that α-Fe_2O_3 shows weak ferromagnetism, whose mechanism was explained by
Dzyaloshinsky[4] and Moriya[5]. Therefore, difference of magnetization between the sample and
buffer layer corresponds to the magnetization of α-$(InFe)_2O_3$ alloy film.

CONCLUSIONS

We successfully fabricated the α-$(InFe)_2O_3$ alloy films on sapphire substrates by inserting
buffer layer. The iron compositions in the α-$(In_{1-x}Fe_x)_2O_3$ films, x, were artificially tuned for the

entire range of 0 to 1 by changing the iron precursor composition in source solution. Magnetic measurements revealed that α-$(In_{1-x}Fe_x)_2O_3$ (x = 0.22) alloy film shows ferromagnetism at 5 K.

ACKNOWLEDGMENTS

The authors deeply acknowledge Profs. M. Suzuki and I. Kakeya for supporting the magnetic measurements and fruitful discussions.

REFERENCES

1. H. Munekata, H. Ohno, S. von Molnar, A. Segmüller, L. L. Chang, and L. Esaki, Phys. Rev. Lett. **63,** 1849 (1989).
2. H. Ohno, A. Shen, F. Matsukura, A. Oiwa, A. Endo, S. Katsumoto, and Y. Iye, Appl. Phys. Lett. **69,** 363 (1996).
3. K. Kaneko, T. Nomura. I. Kakeya, and S. Fujita, Appl. Phys. Express **2,** 075501 (2009).
4. I. Dzyaloshinsky, J. Phys. Chem. Solids **4,** 241 (1958).
5. T. Moriya, Phys. Rev. Lett. **4,** 228 (1960).

Mater. Res. Soc. Symp. Proc. Vol. 1494 © 2013 Materials Research Society
DOI: 10.1557/opl.2013.410

Epitaxial Growth of (Na,K)NbO$_3$ Films by Pulsed Laser Deposition

K. Sakurai[1,2], T. Hanawa[1,2], N. Kikuchi[1,2], K. Nishio[2], K. Tonooka[1], R. Wang[1], H. Bando[1], and H. Takashima[1]

[1]National Institute of Industrial Science and Technology (AIST), 1-1-1 Umezono, Tsukuba-shi, Ibaraki 305-8568, JAPAN
[2]Department of Materials Science and Technology, Tokyo University of Science, 2641 Yamazaki, Noda-shi, Chiba 278-8510, JAPAN

ABSTRACT

(Na,K)NbO$_3$ is a promising candidate for lead-free piezoelectric materials. (Na$_{1-x}$K$_x$)NbO$_3$ films (x = 0.3–0.7) were epitaxially grown on a (100)SrTiO$_3$ substrate via pulsed laser deposition. The effects of substrate temperature and oxygen pressure during deposition on the crystallinity of the films were examined: both parameters affected the mosaic spread of the crystallites and the formation of an impurity phase. In this study, the optimum conditions for the preparation of highly crystalline films were a substrate temperature of 800 °C and oxygen pressure of ~60 Pa. The lattice constants parallel and perpendicular to the substrate surface responded differently to changes in x: the constant parallel to the surface increased with increasing x, while the constant perpendicular to the surface was maximized at x = 0.5. The difference in the dependence of the lattice constants could be explained by the elastic distortion of the lattice.

INTRODUCTION

Piezoelectric materials are used for various electronic devices such as ultrasonic sensors, gyro sensors, fuel injectors for diesel engines, print-heads for ink-jet printers, and so on. Recently, harvesting of mechanical energy, such as vibration, which can be converted to electrical energy via piezoelectric actuators, has attracted a lot of attention because of energy conservation measures [1,2]. Typical piezoelectric materials are lead-containing perovskites such as Pb(Zr,Ti)O$_3$ (PZT). However, it is likely that the use of these lead-containing materials in electrical devices and vehicles will be restricted in the near future because of their toxicity [3]. Although BaTiO$_3$ is a known lead-free piezoelectric material, its lower Curie temperature (T$_c$) than that of PZT is a problem for industrial applications. Alkaline niobate-based ceramics, in particular the (Na,K)NbO$_3$ series, are promising candidates for lead-free piezoelectric materials because they have a high T$_c$ and piezoelectricity comparable with those of PZT [4,5]. (Na,K)NbO$_3$-LiTaO$_3$, which has a T$_c$ of ~250 °C, has been reported to have high piezoelectric constants (d$_{33}$) of ~300 pC/N and 416 pC/N for ceramics without and with texturing, respectively [6]. A (Na,K)NbO$_3$-BaZrO$_3$-(Bi,Li)TiO$_3$ system (T$_c$ = 243 °C) has been reported to exhibit a higher d$_{33}$ of 420 pC/N without any texturing [7].

In this study, (Na$_{1-x}$K$_x$)NbO$_3$ (NKN) films were deposited on (100)SrTiO$_3$ (STO) substrates via pulsed laser deposition (PLD) as the first step to the deposition of films with a complex composition of BaZrO$_3$ and (Bi,Li)TiO$_3$ containing (Na,K)NbO$_3$. Several reports on (Na,K)NbO$_3$ films have been published [4,8,9]; however, the effect of the alkaline composition on the crystallographic parameters has not been elucidated. The effect of the composition of the films on the crystal phase and crystallinity was examined using out-of-plane and in-plane X-ray diffraction measurements.

EXPERIMENT

$(Na_{1-x}K_x)NbO_3(x = 0.3–0.7)$ targets were prepared via a solid-state reaction. Weighed samples of Na_2CO_3 (purity: 2N), K_2CO_3 (2N), and Nb_2O_5 (3N) were ball-milled in ethanol for 12 h with a ZrO_2 ceramic ball. The ball-milled slurry was dried and pressed into disks (22 mm diameter) at 80 MPa for 2 min. These disks were calcined at 900 °C for 4 h and then ground in a mortar. The resultant ground powders were then ball-milled in ethanol for a further 12 h. The ball-milled slurry was dried and pressed into disks (22 mm diameter) at 150 MPa for 2 min. These disks were then calcined at 1030 °C for 2 h. The resultant NKN disks were identified to have a perovskite structure by X-ray diffraction measurements and were used as targets for pulsed laser deposition.

NKN films were prepared using the PLD process. The substrates comprised 10 mm × 10 mm squares of 0.5 mm thick (100)-oriented single-crystal $SrTiO_3$ (STO). STO has a cubic perovskite structure with a lattice constant of a = 0.3905 nm, which is close to that of NKN. A preparation chamber was evacuated to ~2 × 10^{-5} Pa using a turbomolecular pump. Ablation was induced using a frequency-quadrupled Nd-YAG laser operating at λ = 266 nm. The fixed PLD conditions were as follows: Laser repetition rate of 16 Hz, distance of 40 mm between the target and substrate, energy density of 10 W/cm^2, and 67200 laser shots. The substrate temperature (T_s) and oxygen pressure (P_{O2}) during deposition were controlled in the ranges 500–800 °C and 1.0–90 Pa, respectively.

The crystal phase of the films was investigated via CuKα X-ray diffraction (XRD, X' Pert Pro, PANalytical) analysis using Bragg-Brentano measurements. XRD rocking curves ω scans were measured to evaluate the mosaic spread of the crystallites in the NKN films. In order to estimate the lattice constant of the films, a reciprocal space map was determined. The chemical composition of the films was estimated using wavelength-dispersive X-ray fluorescence analysis (ZSX, Rigaku).

RESULTS AND DISCUSSION

Figure 1(a) shows XRD patterns of the Bragg-Brentano measurement of $(Na_{0.7}K_{0.3})NbO_3$ films deposited at various substrate temperatures at an oxygen pressure of 1.0 Pa. Two intense peaks at 2θ = 23.0° and 47.0° were assigned to $SrTiO_3$ (001) and (002), respectively, whereas weak and sharp diffraction peaks at 2θ = 20–21° and 37–42° were assigned to spectral lines of STO (001) and (002) such as Kβ and tungsten Kα$_1$, respectively; these peaks were present because the XRD measurements were taken without a monochromator. Two intense peaks at 2θ = 22.5° and 46.0° were found for the films deposited at ≥500 °C and were assigned to NKN (001) and (002), respectively. There were no significant differences in the patterns of the films with different x values. Therefore, the NKN films deposited on a (100)STO substrate at ≥500 °C were found to grow along the c-axis on the substrate. Several weak peaks that were assigned to a pyrochlore phase with an alkaline-deficient structure such as $Na_2Nb_4O_{11}$ were also found for the films deposited at ≥600 °C [10]. Figure 1(b) shows the full width at half maximum (FWHM) values of the rocking curve for the films deposited at various T_s values at a P_{O2} of 1.0 Pa. The FWHM monotonically decreased with increasing T_s, which indicates that increasing the T_s

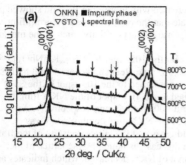
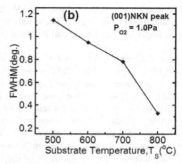

Fig. 1 $(Na_{0.7}K_{0.3})NbO_3$ films deposited at various T_s values (P_{O2} = 1.0 Pa).
    (a) XRD patterns of the Bragg-Brentano measurements
    (b) FWHM of the rocking curve for the films

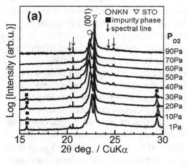
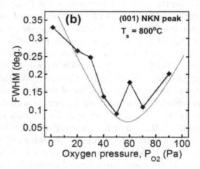

Fig. 2 $(Na_{0.7}K_{0.3})NbO_3$ films deposited at various P_{O2} values (T_s = 800 °C).
    (a) XRD patterns of the Bragg-Brentano measurements
    (b) FWHM of the rocking curve for the films (dotted curve is provided as a guide).

reduces the mosaic spread of crystallites in NKN films leading to higher crystallinity. Although heating the substrate above 800 °C may further reduce the mosaic spread, unfortunately our PLD equipment could not achieve these temperatures so the effect could not be investigated.

Figure 2(a) shows XRD patterns of the Bragg-Brentano measurements of the films deposited at a T_s of 800 °C under various P_{O2} values. Two intense peaks assigned to STO (001) and NKN (001) and associated peaks at 2θ = 20–21° were found for all the films in this figure. Weak diffraction peaks at 2θ = 16° and 29° were assigned to a pyrochlore phase; these peaks decreased with increasing P_{O2} and disappeared for the films deposited at $P_{O2} \geq 40$ Pa. Another peaks that were found at 2θ = 24–25° in the patterns of the films deposited at $P_{O2} \geq 20$ Pa were thought to be the spectral line since these peaks were not evident in the pattern obtained with a monochromator. Figure 2(b) shows the FWHM of the rocking curve for the films deposited at a

T_s of 800 °C under various P_{O2} values. The FWHM decreased with increasing P_{O2} and showed a minimum value of ~0.1° at ~60 Pa. These results indicate that the oxygen pressure during film deposition strongly affects the formation of the crystal phase and crystallinity, such as the mosaic spread.

Figure 3 shows XRD reciprocal space maps of near the 103^* spot for $(Na_{1-x}K_x)NbO_3$ (x = 0.3, 0.5, 0.7) films deposited at 800 °C under a P_{O2} of 70 Pa. Two clear spots were found in these figures. The intense spot in the upper part of the figures is assigned to the 103^* spot of the STO substrate and the weak spot (in the lower part) is assigned to the 103^* spot of the NKN film. In this figure, the x-axis and y-axis represent the <100> and <001> directions, respectively. As shown in Figure 3(a), the Q_x value of the spot assigned to the NKN films with x = 0.3 was almost the same as that of the 103^* spot of STO; this indicates that the lattice constant of the film parallel to the substrate surface, i.e., along the a-axis in this case, is bound by the lattice of the substrate. With increasing x, the Q_x value of the 103^* spot of NKN decreased, which indicates an increase in the lattice constant of the a-axis.

Parallel (i.e., in-plane) and perpendicular (i.e., out-of-plane) film lattice constants were estimated from the reciprocal space maps. Figure 4(a) shows the lattice constants of the a-, b- and c-axes for the films as a function of x in $(Na_{1-x}K_x)NbO_3$; the a- and b-axes are in-plane and the c-axis is out-of-plane. The lattice constants of the a- and b-axes showed similar trends with respect to x, but their values differed. It is known that $(Na_{1-x}K_x)NbO_3$ sequentially forms rhombohedral, orthorhombic, tetragonal, and cubic phases as the temperature increases. At room temperature (i.e., 25 °C), $(Na_{1-x}K_x)NbO_3$ forms an orthorhombic phase [11]. The difference in lattice constants of the a- and b-axes is regarded as significant in consideration of the experimental error (±0.0002 nm). Therefore, like the bulk specimen, the epitaxial films were considered to form an orthorhombic phase. It was also found that the lattice constants parallel and perpendicular to the substrate surface showed different trends with respect to x. The lattice constants of the a- and b-axes increased with increasing x, while that of the c-axis showed a maximum at x = 0.5. Since the ionic radius of K^+ is larger than that of Na^+, the lattice constants of the film are expected to increase with increasing x. Figure 4(b) shows the ratios of the lattice constants, i.e., c/a and c/b, and volume of the primitive unit cell for the NKN films (V_{NKN}) with respect to x. Both ratios also indicated a maximum at x = 0.5. In contrast, the volume increased with increasing x, which is likely due to the difference in the ionic radii of Na^+ and K^+.

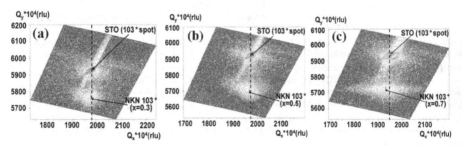

Fig. 3 XRD reciprocal space maps of the $(Na_{1-x}K_x)NbO_3$ films near the 103^* spot of the STO substrate. The films were deposited at T_s = 800 °C in P_{O2} = 70 Pa: (a) x = 0.3, (b) x = 0.5, and (c) x = 0.7.

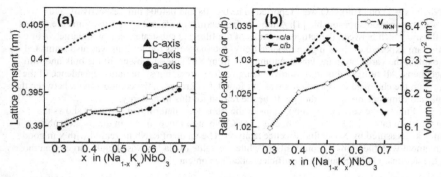

Fig. 4 (a) Lattice constants of the a-, b-, and c-axes for the $(Na_{1-x}K_x)NbO_3$ films. (b) Ratios of the lattice constants of the c-axis to those of the a- and b-axes and volume of the primitive unit cell for the NKN films (V_{NKN}).

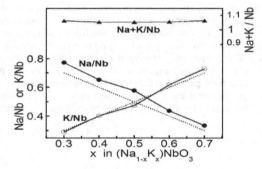

Fig. 5 Chemical composition ratios of the resulting $(Na_{1-x}K_x)NbO_3$ films as a function of x. The solid and open circles represent Na/Nb and K/Nb, respectively, while the dotted lines represent the nominal values, and the solid triangles show the resulting (Na+K)Nb.

It is well known that an epitaxial film grown on a substrate matches its in-plane lattice constant to that of the substrate, resulting in elastic distortion of the lattice of the film. It is considered that the increase in the lattice constant of the c-axis with increasing x value up to 0.5 is due to this elastic distortion. The lattice constants of the a- and b-axes for the films with $x \leq 0.5$ were bound by the lattice of the substrate because the mismatch of the film and substrate lattice constants may be small and the lattice constant of the c-axis monotonically increases with increasing x. On the other hand, a slight decrease in the lattice constant of the c-axis at $x > 0.5$ was likely due to the decrease in the elastic distortion of the lattice. As shown in Figure 4(a), the slopes of plots of the lattice constants of the a- and b-axes versus x were steeper for the films with $x > 0.5$; this means that increasing x above 0.5 increased the difference between the in-plane lattices of the film and substrate, resulting in a decrease of the lattice constant of the c-axis due to the decreased elastic distortion. Since the lattice constants of the a-axis for epitaxial $NaNbO_3$ (x = 0) and

$KNbO_3$ ($x = 1$) on (100)STO were reported to be 0.3904 and 0.4004 nm, respectively, our consideration seems reasonable [4]. Although the maxima of c/a and c/b for the films are related to the mismatch between the lattice constants of the film and substrate, films with large c/a or c/b ratios are expected to show high piezoelectricity. Internal stress of the films was also estimated to explain the variation in the lattice parameters, from XRD measurements of the bulk and film specimens. All films in this study showed compressive stress, but systematic dependence of the stress on the chemical composition x was not found. Difference in the composition x between the bulk and film specimens was thought to be the reason for this irregularity.

Finally, the chemical compositions of the films were determined. Figure 5 shows the composition ratio, i.e., K/Nb, Na/Nb, and (Na+K)/Nb, as a function of x in the $(Na_{1-x}K_x)NbO_3$ films, as measured by X-ray fluorescence analysis. The value of K/Nb increased with x in good agreement with the nominal composition, while the value of Na/Nb was slightly larger than those of the nominal Na/Nb resulting in sufficient alkaline content.

CONCLUSIONS

Lead-free piezoelectric materials, i.e., $(Na_{1-x}K_x)NbO_3$ films, were deposited on $(100)SrTiO_3$ substrates using pulsed laser deposition. The effect of the composition of the films on the crystal phase and crystallinity was examined via out-of-plane and in-plane X-ray diffraction measurements. NKN films were epitaxially grown on the substrate at substrate temperatures of 500–800 °C in the presence of oxygen. Optimum conditions for the preparation of the films in this study were a substrate temperature of 800 °C and an oxygen pressure of ~60 Pa. The lattice constant of the films parallel to the substrate surface increased with increasing x, while that of the films perpendicular to the substrate surface showed a maximum at $x = 0.5$. The difference in the dependence of the lattice constants could be explained by the elastic distortion of the lattice.

REFERENCES

1. G. Poulin, E. Sarraute and F. Costa, *Sensors and Actuators A* **116**, 461 (2004)
2. S. P. Beeby, M. J. Tudor and N. M. White, *Meas. Sci. Technol.* **17**, R175 (2006)
3. "Adaptation to scientific and technical progress of Annex II to Directive 200/53/EC (ELV) and of the Annex to Directive 2002/95/EC (RoHS) Final Report", Öko-Institute e.V.-Institute for Applied Ecology, Germany
4. T. Saito, T. Wada, H. Adachi and I. Kanno, *Jpn. J. Appl. Phys.* **43**, 6627 (2004)
5. Y. Nakashima, W. Sakamoto, H. Maiwa, T. Shimura and T. Yogo, *Jpn. J. Appl. Phys.* **46**, L311 (2004)
6. Y. Saito, H. Takao, T. Tani, T. Nonoyama, K. Takatori, T. Homma, T. Nagaya and M. Nakamura, *Nature* **432**, 84 (2004)
7. R. Wang, H. Bando and M. Itoh, IMF-ISAF 2009 Conf. Xi'an, DO-033, (2009)
8. K. Shibata, K. Suenaga, A. Nomoto and T. Mishima, *Jpn. J. Appl. Phys.* **48**, 121408 (2009)
9. Y. Wakasa, I. Kanno, R. Yokokawa, H. Kotera, K. Shibata and T. Mishima, *Sensors and Actuators A* **171**, 223 (2011)
10. T. Saito, T. Adachi, T. Wada and H. Adachi, *Jpn. J. Appl. Phys.* **44**, 6969 (2005)
11. L. Egerton and D. M. Dillon, *J. Am. Ceram. Soc.* **42**, 438 (1959)

Mater. Res. Soc. Symp. Proc. Vol. 1494 © 2013 Materials Research Society
DOI: 10.1557/opl.2013.137

Thickness Dependent Optical Properties of WO₃ Thin Film using Surface Plasmon Resonance

Ayushi Paliwal[1], Monika Tomar[2] and Vinay Gupta[1]
[1]Department of Physics and Astrophysics, University of Delhi, Delhi 110007, INDIA
[2]Department of Physics, Miranda House, University of Delhi, Delhi 110007, INDIA
[1]**Email id:** drguptavinay@gmail.com; vgupta@physics.du.ac.in
Contact no: +91 9811563101

ABSTRACT

The effect of tungsten oxide (WO₃) thin film thickness on the surface plasmon resonance (SPR) properties have been investigated. WO₃ films of varying the thickness (36 nm, 60 nm, 80 nm, 100 nm, 150 nm and 200nm) have been deposited onto Au coated prism (Au/prism) by radio frequency (RF) magnetron sputtering technique. The SPR responses of bilayer films were fitted with the Fresnel's equations in order to calculate the dielectric constant of WO₃ thin film. The variation of complex dielectric constant and refractive index with the thickness of WO₃ thin film was studied.

INTRODUCTION

Surface plasmon resonance (SPR) technique has been widely recognized as a valuable tools for investigating surface interactions [1]. Since the surface plasmon (SP) waves are excited at metal-dielectric interface, thus any change in the vicinity of the interface can be monitored. SP wave at the metal-dielectric interface can be excited in a resonant manner by a visible or infrared light beam. In the Kretschmann configuration a light wave is totally reflected at the interface between the prism coupler and the thin metal layer (deposited on the prism surface) and excites a SP wave at the outer boundary of the metal by evanescently tunnelling through the thin metal layer. Thin films of noble metal such as Gold (Au) and Silver (Ag) are capable of supporting SP waves [2]. Any physical factor that causes variations in the refractive indices, surface inhomogeneity and thickness of the sensing film forming the SPR configurations will generate significant changes in the plasmon dispersion relations and hence change the SPR curve [3]. The changes in the SPR curve can be used to calculate the dielectric constant of the sensing film.

Tungsten trioxide is a wide band gap semiconductor having monoclinic structure that has good electrochromic properties. WO₃ thin films have been extensively used for the semiconductor gas sensors which require high operating temperatures. SPR technique on the other hand, is best suited at room temperature as well. Dielectric properties of the WO₃ film change appreciably on interaction with target even at room temperature , however, detection of dielectric properties using conventional electrical measurement techniques is not easy for WO₃ because of high losses. SPR is a very efficient tool to measure the dielectric properties of rich layers. Hence, in the present work an effort has been made to study the dielectric properties of WO₃ thin film of varying thickness using SPR technique for their subsequent use in gas sensing applications.

EXPERIMENT

The SP modes have been excited at the WO_3-gold interface in the Kretschmann configuration using a right angled BK7 glass prism (n_p =1.517) with the help of a laboratory assembled system. The optimized thin film of gold (thickness ~40nm) was deposited on the hypotenuse face of glass prism (BK-7) by thermal evaporation technique. The gold plated prisms were annealed at 300°C in air for 1 hour to get the stabilised SPR mode. The Au thin film is also deposited on BK7 glass slide under similar growth conditions for structural and optical characterization. WO_3 thin film was deposited on gold plated glass prisms by rf-magnetron sputtering technique under varying deposition conditions using a 2″ diameter metallic tungsten target (99.999% pure). The thickness of WO_3 thin film is varied by varying the deposition time. WO_3 thin films were deposited with 60%Ar and 40%O_2 as the sputtering gas at a sputtering pressure of 20mTorr and substrate temperature of 300°C by applying a power of 50W. For optical studies, WO_3 films were deposited on the BK-7 glass substrate under identical deposition conditions.

Crystallographic properties of WO_3 thin films were studied using a Bruker D 80 X Ray diffractometer. UV-visible spectra of WO_3 thin films was taken within the wavelength range 190 – 1100 nm using a computer interfaced Perkin Elmer (lambda 35) UV-visible dual beam spectrophotometer.

A low cost table top SPR system based on prism coupling is laboratory made. The complete SPR set up consists of a p-polarized He-Ne laser (λ=633 nm), a prism table fixed on XYZ precision rotation stage, and a photo detector. The XYZ rotational stage has fine linear movements along the X, Y and Z directions with an accuracy of 0.001 cm. The prism table mounted on XYZ stage could be rotated precisely about its axis using a stepper motor, with an accuracy of 0.03°. The photo detector is fixed on a rigid stand which can be revolved freely about an axis passing through centre of prism table (i.e. concentric movement with respect to prism table). The photo detector is attached with a suitable optical power meter (Newport make) which is very sensitive to fine change in the intensity laser beam and the intensity of the reflected laser beam may be recorded in Watt (W), decibel (dB) or dBm upto a lower detection limit of nW. To measure the SPR response, a p-polarized He-Ne laser (λ=633 nm) was made to fall on the surface of WO_3/Au/prism film and the intensity of reflected light was measured as the function of incident angle. The SPR responses of WO_3/Au bi-layer films for varying thickness of WO_3 were measured.

DISCUSSION

The XRD pattern of the WO_3 thin film for different thickness is shown in figure 1.The broad reflection peak was observed at 2θ = 23.5° corresponds to (020) plane of WO_3 which confirms the formation of monoclinic structure of WO_3 [5]. It may be seen that the intensity of (020) plane keeps on increasing with WO_3 film thickness, however, there was no shift in the peak position. The interplanar spacing is calculated using Bragg's diffraction condition i.e. 3.83 A° which is found to be constant with the thickness of tungsten oxide thin film. The grain size is estimated using the Scherrer's formula and is found to be in the range of 0.8-1.3A°. The value of the grain size increases with the increase in thickness of WO_3 thin film. As the film thickness increases, complete nucleation of WO$_3$ grains takes place increasing the grain size and hence the intensity of dominant (020) diffraction peak (figure 1).

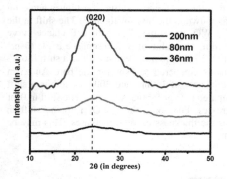

Figure 1. XRD spectra of WO$_3$ thin film as a function of thickness

The optical transmission spectra of the WO$_3$ thin films deposited on BK-7 glass substrate for varying thickness is shown in figure 2(a). The films are found to be highly transparent (> 80%) in visible region. As the thickness increases, a well defined fringe pattern appear indicating the deposition of a homogeneous and uniform thin film. Optical band gap of the WO$_3$ thin film was calculated from the Taue plot. Estimated value of bandgap for WO$_3$ thin film is found to be in the range of 3.59-3.91 eV which is observed to be higher than the values reported by other workers[6].The bandgap increases with the increase in thickness of tungsten oxide thin film as shown in figure 2(b).

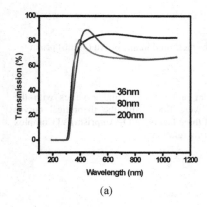

(a)

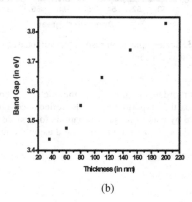

(b)

Figure 2. (a) Transmittance spectra of WO_3 thin films for varying thickness (b) Variation of Band gap of WO_3 thin films as a function of film thickness

Figure 3 shows the SPR reflectance curves for WO_3/Au/prism interface for WO_3 film of 36nm thickness. It may be seen that the SPR reflectance curve shifts to a higher angle on incorporating WO_3 film compared to Au/prism as shown in the inset of figure 3. The shift in the SPR reflectance curve is due to the presence of WO_3 thin films. The SPR reflectance curve further shifts for varying thickness of WO_3 thin film. The value of SPR resonance angle (θspr) for the WO_3 thin films with the thickness of 36 nm, 60 nm, 80 nm, 100nm exhibit a shift (56.8°, 54°, 49.1° and 44.6° respectively) to higher angles compared to θspr for the bare Au/prism (43.8°). However, the WO_3 thin film with the thickness of 150nm and 200nm exhibits a shift (42.3° and 40.8° respectively) to lower angle compared to θspr for the Au/prism. The shift in θspr angle with increasing WO_3 thin thickness decreases. For 36nm thick WO_3 film the shift in resonance angle is maximum which decreases as the thickness of WO_3 increases. This may be attributed to decrease in the roughness of the film.

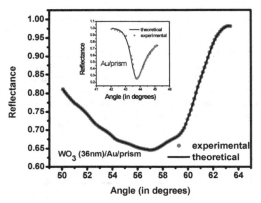

Figure 3. Surface plasmon resonance reflectance curves measured in air for Au (40nm)/prism (inset) and WO_3(36nm)/Au/prism

In order to know how the dielectric constant and refractive index varies with the thickness, the experimental SPR reflectance curve were fitted with the theoretical curves simulated by using Fresnel's equations for a system of three layers (WO_3/Au/prism)[9].Complex refractive index (n_i+ik_i) was evaluated using the following relation:

$$n_i = \left[\frac{(\varepsilon_i'^2 + \varepsilon_i''^2)^{1/2} + \varepsilon_i'}{2} \right]^{1/2} \tag{1}$$

$$k_i = \left[\frac{(\varepsilon_i'^2 + \varepsilon_i''^2)^{1/2} - \varepsilon_i'}{2} \right]^{1/2}. \tag{2}$$

where ε_i' and ε_i'' are the real and imaginary part of dielectric constant respectively

ε_i' and ε_i'' are varied until entire SPR reflectance curve fitted well to the experimental data and the best fitting is shown in the inset of Fig. 3 for the Au/prism having Au thickness of 40nm. The estimated value of dielectric constant of Au was about $-11.9+0.7i$, which is close to reported values [10]. Now estimated value of the dielectric constant of Au is used to determine the dielectric constant of WO_3 thin film using the Fresnel's equations for a three layer system. The dielectric constants and refractive indices are estimated for varying thickness of WO_3 thin film and summarized in table I. The estimated values of dielectric constant and refractive index for the WO_3 thin film lie in the range 3.05 to 7.44 and 1.83 to 2.75 respectively which matches well with the values reported in literature [11].

Table I. Values of complex dielectric constant and refractive index

Thickness (nm)	Dielectric constant(real part)	Dielectric constant (imaginary part)	Refractive index	Extinction coefficient
36	3.05	0.005	1.83	1.19
60	4.07	0.004	2.01	0.99
80	4.89	0.0031	2.21	0.85
100	5.97	0.0026	2.34	0.68
150	6.59	0.0017	2.56	0.32
200	7.44	0.0013	2.75	0.13

It may be seen from table I that the real part of dielectric constant increases with the increase in thickness of WO_3 thin film whereas imaginary part decreases. Similarly, refractive index increases and extinction coefficient decreases with increase in thickness. This may be related to the fact that increasing thickness, reduces the film roughness, enhances the grain size and hence reduces the losses.

CONCLUSIONS

The structural and optical characteristics of WO_3 thin film have been studied as a function of its thickness. The grain size and optical band gap of the film increases with its thickness. The SPR reflectance curve of WO₃ dielectric layer with different thickness have been studied by surface plasmon resonance technique. The dielectric constant and refractive index of WO_3 thin film increase with increasing the film thickness whereas the dielectric loss as well as extinction coefficient decreases with film thickness.

ACKNOWLEDGMENTS

Authors are thankful to DST, Govt. of India for financial support. One of the authors (AP) is grateful to CSIR for research fellowship and University of Delhi for Teaching Assistantship.

REFERENCES

1. X.C. Yuan, B. Hong, Y.G. Tan, D.W. Zhang, R. Irawan, and S.C. Tjin, *J. Opt. A: Pure Appl. Opt.* **8,** 959-963 (2006).
2. D.Yang, H.H. Lu, B.Chen, and C.W. Lin, *Sensors and Actuators B* **145**, 832-838 (2010).
3. Z. Bao, G.L.D. Jiang, W. Cheng, and X. Ma, *Materials Science and Engineering B* **171**, 155-158 (2010).
4. M. Acosta, D. González, and I. Riech, *Thin Solid Films* **517**, 5442-5445 (2009).
5. Y. Yamada, K. Tabata, and T. Yashima, *Solar Energy Materials and Solar Cells* **91**, 29-37 (2007).
6. M. G. Hutchins, O. Abu-Alkhair, M.M. El-Nahass, and K. Abd El-Hady, *Materials Chemistry and Physics* **98**, 401-405 (2006).
7. M. Bao, G. Li, D. Jiang, W. Cheng, and X. Ma, *Materials Science and Engineering B* **171,** 155-158 (2010).
8. H.Raether, "Surface Plasmons on smooth and rough surfaces and on Gratings", Springer Verlag, Berlin Hiedelberg, Tokyo.
9. S. K. Özdemir , and G. Turhan-Sayan, *J. Lightwave Technol.* **21**, 805-814 (2003).
10. N. Mehan, V. Gupta, K. Sreenivas, and A. Mansingh, *J. Appl. Phys.* **96**, 3134 (2004).
11. H. Deng, D.Yang, B.Chen, and C.W. Lin, *Sensors and Actuators B* **134**, 502-509 (2008).

Mater. Res. Soc. Symp. Proc. Vol. 1494 © 2012 Materials Research Society
DOI: 10.1557/opl.2012.1582

Optical, Structural, and Electrical Properties of Vanadium Dioxide Grown on Sapphire Substrates with Different Crystallographic Orientations

M. Nazari,[1,2] Y. Zhao,[2,3] Y. Zhu,[2,3] V. V. Kuryatkov,[2,3]
A. A. Bernussi,[2,3] Z. Fan,[2,3] and M. Holtz[1,2]

[1]Department of Physics and Nano Tech Center, Texas Tech University, Lubbock, TX 79409
[2]Nano Tech Center, Texas Tech University, Lubbock, TX 79409
[3]Department of Electrical Engineering, Texas Tech University, Lubbock, TX 79409

ABSTRACT

The phase transition of VO_2 grown on sapphire having different crystallographic growth planes is examined experimentally. Measurements of electrical resistivity are compared with spectroscopic ellipsometry studies, to obtain complex index of refraction and plasma frequency, and transmission in the terahertz frequency range, each as a function of temperature.

INTRODUCTION

The metal-insulator transition (MIT) of vanadium dioxide (VO_2) has drawn intense scientific interest for understanding fundamental mechanisms of these transformations.[1] Single crystal VO_2 exhibits an abrupt decrease in electrical resistivity, by up to five orders of magnitude, within 0.1 °C upon heating through the MIT temperature of $T_{MIT} \sim 68$ °C. This phase transition is accompanied by concomitant changes in the optical properties in the terahertz (THz), far infrared (IR), and visible spectral ranges.[2-5] The MIT is accompanied by a structural phase transition from monoclinic (M_1) crystal structure at low-temperature to tetragonal rutile (R) in the high-temperature metallic phase.[6,7] These properties render VO_2 a promising candidate for a number of interesting applications in electronics, photonics, and sensors.

It is anticipated that VO_2 films grown on the different orientation of sapphire substrates will exhibit different crystal orientation and quality, including grain boundary, grain size, built-in strain, and V-V bond orientation.[8] These differences may impact free carrier density, T_{MIT}, hysteresis, and abruptness of the transition, thus affecting the device modulation performance. Recent investigations of the effect of stress on VO_2 have resulted in new understanding of the material phase diagram.[9-14] Stress along the M_1 (011) crystal axis, corresponding to $(110)_R$,[15] has been shown to induce a phase transition from the M_1 phase to M_2, in which alternating V chains pair without twisting, while the others twist without pairing.[16,17] The resulting diversity of phenomena has not yet been explored, particularly for thin films deposited on substrates which are generally more amenable to device applications.

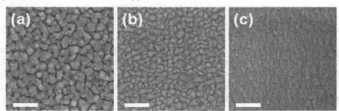

Figure 1. SEM images of VO_2 films grown (a) c-plane, (b) m-plane, and (c) r-plane sapphire substrates. The scale bars are 1 μm.

We report electrical and optical studies of VO_2 deposited on sapphire substrates having different crystallographic orientations for the planar growth surfaces. The VO_2 crystallographic orientation is found to depend on the substrate used even under identical deposition conditions. The structural variation has an impact on the electrical and optical properties.

EXPERIMENT

VO_2 thin films ~120 nm in thickness were grown on c-, r-, and m-plane sapphire substrates (samples denoted c, r, and m, respectively) under identical conditions by computer controlled pulsed DC reactive sputtering deposition from a high purity vanadium target (99.95%). The growth temperature and sputtering gas ambient were previously optimized[4,18,19] and precisely controlled to yield high-purity VO_2. The sputtering pressure was set at 3 mTorr

with an O_2/Ar flow ratio of 11%. All three samples were grown at 575 °C with thickness ~ 130 nm. According to our prior x-ray diffraction and transmission electron microscopy studies of these samples,[8] deposition on c-plane sapphire results in VO_2 with the V—V chains of the R phase oriented in the plane of the sapphire. In contrast, the m and r samples have C_R axis directed out of (but not normal to) the sapphire substrate plane. Based on this, the substrates are expected to have different stress effects on the VO_2.

The resistivity dependence on sample temperature in heating and cooling cycles were measured using the van der Pauw configuration. For the electrical measurements, Ti (50 nm)/Au (120 nm) metal stacks deposited by e-beam evaporation and lift-off process were used as electrodes. Spectroscopic ellipsometry (SE) measurements were carried out between photon energies of 0.6 and 6.5 eV at angle of incidence 70°.[20] Results were modeled using commercial software. THz time-domain transmission measurements were conducted to obtain the electric field amplitude transmitted through the VO_2 samples at normal incidence,[2] these were then converted to frequency-domain spectra. The sample temperatures were controlled during each measurement using a thermoelectric heater/cooler stage equipped with a thermistor.

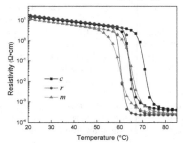

Figure 2. Resistivity vs. temperature for the three samples through the MIT process.

DISCUSSION

Scanning electron microscope (SEM) images of the as-grown surfaces are shown in Fig. 1. All samples exhibited crystalline morphology but the nano-grain sizes vary with the substrate orientation. VO_2 films grown on the c-plane substrate revealed the presence of voids, which resulted in a semi-continuous film. In contrast, the SEM images of the m and r samples exhibited dense nano-grains forming a continuous film.

Resistivity (ρ) measurements show the expected abrupt drop near T_{MIT}. This is illustrated in Fig. 2 for each sample during both heating and cooling processes. Resistivity is seen to decrease by approximately four orders of magnitude when the material transforms from the insulating to conducting phase. The hysteresis loop is observed in each sample. For the temperature upswing, we obtain T_{MIT} values. These are summarized in Table 1, along with T_{onset} corresponding to the temperature at which the decrease in ρ deviates from the exponential dependence as T increases. The dependences observed in ρ are confirmed by the optical measurements and may attributed to the superior film morphology seen in Fig. 1(c).

Table 1. Summary of onset and MIT temperatures based on independent measurements.

Sample	Resistivity		Plasma Frequency		THz Transmission	
	T_{onset} (°C)	T_{MIT} (°C)	T_{onset} (°C)	T_{MIT} (°C)	T_{onset} (°C)	T_{MIT} (°C)
c	62±2	71±1	64±2	72±2	72±2	75±2
r	62±2	64±1	58±2	63±2	60±2	65±2
m	57±2	65±1	56±2	64±2	64±2	67±2

SE data were fitted using a three-layer model. The sapphire substrate layer utilizes measurements obtained directly from identical bare substrates oriented in-plane the same as during the VO$_2$ SE measurements. The second layer is the ~ 120 nm thick VO$_2$ which we describe using a frequency (ω) dependent, complex dielectric function

$$\varepsilon(\omega) = \varepsilon_{\infty} + \sum_{j} \varepsilon_{TL,j}(\omega) + \varepsilon_D(\omega) \qquad (1)$$

where ε_{∞} is the high-frequency dielectric constant. The Tauc-Lorentz dielectric function $\varepsilon_{TL,j}$ is described in Ref. [21]; the imaginary component is non-zero in the photon energy range $> E_g$,

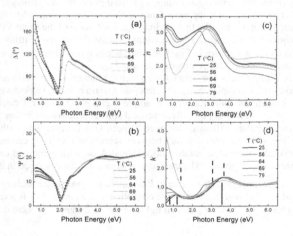

Figure 3. Measured (a) Δ and (b) Ψ from SE and fit results using the three-layer model for sample c at room temperature and near the phase transition. (c) n and (d) k for VO$_2$ (only) layer of sample c.

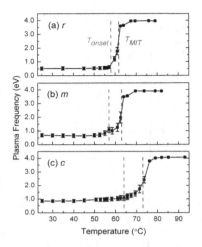

Figure 4. Temperature dependence of the plasma frequency (expressed in eV) for the three samples investigated here. Shown is the upstroke in temperature only. The solid vertical lines denote the onset temperature of the phase transition and T_{MIT}.

where the material absorbs light, thus identifying the optical gap. ε_D is the Drude function. Below the phase transition temperature ε_D accounts for the presence of free-carriers due to unintentional doping. In the high temperature range ε_D describes the free-electron gas of the conductor. The third layer accounts for the top ~ 6 nm of VO_2, which is found to be rough from atomic force microscopy. We use an effective medium approximation of the same VO_2 described above and vacuum, with no attempt to take into account in-plane feature size. Fitting results for ellipsometric angles Δ and Ψ are shown in Fig. 3 for sample c at room temperature, well above the phase transition, and at intermediate temperatures near T_{MIT}.

Based on these fitting results, we obtain the dielectric function of the VO_2 material. Shown in Fig. 3(c) and (d) are complex index of refraction spectra n and k, respectively, calculated from the dielectric function. Results for samples r and m are similar. Optical transition energies from the fitting are shown in Fig. 3(d) as vertical solid (dashed) lines below (above) the phase transformation temperature. Although the overall spectra are in agreement with previously published work,[6,22,23] the transition energies vary between reports and with our current results. This may be due to a combination of substrate induced stresses, deposition method and conditions, and measurement technique. The transition energies will be described in detail in a separate publication.[24] The most notable change in k occurs in the low photon-energy range, primarily as a result of the elevated free carrier concentration in the conducting R phase, and the resulting increase in plasma frequency ω_P shown in Fig. 4 for each sample (T increasing only).

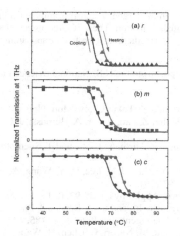

Figure 5. THz field amplitude transmission at 1.0 THz as a function of the temperature (symbols) for VO$_2$ films grown on (a) c-plane, (b) r-plane, and (c) m-plane sapphire substrates. Values are normalized to transmitted field amplitude at $T = 40$ °C for each sample.

Since the free-carrier concentration is proportional to ω_P^2, we estimate T_{onset} where the plasma frequency begins to rise, and T_{MIT} near the midpoint of the rise in ω_P^2 corresponding to the maximum in $d\ln(\omega_P^2)/dT$. These are indicated in each panel of Fig. 4 and summarized in Table 1. Reasonable agreement is seen between the SE and resistivity measurements. The sudden increase in free-carrier concentration is also expected to affect the transmission properties at low photon energy. Shown in Fig. 5 is the temperature dependence of the transmitted field amplitude at frequency 1.0 THz. Data are normalized to the amplitude at $T = 40$ °C for each VO$_2$ sample. The typical thermal hysteresis loop is observed for all three cases. The values obtained from the terahertz studies for T_{MIT} and T_{onset} of sample r are in good agreement with what is obtained from the resistivity or SE measurements, as summarized in Table 1. Sample r exhibits an abrupt transition so that any deviations between techniques are expected to be smaller than the total error. In contrast, the T_{MIT} and T_{onset} values for samples c and m are noticeably higher than what was obtained from the other two approaches. This is most notable in T_{onset}. One possible explanation is the formation of M$_2$-phase VO$_2$, which is expected to be transparent to terahertz light. However, this possibility is only valid for sample c, due to the crystal orientation with the C_R axis in the substrate plane.[8] For sample m we do not anticipate formation of the M$_2$ phase. Our results suggest that the higher free-carrier concentrations implied by ρ or ω_P are needed to absorb the terahertz light in VO$_2$, thereby producing a shift in the observed dependence to higher temperature. This conclusion has implications for optical modulation experiments based on partial transformation of the VO$_2$.

CONCLUSIONS

We observe overall agreement of the phase transition properties of VO$_2$ based on electrical and optical measurements, including in the terahertz range. Agreement is observed for T_{MIT} and T_{onset} between resistivity and plasma frequency, obtained from SE measurements. The corresponding quantities differ in the terahertz transmission measurements, particularly for samples which exhibit a sluggish transition. These results suggest that higher carrier concentrations are needed to observe the transition from transparent to opaque at ~ 1 THz.

ACKNOWLEDGMENTS

The authors acknowledge support from the U.S. National Science Foundation ECCS-1128644.

REFERENCES

[1] M. Imada, A. Fujimori, and Y. Tokura, Rev. Mod. Phys. **70,** 1039 (1998).

[2] Y. Zhu, Y. Zhao, M. Holtz, Z. Fan, and A. A. Bernussi, J. Opt. Soc. Am. B **29,** 2373 (2012).

[3] M. Nakajima, N. Takubo, Z. Hiroi, Y. Ueda, and T. Suemoto, Appl. Phys. Lett. **92,** 011907 (2008).

[4] C. H. Chen, Y. H. Zhu, Y. Zhao, J. H. Lee, H. Y. Wang, A. Bernussi, M. Holtz, and Z. Y. Fan, Appl. Phys. Lett. **97** (2010).

[5] R. Balu and P. V. Ashrit, Appl. Phys. Lett. **92,** 021904 (2008).

[6] H. W. Verleur, A. S. Barker, Jr., and C. N. Berglund, Phys. Rev. **172,** 788 (1968).

[7] C. N. Berglund and H. J. Guggenheim, Phys. Rev. **185,** 1022 (1969).

[8] Y. Zhao, J. H. Lee, Y. Zhu, M. Nazari, C. Chen, H. Wang, A. Bernussi, M. Holtz, and Z. Fan, J. Appl. Phys. **111,** 053533 (2012).

[9] J. Wu, Q. Gu, B. S. Guiton, N. P. de Leon, L. Ouyang, and H. Park, Nano Lett. **6,** 2313 (2006).

[10] J. I. Sohn, H. J. Joo, D. Ahn, H. H. Lee, A. E. Porter, K. Kim, D. J. Kang, and M. E. Welland, Nano Lett. **9,** 3392 (2009).

[11] J. Cao, Y. Gu, W. Fan, L. Q. Chen, D. F. Ogletree, K. Chen, N. Tamura, M. Kunz, C. Barrett, J. Seidel, and J. Wu, Nano Lett. **10,** 2667 (2010).

[12] A. C. Jones, S. Berweger, J. Wei, D. Cobden, and M. B. Raschke, Nano Lett. **10,** 1574 (2010).

[13] Y. Gu, J. Cao, J. Wu, and L.-Q. Chen, J. Appl. Phys. **108,** 083517 (2010).

[14] J. M. Atkin, S. Berweger, E. K. Chavez, M. B. Raschke, J. Cao, W. Fan, and J. Wu, Phys. Rev. B **85,** 020101 (2012).

[15] M. Marezio, D. B. McWhan, J. P. Remeika, and P. D. Dernier, Phys. Rev. B **5,** 2541 (1972).

[16] J. P. Pouget, H. Launois, J. P. D'Haenens, P. Merenda, and T. M. Rice, Phys. Rev. Lett. **35,** 873 (1975).

[17] T. M. Rice, H. Launois, and J. P. Pouget, Phys. Rev. Lett. **73,** 3042 (1994).

[18] C. H. Chen and Z. Y. Fan, Appl. Phys. Lett. **95** (2009).

[19] C. Chen, Y. Zhao, X. Pan, V. Kuryatkov, A. Bernussi, M. Holtz, and Z. Fan, J. Appl. Phys. **110,** 023707 (2011).

[20] A. Majumdar, R. D. Bogdanowicz, and R. Hippler, Photonic Lett. Poland, **37,** 70 (2011).

[21] J. G. E. Jellison and F. A. Modine, Appl. Phys. Lett. **69,** 371 (1996).

[22] M. M. Qazilbash, A. A. Schafgans, K. S. Burch, S. J. Yun, B. G. Chae, B. J. Kim, H. T. Kim, and D. N. Basov, Phys. Rev. B **77,** 115121 (2008).

[23] K. Okazaki, S. Sugai, Y. Muraoka, and Z. Hiroi, Phys. Rev. B **73,** 165116 (2006).

[24] M. Nazari, Y. Zhao, V. V. Kuryatkov, Z. Y. Fan, A. A. Bernussi, and M. Holtz, (Submitted to Phys. Rev. B.

Mater. Res. Soc. Symp. Proc. Vol. 1494 © 2013 Materials Research Society
DOI: 10.1557/opl.2013.593

Dynamic Properties of Spectrally Selective Reactively Sputtered Metal Oxides

A.V. Adedeji*[1], S.D. Worsley[1], T.L. Baker[1], R. Mundle[2], A.K. Pradhan[2], A.C. Ahyi[3] and T. Isaacs-Smith[3]

[1]Department of Chemistry, Geology & Physics, Elizabeth City State University, 1704 Weeksville Road, Elizabeth City, NC 27909, USA
[2]Center for Materials Research, Norfolk State University, 700 Park Avenue, Norfolk, VA 23504, USA
[3]Department of Physics, Auburn University, Auburn, AL 36849, USA

ABSTRACT

Thin films of Transition Metal Oxides (TMOs) were deposited by reactive sputtering of pure transition metal targets in Argon-Oxygen gas mixture at elevated substrate temperature for efficient energy consumption. The atomic composition and thickness of the TMO films was determined by Rutherford Backscattering Spectroscopy (RBS). Optical transmittance and reflectance spectrum of the films on quartz substrate was measured with thin film measuring system at room temperature and slightly elevated temperature. The surface morphology and structure of the TMO films was determined with Atomic Force Microscope (AFM).

INTRODUCTION

In the 21st century, energy conservation and efficient energy consumption are extremely important in order to prevent looming energy crisis. Smart window technology that uses chromogenic materials to vary the throughput of solar radiation in buildings and automobile and similar applications may become essential part of the energy conservation model of the near future. Most of the known chromogenic materials are based on Transition Metal Oxides (TMOs) and their alloys [1]. Transition metal oxides are complex oxides with very rich physics and many potential applications. The complexity of TMOs arises from competing interactions and coupling between charge, spin and orbital wave functions of electrons [2,3]. Couplings and interactions lead to important changes that drive the system into new states with very different properties when external stimuli are applied. For example, temperature can induce metal-insulator, charge and magnetic ordering, or superconducting transitions [3]. Chemical substitution is another method used to tune materials properties. Apart from advancing the fundamental understanding of low-dimensional TMO, basic research on the optical properties can lead to practical applications, especially in energy-related technology [2,4,5].

Many TMOs are known to be chromogenic materials. They allow the transmittance of visible light and solar energy to be varied under the action of an external stimulus [6]. In warm climate, windows can be coated with spectrally selective thin films that will allow the transmittance of visible light and reflect the infrared. In climates where the ambient temperature is not constant, thin films with dynamic optical properties will be the solution solar energy control [7,8]. Thin films that respond to light stimulus are called photochromic (e.g TiO_2 and

MoO$_3$ based films) [9]; thin films that respond to temperature are called thermochromic (e.g. VO$_2$ and VO$_2$ doped with other transition metals) [3,10]; thin films that respond to applied potential are called electrochromic (e.g. NiO, WO$_3$) [11,12].

In this article we describe the deposition of TMOs and mixed TMOs thin films by reactive magnetron sputtering. The oxides of the following transition metals and some of their alloys were sputter-deposited: Ti, V, Cr, Ni, Cu, Mo, Ta and W. We investigated the optical transmittance and reflectance as a function of wavelength and temperature. The composition and morphology of the films were also reported.

EXPERIMENTAL PROCEDURE

All the films were deposited on fused quartz and Si\CVD diamond substrates and that were pre-cleaned using standard organic cleaning process. Metal targets were reactively sputtered in an ultra high vacuum magnetron sputtering system. The ratio of O$_2$/Ar in the sputtering gas is 1/10 for all depositions. During all depositions the substrate was kept at 300°C and rotated. The base pressure in the sputtering chamber was about 5 x 10^{-8} Torr. The sputtering target and purity, power, pressure and time are shown in the table below. For the V-Ti-O samples, vanadium and titanium targets were reactively co-sputtered in O$_2$/Ar gas at 300°C. Other titanium oxide sample was deposited from TiO$_2$ target. The V-Ti-O sample was deposited by co-sputtering V and Ti targets. All the TMOs films were sputtered on quartz and Si\CVD diamond substrates.

Table 1: Sputter targets and deposition conditions

Target sputtered	% purity	Sputtering Power (W)	Chamber Pressure (mtorr)	Sputtering Time (mins.)	Thin Films Samples
Ti	99.5	DC. 100	10	30	Ti-O
V	99.5	DC. 100	10	30	V-O
Cr	99.95	DC. 50	3	30	Cr-O
Ni$_{93}$V$_7$	99.95	DC. 50	3	30	Ni-V-O
Cu	99.99	DC. 30	3	25	Cu-O
Mo	99.95	DC. 50	3	30	Mo-O
Ta	99.95	DC. 50	3	30	Ta-O
W	99.95	DC. 50	3	30	W-O
V & Ti co-sputtered	-	V: DC. 100 Ti: RF. 30	10	30	V-Ti-O
TiO$_2$	99.9	RF. 100	10	30	TiO$_2$

The accelerator at Auburn University produced 2 MeV Helium ions for RBS analysis of samples and Rutherford Universal Manipulation Program (RUMP) was used to fit the experimental data and determine the samples' composition and thickness. Thin films measuring system was used to measure the transmittance and reflectance spectrum from 200 – 1100 nm as a function of temperature. The samples surface morphology was determined with Atomic Force Microscope.

RESULTS AND DISCUSSION

Samples Composition and Thickness

Figure 1 shows the RBS data for TMOs films on Si\2 μm CVD diamond substrates. The wide window from channel 500 to about 100 allows easy observation and quantification of the composition and thickness. The RUMP simulation lines and the parameters are shown in the figures as well. The thickness of each sample and the metal-to-oxygen atomic ratios are shown in table 2.

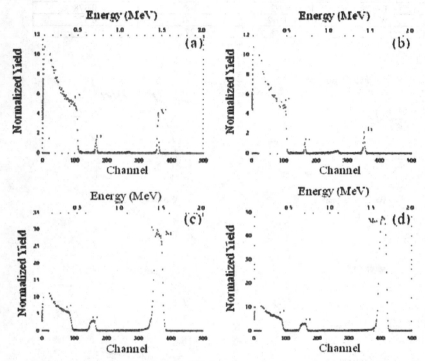

Fig. 1: RBS spectra of (a) Si\C\V-O (b) Si\C\Ti-O (c) Si\C\Ni-V-O and (d) Si\C\Mo-O. The red lines are the RUMP simulation fits.

Optical Properties of TMOs Films

The plots of transmittance and reflectance versus wavelength obtained from thin film measuring system are shown in figure 2. In figure 2a, the transmittance edge of the V-O sample is at longer wavelength compared with the Ti-O sample. Both samples and their alloys have transparency close to 100 % in the visible region of the spectrum.

Table 2: Shows the sputter targets, the TMOs atomic composition, thickness and roughness.

Target sputtered	Atomic composition	Sample thickness (nm)	AFM Surface Roughness (nm)
Ti	Ti/O = 1/3.5	14	0.293
V	V/O = 1/3	18	0.518
Cr	Cr/O = 1/2.5	15	0.422
$Ni_{93}V_7$	Ni/V/O = 1/0.2/1.0	165	1.06
Cu	Cu/O = 1/0.1	80	38.0
Mo	Mo/O = 1/1.8	180	0.731
Ta	Ta/O = 1/2	160	0.378
W	W/O = 1/1.2	160	0.325
TiO_2	Ti/O = 1/3	27	-

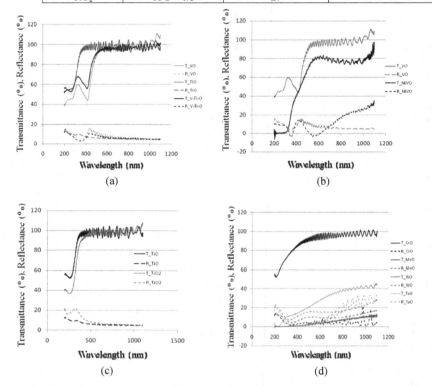

(a)

(b)

(c)

(d)

Fig. 2: Transmittance and Reflectance spectrum of TMO films on quartz substrates: (a) V-O, Ti-O and V-Ti-O (b) V-O and Ni-V-O (c) Ti-O and TiO (sputtered from TiO_2 target) and (d) Cr-O, Mo-O, W-O and Ta-O.

In figure 2b, the transmittance of the Ni-V-O sample drop to almost zero in the uv while the transmittance in the visible region remain high, about 80%. Figure 2c shows that there is no significant difference between Ti-O that was reactively sputtered from pure titanium target and Ti-O deposited from TiO_2 target. The samples in figure 2d do not have the same thickness. Chromium oxide is about a factor of 10 thinner than Mo, W and Ta oxides.

Figure 3 shows the reflectance of the various TMO samples as a function of temperature at 900 nm. All the TMOs investigated do not show any thermochromic properties between room temperaure and about 100°C in the wavelegth region investigated.

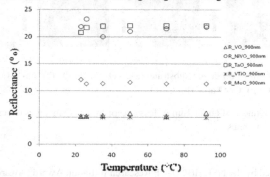

Fig. 3: Reflectance versus Temperature of reactively sputtered TMOs at 900 nm wavelength.

The atomic composition and structure of the deposited TMO films determine the thermochromic propoerty. For example, VO_2 films are known to show electrical and optical transitions at about 68°C, but the ratio of V/O is the reactively sputtered sample is 1/3 and the films are mostly amorphous.

Surface Morphology of TMO Films

The AFM images of some of the TMO films on quartz are shown below. The samples are generally very smooth. The surface roughness is 1.06 nm or less except for Cu-O sample that is very rough (38 nm).

CONCLUSIONS

Transition metal oxides thin films of Ti, V, Cr, Ni-V, Cu, Mo, W and Ta were deposited by reactive sputtering in O_2/Ar gas at 300°C substrate temperature. The composition and thickness of the films were obtained from RBS data analysis using RUMP. The transmittance and reflectance spectrum of the films was obtained as a function of temperature from room temperature to less than 100°C. No thermochromic characteristic was observed for all the films deposited on quartz substrates in this temperature range and wavelength region. It is known that thermochromism is sensitive to atomic composition and surface structure. The TMO films were

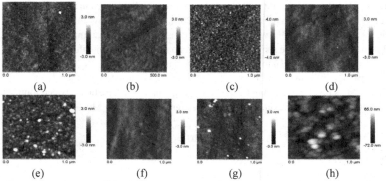

Fig. 4: AFM images of TMO films on quartz substrate (a) V-O (b) Ti-O (c) Ni-V-O (d) Ta-O (e) Mo-O (f) W-O (g) Cr-O and (h) Cu-O.

very smooth with surface roughness of about 1nm or less except for Cu-O that was observed to be very rough (about 38 nm roughness).

ACKNOWLEDGMENTS

This work is supported by the NSF HBCU-UP Research Initiation Award (Award No. 1137470) and NSF Major Research Instrumentation (Award No. 1040200) grants.

REFERENCES

1. S.S. Kanu, R. Binions, Thin Films for Solar Control Applications, Proc. R. Soc. A **466** (2010) 19-44
2. Jinbo Cao and Junqiao Wu, Strain effect in low-dimensional transition metal oxides, Materials Science and Engineering R **71** (2011) 35-52
3. J.L. Musfeldt, Functional Metal Oxide Nanostructures, Springer Series in Materials Science 149, DOI 10.1007/978-1-4419-9931-3_5, J. Wu et al (eds.): Chapter 5 - Optical Properties of Nanoscale Transition Metal Oxides.
4. J. Oensstein and A.J. Millis, Advances in the Physics of high-temperature superconductivity, Science **288** (2000) 468-474
5. D.N. Basov and T. Timusk, Electrodynamics of high-T_c superconductors, Rev. Mod. Phys. **77** (2005) 721.
6. C.G. Granqvist, S. Green, G.A. Niklasson, N.R. Mlyuka, S. von Kraemer and P. Georen, Advances in Chromogenic materials and devices, Thin Solid Films **518** (2010) 3046-3053.
7. C.M. Lampert, Smart switchable glazing for solar energy and day light control, Solar Energy Mater., Solar Cells **52** (1998) 207.
8. C.G. Granqvist, Solar energy materials, Adv. Mater. **15** (2003) 1789-1803.

9. C.G. Granqvist, Radiative heating and cooling with spectrally selective surfaces, Appl. Opt. **20** (1981) 2606

10. T.D. Manning, I.P. Parkin, R.J.H. Clark, D. Sheel, M.E. Pemble and D. Vernadou, Intelligent window coatings: atmospheric pressure chemical vapour deposition of vanadium oxides, J. Mater. Chem. **12** (2002) 2936-2939

11. C.M. Lampert, Optical switching technology for glazings, Thin Solid Films **236** (1993) 6

12. I. Bouessay, A. Rougier, P. Poizot, J. Moscovici, A. Michalowicz and J.M. Tarascon, Electrochromic degradation in nickel oxide thin films: a self-discharge and dissolution phenomenon, Electrochim. Acta **50** (2005) 3737

Mater. Res. Soc. Symp. Proc. Vol. 1494 © 2012 Materials Research Society
DOI: 10.1557/opl.2012.1647

Effects of La$_{0.5}$Sr$_{0.5}$CoO$_3$ sol concentration on the microstructure and dielectric properties of Ba$_{0.6}$Sr$_{0.4}$TiO$_3$ films prepared by sol-gel method on Ti substrate

Dan Jiang[1], Songwei Han[1], Xuelian Zhao[1] and Jinrong Cheng[1]*
School of Materials Science and Engineering, Shanghai University, Shanghai 200072
*corresponding author: jrcheng@shu.edu.cn

ABSTRACT

Ba$_{0.6}$Sr$_{0.4}$TiO$_3$ (BST) thin films were deposited on La$_{0.5}$Sr$_{0.5}$CoO$_3$ (LSCO) buffered Ti substrates. Both BST and LSCO were prepared by sol-gel method. X-ray diffraction (XRD) and scanning electron microscopy (SEM) analysis were used to investigate the effect of LSCO sol concentration on the crystallinity and surface morphology of the films. The results show that with the increase of LSCO sol concentration, BST films show variation of the structure and dielectric properties. BST films for LSCO of 0.2 mol/L exhibit a better crystallinity and improved dielectric properties, with the tunability, dielectric constant and tanδ of 30%, 420 and 0.028 respectively.

INTRODUCTION

Ba$_{1-x}$Sr$_x$TiO$_3$ (BST) is one of the candidates for tunable microwave devices for wireless telecommunications [1,2] for its highly nonlinear dielectric response and high tunability in the vicinity of the paraelectric to ferroelectric phase transformation temperature Tc [3], Usually, BST thin films have been deposited by radio frequency (RF) sputtering [4], pulsed laser deposition (PLD) [5], metal organic chemical vapour deposition (MOCVD) [6] and sol-gel method [7]. Compared with other methods, the sol-gel method has several advantages due to its simplicity, easy control of the film composition, low cost of the apparatus and raw material [8].

Usually, platinized silicon is employed as substrate for BST thin films. However, some problems existed, such as the formation of hillocks in Pt electrode at high temperature [9] and the diffusion between electrode and Si, which greatly undermine the dielectric properties and hinder practical applications of BST thin films [10]. Recently, depositing BST thin films directly on base metal substrate have attracted much attention for their advantages of mechanical flexibility of the substrate and low cost. However, the low oxidation resistance of base metal substrates is one of the principal limitations in preparing BST films on metal substrates, which caused diffusion or reaction between the films and the substrates in the process of heat treatment [11]. LSCO, which has a pseudocubic perovskite structure with a lattice constant of 0.3835 nm [12] has been used as buffer layer. There are high lattice match and structural compatibility for the BST/LSCO structures, which are potential to improve dielectric properties of the films on base metal substrates. Recently, BST thin films on Ti substrates with LSCO buffer layers have been few reports. In this study, BST films were prepared on Ti substrates by sol-gel method with LSCO buffer layers. The effect of the LSCO sol concentration on dielectric properties of BST films was investigated.

EXPERIMENT

The LSCO sol were prepared by using lanthanum nitrate, strontium nitrate and cobaltous acetate as starting materials, which were mixed in a molar ratio of 1:1:2 dissolved in the mixed solvents of acetic acid and deionized water to obtain LSCO sol of 0.1, 0.2 and 0.3 mol/L , respectively. The 7 wt% polyvinyl alcohol (PVA) was employed as the stabilizing agent. Barium acetate, Strontium acetate, and tetrabutyl titanate were used as precursor materials when preparing Ba$_{0.6}$Sr$_{0.4}$TiO$_3$ sol of 0.4 mol/L.

The LSCO sol of 0.1, 0.2, 0.3 mol/L was separately spin coated onto Ti substrates to about 150 nm at 4000 rpm for about 30 seconds, followed by heat treatment at 350 °C for 3 minutes. And the LSCO were crystallized at 700 °C for 10 minutes. Then, BST thin films with thickness of about 500 nm were deposited on LSCO followed by the crystallization at 750 °C for 15 minutes.

For dielectric measurements, the metal-insulator-metal (MIM) capacitor configuration was fabricated. The gold electrode with diameter of 0.4 mm was evaporated through a shadow mask. The crystal structure of BST thin films were obtained by X-ray diffraction (XRD). The morphology of surface was observed by a scanning electron microscopy (SEM). Capacitance and dielectric loss were measured using an Aglient 4294A impedance analyzer.

RESULTS AND DISCUSSION

XRD patterns of the LSCO films on Ti and BST films on the LSCO coated Ti substrates were shown in Fig. 1. It can be seen that LSCO films for different sol concentrations were identified to be perovskite structure with random orientation. With the increase of LSCO sol concentration, the (111) peak getting stronger indicating that the crystallinity of LSCO films is enhanced with increasing the LSCO sol concentration. Fig. 1(b) reveals that BST films exhibit the pure perovskite structure with no evidence of secondary phase formation. The (110) and (111) diffraction peaks of BST films for LSCO sol of 0.2 mol/L indicated the better crystallinity of the films. Upon using the LSCO buffer layer, BST thin films can be well crystallized at 750 °C which is lower than that of BST films on platinized Si substrates [13].

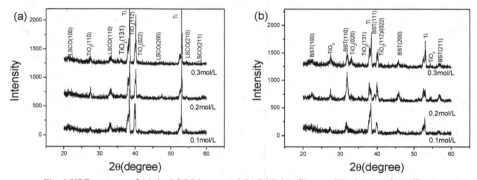

Fig. 1 XRD patterns of (a) the LSCO layers and (b) BST thin films on Ti substrates for different concentrations of LSCO sol

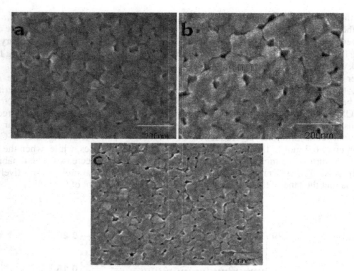

Fig. 2 SEM images of the surface of BST thin films for LSCO sol of (a) 0.1, (b) 0.2 and (c) 0.3 mol/L

Fig. 2 show SEM images of BST thin films for LSCO sol of 0.1, 0.2 and 0.3 mol/L. BST thin films exhibit a good uniformity of microstructure. The grain size, as averaged from the graphs, increases as the LSCO sol concentration changes from 0.1 to 0.2 mol/L. With further increasing the LSCO sol concentration to 0.3 mol/L, the grain size of BST thin films decreases. The average grain size of BST films for LSCO sol of 0.2 mol/L is approximately 91 nm. Micro-cracks can be observed from SEM images that might be the defects of fast volatile of organic compounds during the process of heat treatment.

The dielectric constants are calculated from the capacitance values using the paralleled plate capacitor formula. Fig. 3 shows dielectric constant and tanδ of BST thin films for LSCO sol of different concentrations as a function of frequency. At low frequencies, the tanδ of all the films is virtually in the similar range, changing slightly with frequency, indicating that the concentration of the space charges in the interfacial layer between BST/LSCO is very small. It can be concluded that the nature of LSCO layers has a direct impact on the dielectric properties of BST films. BST films for LSCO sol of 0.2 mol/L have relatively low tanδ of about 0.028 and dielectric constant of about 420 at 10^4 Hz. The dielectric constant of BST films for LSCO sol of 0.2 mol/L and 0.3 mol/L decrease gently with the frequency up to 10^7 Hz. However, the dielectric constant significantly decreases whilst tanδ increases for the BST films of LSCO sol of 0.1 mol/L with the measurement frequency of above 10^6 Hz.

Fig. 4 exhibits dielectric constant and tanδ of BST thin films for LSCO sol of different concentrations as a function of bias voltage at frequency of 10^6 Hz. The tunability is defined as the ratio of $(C_O - C_V)/C_O$ or $(\varepsilon_{max} - \varepsilon_{min})/\varepsilon_{max}$, where C_O is the capacitance at zero bias, C_V is the capacitance at V bias, ε_{max} and ε_{min} are the maximum and minimum measured dielectric constant. The decrease of dielectric constant was observed with the increase of applied electric field. The rate of change is relative quick for BST thin films with LSCO sol of 0.2 mol/L, which indicating a high tunability. BST thin films for LSCO sol of 0.2 mol/L have relative large grain size, resulting in the enhanced dielectric tenability [14]. The increase in grain size enhances the dipole density, however decreases the amount of grain boundaries, inducing fewer grain boundary defects [15]. The tanδ of the films are reasonable and stable with the electric field varying from -200 to 200 kV/cm.

In order to compare films with varied dielectric properties for use in tunable device applications, the figure of merit (FOM) is a frequently used parameter to characterize the correlations between the tunability and the dielectric loss. High tunability and low dielectric loss can insure FOM defined as the ratio of the tunability/loss. Ideally, the FOM value is desired to be as high as possible. Fig.5 reveals the tunability, tanδ and the FOM of BST thin films for LSCO sol of different concentrations at frequency of 106 Hz. With increasing the LSCO sol concentration up to 0.2 mol/L, the value of tunability and FOM increases, while, when the LSCO sol concentration turn to 0.3 mol/L, the value of tunability and FOM decreases. The tunability of the BST films was 10%, 30% and 22% for LSCO sol of 0.1, 0.2 and 0.3 mol/L respectively. The trend of tanδ is that the tanδ of the BST is the smallest one for LSCO sol of 0.2 mol/L.

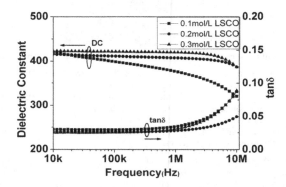

Fig.3 Dielectric constant and tanδ of BST thin films for LSCO sol of different concentrations as a function of frequency

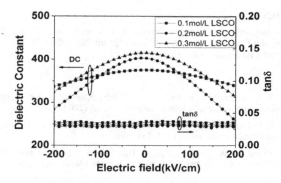

Fig.4 Dielectric constant and tanδ of BST thin films as a function of bias voltage at frequency of 10^6 Hz for LSCO sol of different concentrations

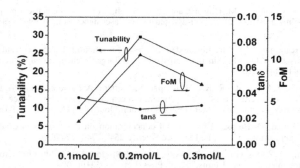

Fig.5 Tunability, tanδ and FOM of BST thin films for LSCO sol of different concentrations at frequency of 10^6 Hz

CONCLUSION

Perovskite BST thin films have been prepared on the Ti substrates at a lower annealing temperature of 750 °C, by using LSCO buffer layers. The LSCO sol of 0.2 mol/L is beneficial to giving the well crystallized LSCO buffer layers, on which BST thin films have the relative large grain size and the homogeneous structure. BST thin films for LSCO sol of 0.2 mol/L have relative high tunability of 30% and the low tanδ of 0.028. The LSCO buffer sol concentration plays an important role in the microstructure and dielectric properties of the BST thin films, which is useful in tailoring the orientation, microstructure, and properties of BST thin films for practical applications.

ACKNOWLEDGMENT

This work was supported by Shanghai Rising Star Program under Grant No. 08QH14008, Shanghai education development foundation under grant No. 08SG41 and National Nature Science Foundation of China under Grant No. 50872080.

REFERENCES

1. A. K. Tagantsev, V. O. Sherman, K. F. Astafiev, J. Venkatesh, and N. Setter, "Ferroelectric materials for microwave tunable applications," Electroceramics. 11,5-66 (2003).

2. O. G. Vendik, E. K. Hollmann, A. B. Kozyrev, and A. M. Prudan, "Ferroelectirc tuning of planar bulk microwave devices," Superconductivity. 12, 325 (1999).

3. S. Zhong, S. P. Alpay, M. W. Cole, E. Ngo, S. Hirsch, and J. D. Demaree, "Highly tunable and temperature insensitive multilayer barium strontium titanate films," Appl. Phys .Lett . 90, 092901 (2007).

4. T. Kikkawa, N. Fujiwara, H. Yamada, S. Miyazaki, F. Nishiyama, and M. Hirose, "Energy band structure of Ru/(Ba, Sr)TiO/Si capacitor deposited by inductively coupled plasma-assisted radio-frequency-magnetron plasma sputtering ,"Appl. Phys. Lett. 81,2821-2823(2002).

5. A. Srivastava, V. Craciun, J. M. Howard, and R. K. Singh, " Enhanced electrical properties of $Ba_{0.5}Sr_{0.5}TiO_3$ thin films grown by ultraviolet-assisted pulsed-laser deposition,"Appl. Phys.Lett. 75,3002-3004(1999).

6. S. Saha, D. Y. Kaufman, S. K. Streiffer, and O. Auciello, "Anomalous leakage current characteristics of $Pt/(Ba_{0.75} Sr_{0.25}) Ti_{1+y}O_{3+z}/Pt$ thin films grown by metalorganic chemical vapor deposition,"Appl. Phys.Lett. 83,1414-1416 (2003).

7. P. Fornasiero, R. D. Monte, G. R. Rao, J. Kaspar, S. Meriani, A.Trovarelli, and M. Graziani, "Rh-loaded CeO_2-ZrO_2 solid-solutions as highly efficient oxygen exchangers: dependence of the reduction behavior and the oxygen storage capacity on the structural,"Journal of Catalysis. 151,168-177 (1995).

8. M. Dutta, S. Mridha, D. Basak, "Effect of sol concentration on the properties of ZnO thin films prepared by sol-gel technique," Applied Surface Science.254 , 2743–2747(2008).

9. H. Han, X. Song, J. Zhong et al, Highly a-axis-oriented Nb-doped PZT thin films grown by sol–gel technique for uncooled infrared dectors," Appl. Phys. Lett. 85, 5310(2004).

10. S.G. Ghonge, E. Goo, R. Ramesh, R. Haakenaasen, D.K. Fork, "Microstructure of epitaxial oxide thin film heterostructures on silicon by pulsed laser deposition" Appl. Phys. Lett. 64, 3407 (1994).

11. Jong-Jin Choi, Jungho Ryu, Byung-Dong Hahn, Woon-Ha Yoon, and Dong-Soo Park, "Effects of a Conducting $LaNiO_3$ Thick Film as a Buffer Layer of a $Pb(Zr,Ti)O_3$ Film on Titanium Substrates," Appl. Ceram. Technol.6 (6) 687–691 (2009).

12. P.W. Chan, W. Wu, K.H. Wong, K.Y. Tong, J.T. Cheung, "Preparation and characterization of epitaxial films and of an a-axis oriented heterostructure on (001) by pulsed laser deposition," Physics D: Applied Physics.30, 957 (1997).

13. Jia Gong.,et al, "Improvement in dielectric and tuanble properties of Fe-doped $Ba_{0.6}Sr_{0.4}TiO_3$ thin films grown by pulsed laser deposition," unpublished.

14. Nayak, M. and Tseng, T.-Y, "Dielectric tunability of barium strontium titanate films prepared by a sol-gel method,"Thin Solid Films. 408, 194–199(2002).

15. Feng Chen, Jinrong Cheng, Shenwen Yu, Zhongyan Meng, "Structural and electrical properties of $Pb(Zr_{0.53}Ti_{0.47})O_3$ films prepared on $La_{0.5}Sr_{0.5}CoO_3$ coated Si substrates,"Eur. Ceram. Soc. (2009).

Mater. Res. Soc. Symp. Proc. Vol. 1494 © 2013 Materials Research Society
DOI: 10.1557/opl.2013.177

Potential Driven Chemical Expansion of $La_{0.6}Sr_{0.4}Co_{1-x}Fe_xO_{3-\delta}$ Thin Films on Yttria Stabilized Zirconia

Kee-Chul Chang[1], Brian J. Ingram[2], E. Mitchell Hopper[2], Miaolei Yan[3], Paul Salvador[3], and Hoydoo You[1]

[1]Materials Science Division, Argonne National Laboratory, 9700 S Cass Ave., Argonne, IL 60439, U.S.A
[2]Chemical Sciences and Engineering Division, Argonne National Laboratory, 9700 S Cass Ave., Argonne, IL 60439, U.S.A
[3]Department of Materials Science and Engineering, Carnegie Mellon University, Pittsburgh, 5000 Forbes Ave., Pittsburgh, PA 15213, U.S.A

ABSTRACT

To better understand the response of oxygen vacancy concentration to applied potential, the lattice parameter of pulsed laser deposited $La_{0.6}Sr_{0.4}Co_{1-x}Fe_xO_{3-\delta}$ thin films was monitored using in situ X-ray diffraction. We demonstrate that the chemical expansion under applied potential depends on the cathode morphology, which determines the contribution of different reaction pathways. We investigated applied potential dependent lattice expansion on $La_{0.6}Sr_{0.4}Co_{1-x}Fe_xO_{3-\delta}$ with 3 different Co:Fe ratios in an attempt to connect bulk chemical expansion data to thin films. We find that the chemical expansion trends in thin films are different than expected from bulk data.

INTRODUCTION

Oxygen vacancies play a prominent role in the function of ionic devices such as solid oxide fuel cells (SOFC), oxygen ion transport membranes and oxygen sensors. Therefore, understanding the oxygen vacancy distribution inside such devices during operation, which is related to the reaction pathways, will promote fundamental understanding and lead to better performance.

Chemical expansion, the change in lattice parameter due to changes in the defect composition in the film, provide a convenient, indirect method of measuring the oxygen vacancy concentration. Although we can measure relative changes in the oxygen vacancy concentration using the lattice constants, putting these values into an absolute scale is challenging due to in-plane stresses and other thin film effects. In this paper, we measured the lattice expansion behavior of $La_{0.6}Sr_{0.4}Co_{1-x}Fe_xO_{3-\delta}$ thin films under applied potential to see if the thin film oxygen non-stoichiometry can be related to the recently published[1,2,3] bulk values.

EXPERIMENTAL DETAILS

We studied three thin films of $La_{0.6}Sr_{0.4}Co_{1-x}Fe_xO_{3-\delta}$ (x=0.05, 0.8, and 1, hereafter referred to as LSC, LSCF, and LSF, respectively in this paper) on yttria stabilized zirconia (YSZ) single crystal substrates. The cathode film samples were grown on YSZ(111) by pulse laser deposition (PLD). To run the system as a half-cell, we used screen-printed porous Au on the film surface and Pt mesh on the YSZ as current collectors. The in situ X-ray experiments were performed at beamline 12-BM at the Advanced Photon Source, Argonne National Lab. A schematic illustration of the experimental setup, which allows us to simultaneously collect X-ray diffraction, X-ray fluorescence and electrochemical data, is shown in figure 1. Since the effective potential will depend on the distance from the Au current collector, we aligned the sample so that the beam is ~0.5mm away from the Au. A series of X-ray and electrochemical measurements were taken during heat up to ~890K and cool down to room temperature. The sample was held at ~890K for more than 12 hours to induce Sr segregation so that we could correlate electrochemical data to the X-ray measurements.

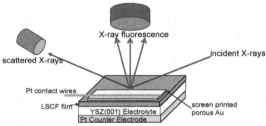

Figure 1. A schematic illustration of the experimental setup

RESULTS AND DISCUSSION

Effect of epitaxy and cathode morphology on lattice expansion

We found that the three different $La_{0.6}Sr_{0.4}Co_{1-x}Fe_xO_{3-\delta}$ samples had different epitaxial relationships to YSZ(111) substrates as shown in figure 2a. LSC had the best epitaxial property, showing uniform thickness evidenced by fringes, while LSCF had no fringes, indicating worse epitaxy. LSF film was found to have no epitaxial relation with the substrate, so we measured (011) peaks at fixed grazing incidence. Despite the different epitaxy, we found that chemical expansion occurred under cathodic potentials in all of the films.

To investigate whether the lattice parameter is due to the film morphology, we also looked at screen printed LSCF powders on YSZ. As shown in figure 2b, lattice expansion was barely observable at -5V applied potential on LSCF powders, in contrast to the much larger expansion at -1.5V observable on thin film LSCF. We hypothesize that this difference is due to the greater contribution of surface reaction pathways to the oxygen reduction reaction in the power samples.

Sr segregation

Total reflection X-ray fluorescence (TXRF) was used to monitor the Sr segregation when the samples were annealed at ~890K. In TXRF, the X-ray fluorescence is measured as a function of X-ray incidence angle and particles on the surface enhance the fluorescence signal below the critical angle (around the peak position of the fluorescence at ~0.17° at 16.5keV X-ray energy). We found evidence of Sr segregation progressing on the LSC and LSCF sample as shown in figure 3. On the LSF sample, shown later in figure 6b, Sr segregation has already occurred at the first TXRF measurement at 840K. The Sr segregation will be correlated to changes in the lattice parameter and electrochemical measurements later in this paper.

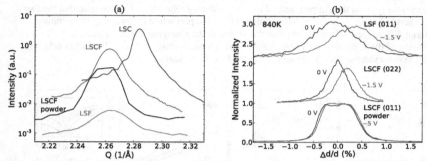

Figure 2. (a) XRD of cathode films around the (011) peak at room temperature (b) Relative shifts in the Bragg peak positions in LSCF, LSF film and LSCF powder under applied potential.

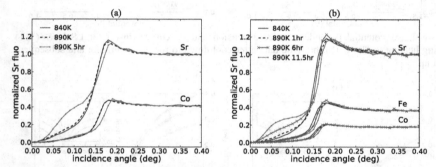

Figure 3. TXRF of (a) LSC and (b) LSCF film at 16.5keV. The low angle Sr fluorescence increases over time when the sample is annealed at 890K over 12 hours, showing that Sr segregation is occurring.

$La_{0.6}Sr_{0.4}Co_{0.95}Fe_{0.05}O_{3-\delta}(LSC)/YSZ(111)$

LSC film was epitaxial with the substrate, so the specular (022) peak was used to monitor the lattice expansion. The potential dependent lattice expansions and currents during heat up at four temperatures, 680, 750, 840, and 890K are shown in figure 4a, with the lattice expansion normalized to the initial value at 0V. 680 to 840K data are compiled from step scans in 100mV increments from 0 to -1V and 0 to 0.5V with a settling time over ~1 minute at each potential while the 890K data is obtained by a continuous scan from 0 to -2V, -2 to 1V, and 1 to 0V at ~25mV/sec rate without settling time. A stepping scan at 890K yielded the results similar to the continuous scan so only the continuous scan will be shown onward. We did not collect any cool down data from this sample, so the sample was subsequently heated again to 910K as shown in figure 4b. We note that the electrochemical data is similar for these two runs but the lattice expansion behavior is different. This result could be due to changes in the beam position, which was 0.8mm away from the Au pad instead of 0.5mm during reheating. We also note that the two lattice expansion measurements at 910K, taken 11 hours apart while maintaining temperature, show large differences. We speculate that this is due to the changing surface and bulk composition of the films due to Sr segregation over this time scale, which will result in different surface oxygen exchange constants and affect the lattice expansion.

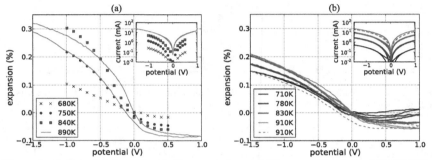

Figure 4. (a) Potential dependent lattice expansion and IV curves (inset) for LSC/YSZ(111) during heating to 890K. (b) the same data after cool down to room temperature and subsequent heat up.

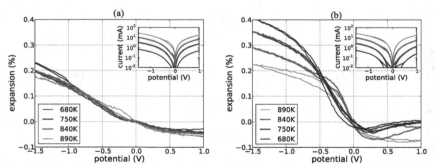

Figure 5. Potential dependent lattice expansion and IV curves (inset) for LSCF/YSZ(111) during (a) heat-up and (b) cool down. The sample was held at 890K for 9 hours between the 2 measurements at 890K.

La$_{0.6}$Sr$_{0.4}$Co$_{0.2}$Fe$_{0.8}$O$_{3-\delta}$(LSCF)/YSZ(111)

The specular (022) peak was used to monitor the lattice expansion of the epitaxial LSCF sample. Potential scans during heat up and cool down (figure 5) show different lattice expansion behavior. After Sr segregation at 890K, there was a slight increase in the lattice expansion response to applied potential, but the current showed almost no change. The lattice expansion was also found to increase significantly at lower temperature after Sr segregation. This trend is opposite of what is expected from bulk data, where it is found that a larger pO2 difference is needed to change the oxygen vacancy concentration by the same amount at lower temperature. One possible explanation for this effect is that the effective potential at the beam position may be different[4] due to changing electronic/ionic conductivity ratio at different temperature. Experimental values[5] suggest that the effective potential will increase at lower temperature due to increasing electronic/ionic conductivity ratios but this may not be observable in LSC due to a sharper decrease in effective potential with distance.

La$_{0.6}$Sr$_{0.4}$FeO$_{3-\delta}$(LSF)/YSZ(111)

LSF film was not epitaxial so we used grazing incidence geometry to minimize the background and monitored the (011) peak. Due to the weak peak intensity, we used wider potential intervals to get 20mV/sec sweep rate as shown in figure 6a. The lattice expansion slows at around -0.5V but does not reach saturation values at -1.5V. This sample was determined to be Sr segregated at 840K as shown in figure 6b, and the polycrystalline nature of the LSF, resulting in many grain boundaries, may be a factor in the enhanced segregation. The lattice expansion at different temperature shows similar trends to Sr segregated LSCF.

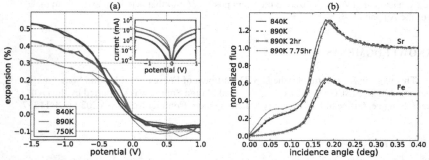

Figure 6. (a) Potential dependent lattice expansion and IV curves (inset) for LSF/YSZ(111). Scan rate was 20 mV/sec. (b) TXRF of LSF at 840K shows that Sr segregation has already occurred.

One unexpected result is that for bulk LSF, the maximum to minimum difference of the lattice expansion should be ~0.4% since δ should reach a limiting value of 0.2 before decomposition occurs[1]. Our data shows expansion range of up to ~0.6% at 750K. We speculate that the thin film may be more stable than bulk at high δ[6]. Another possibility is that the lattice constant may be locked in-plane for this sample, increasing the out-of-plane expansion.

263

CONCLUSIONS

To summarize the data, we also looked at changes in the lattice constant as a function of the distance away from the Au pad to correct for the effective potential at the X-ray beam position at 890K as shown in figure 7. LSCF and LSF lattice expansion fall off faster from the Au pad than LSC. The corrected lattice expansion data shows that the response to potential follows LSF > LSCF > LSC which is opposite of the trend expected from bulk measurements. This suggests that other factors, such as surface or film/electrolyte oxygen exchange coefficients may play a greater role in determining the limiting δ value at high applied potentials.

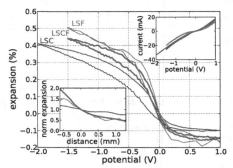

Figure 7. Comparison of lattice expansion for LSC, LSF and LSCF at 890K. The inset shows the lattice expansion at -1V, normalized to the beam position (at 0) that is 0.5mm away from the edge of the Au pad (at -0.5).

ACKNOWLEDGMENTS

This work was supported by the U.S. Department of Energy (DOE), Solid State Energy Conversion Alliance (SECA). Use of the Advanced Photon Source was supported by the DOE, Office of Science, Office of Basic Energy Sciences, under Contract No. DE-AC02-06CH11357.

REFERENCES

1. M. Kuhn, S. Hashimoto, K. Sato, K. Yashiro, J. Mizusaki, Solid State Ionics, **195**, 7 (2011)
2. S. Hashimoto, Y. Fukuda, M. Kuhn, K. Sato, K. Yashiro, J. Mizusaki, Solid State Ionics, **186**, 37 (2011)
3. S.R. Bishop, K.I. Duncan, E.D. Wachsman, J. Electrochem. Soc., **156**, B1242 (2009)
4. M. E. Lynch and M. Liu, J. Power Sources, **195**, 5155 (2010)
5. H. Ullmann, N. Trofimenko, F. Tietz, D. Stöver, and A. Ahmad-Khanlou, Solid State Ionics, **138**, 79 (2000)
6. M. Backhaus-Ricoult et al., Solid State Ionics, **179**, 891 (2008)

Mater. Res. Soc. Symp. Proc. Vol. 1494 © 2013 Materials Research Society
DOI: 10.1557/opl.2013.237

Ruthenium Oxide Electrodeposition on Titanium Interdigitated Microarrays for Energy Storage

K. Armstrong [1], T.M. Dinh [2], D. Pech [2], M. Brunet [2], J. Gaudet[1] and D. Guay[1]

[1] Institut National de la Recherche Scientifique, 1650 Blvd Lionel-Boulet, Varennes, QC, Canada J3X 1S2

[2] Laboratoire d'Analyse et d'Architecture des Systèmes, LAAS-CNRS, 7 av du Colonel Roche, 31077 Toulouse Cedex 4, France.

ABSTRACT

The electrodeposition of hydrated ruthenium dioxide (hRuO$_2$) on Ti interdigitated current collectors deposited onto silicon substrate has been investigated with the objective of preparing a high capacitance and high power micro-supercapacitor (μ-SC) device. Ti current collectors were synthesised by typical photolithography processes, and hRuO$_2$ thin films were electrodeposited from ruthenium chloride precursors. Device specific capacitances exceeding 20 mF·cm^{-2} were obtained, and more than 80 % of that value is retained even at scan rate as high as 1 V·s^{-1} in 0.5 M H$_2$SO$_4$. The mean specific power per active surface area of the device is 368 mW·cm^{-2}. The device is stable and 90% of the initial capacity is retained after 10^5 cycles (1 V potential window). The characteristic response time of the hRuO$_2$ μ-SC is 250 ms, with low ESR (0.61 Ω cm^{-2}) and EDR (0.07 Ω cm^{-2}) values. All these characteristics demonstrate the potential of such μ-SC devices to be part of the next generation of micro-supercapacitors.

INTRODUCTION

The development of low-power integrated circuits (such as sensors, microprocessors or wireless communication chips) has made wireless sensor networks increasingly popular in numerous applications such as transportation, medicine, environment and security. Consequently, it has led to a growing demand for compact energy storage elements integrated in the vicinity of the electronic circuit they are powering [1]. Micro-supercapacitor (μ-SC) can achieve this role, with their reversible energy storage capability, fast charge/discharge rates and outstanding lifetime (over a hundred thousands of cycles) [2]. An interesting avenue to fabricate such storage devices is to use planar interdigitated micro-electrodes, fabricated using conventional photolithography techniques [3] onto a Si substrate on which a pseudocapacitive material can be deposited.

In this study, hydrated ruthenium dioxide (RuO$_x$.nH$_2$O, denoted hRuO$_2$ thereafter) has been selected as pseudocapacitive material due to its metallic conductivity, the reversibility of the electrochemical *redox* reactions involved during the charging/discharging processes, and its chemical and thermal stability [2]. The energy density of this material is hundreds of times higher than solid state or electrolytic capacitors [4]. Moreover, hRuO$_2$ is readily deposited onto various substrates of different shapes, since many versatile techniques have already been identified in the literature [5]. The price of ruthenium could be considered as an obstacle to its widespread use; however, the small amounts of active material used in micro-devices result in ruthenium being only a marginal fraction of the total cost of the device. All these characteristics make hRuO$_2$ an excellent candidate to fit the requirements previously mentioned for integrated compact energy storage elements.

The main challenge of next generation μ-SCs is to combine high energy and high power. In that respect, planar interdigitated micro-electrodes can contribute to maximize the power of a μ-SC device by lowering the cell constant [6]. Therefore, this study will describe how hRuO$_2$ can be prepared by pulsed electrodeposition from an aqueous RuCl$_3$.xH$_2$O solution with good spatial resolution on interdigitated Ti microarrays. It will be shown that this μ-SC device is characterized by a combination of high specific capacitance, very low characteristic response time and the ability to cycle at scan rates as high as 1 V·s^{-1}.

EXPERIMENTAL

A hRuO$_2$ based μ-SC was elaborated with a fixed interdigitated pattern, as shown in Fig. 1. The interspace, i, is 100 μm, the width, w, is 100 μm, the length, L, is 1000 μm, and the number of fingers, N, is 10. The microarrays used as substrates were prepared using conventional photolithography/lift-off processes. The current collector are made of 500 nm thick Ti. All other metallic surfaces are covered with an insulating layer (40 nm Si$_3$N$_4$ / 1 μm SiO$_2$). Electrodeposition of hRuO$_2$ was carried out from a deaerated aqueous chloride solution composed of 5 mM RuCl$_3$.xH$_2$O in 10^{-1} M KCl / 10^{-2} M HCl, aged for 15 days or more to allow quasi-equilibrium of the hydrolysis reaction to be reached at room temperature [7]. The solution pH was adjusted to 2.0 with a KOH aqueous solution and cooled down to 15.0 °C prior to electrodeposition. This solution was added to a 3 electrode deposition cell, with a counter-electrode Pt foil, and slowly stirred. A pulsed voltammetry electrodeposition technique has been chosen to synthesize hRuO$_2$. It consists of a 3 steps programmed potential loop (900 ms @ +0.35 V→ 150 ms @-1.00 V→150 ms at +0.90 V, all vs. SCE), repeated 600 times. After deposition, the μ-SC devices are rinsed and stored for 8 hours in an ultra high vacuum chamber (P ≈ 1 x 10^{-9} Torr). The electrochemical characterizations were performed in deaerated 0.5 M H$_2$SO$_4$ electrolyte at 22°C, using a 1480 potentiostat from *Solartron*, in a 2 electrodes configuration (one for each electrode of the microdevice). EIS measurements were carried out (*Solartron* 1255B frequency analyzer) with a ±10 mV oscillation around OCP, still in this 2 electrodes configuration. Frequencies were ranged from 10^5 to 10^{-2} Hz.

DISCUSSION

A typical hRuO$_2$ based μ-SC device is shown on optical microscope picture of Fig. 1a. It can be seen that a black deposit is formed (Fig. 1b), in good agreement with the literature [8]. There is no deposit visible in the interspace, a requirement to avoid current leakage between two interdigitated electrodes. The resistance between the two electrodes is larger than 15 MΩ. The deposit has a mud-cracked appearance (see Fig. 1b) and is adherent to the Ti microelectrodes.

The mud-cracked microstructure creates diffusion pathways for protons through the mesostructure and facilitates their access to the reaction sites [8].

Fig. 2a shows a typical cyclic voltammogram (CV) acquired at 1 V·s^{-1}. The featureless voltamogramm exhibits a highly symmetrical and rectangular shape, a signature of the numerous oxidation states that overlap each other in hRuO$_2$ [2]. The mean specific cell capacity is 20.3 mF·cm^{-2}, which is comparable to the values reported in literature [9, 10].

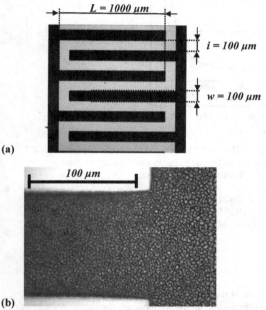

(a)

(b)

Figure 1. Optical microscope pictures of a hRuO$_2$ deposit on a 500 nm Ti interdigitated microarray with i = 100 μm, w = 100 μm, L = 1000 μm and N = 10 (of which only 6 are visible) with **(a)** a 625 X magnification and **(b)** a 6250 X magnification.

Fig. 2b presents the mean specific capacitances as a function of the scan rates obtained from cyclic voltammograms in 0.5 M sulfuric acid solution. It can be seen that over 90 % of the initial capacity is preserved at a scan rate of 1 V·s^{-1}. This indicates that most of the charge is readily accessible and that it is composed of a large fraction of "external" charge (in opposition to an "internal" charge) [11]. This results in a highly reversible electrochemical behavior for these devices, giving them the ability to charge/discharge at high rates.

In order to determine the specific power of the micro-device, it is essential to quantify the equivalent resistances of the devices. A pseudocapacitive material like hRuO$_2$ may be represented by an electrical circuit built of an ideal capacitor with an "equivalent series resistance" (ESR) and an "equivalent distributed resistance" (EDR).

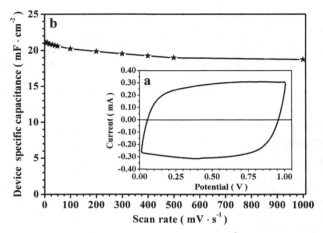

Figure 2. (a) Typical cyclic voltamogramm (1 V·s^{-1}) of a hRuO$_2$ μ-SC device. **(b)** Calculated mean capacitances per active surface area as a function of the scan rates during cyclic voltammetry in 0.5 M H$_2$SO$_4$, at 22°C, in a 2 electrodes configuration.

It was shown elsewhere [12] that a reduction of i reduces significantly the resistive contribution of the electrolyte, and thus the ESR. The maximum specific power, $P_{s,}$ per active surface area, A_{as}, is calculated according to (with ΔU the maximum operation potential window):

$$P_s = \frac{U^2}{4 \cdot (ESR + EDR) \cdot A_{as}} \tag{1}$$

Fig. 3a and 3b shows the *Nyquist* diagram of the device. At high frequencies (>10 kHz), the μ-SC device behave like a resistance. From middle to low frequencies, the imaginary part of the impedance increases and the plot tends to a vertical line characteristic of a capacitive behavior. Fig. 3c corroborated these observations. The ESR and the EDR values are evaluated as described in Fig. 3b. The ESR and EDR sum averages (3 identical cells) is 0.68 Ω·cm^2. Using this value, $A_{as} = 0.014$ cm^2 and a 1 V potential window, the mean cell specific power is 368 mW·cm^{-2}, which is amongst the highest value reported in the literature [13,14,15].

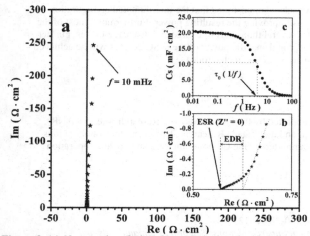

Figure 3. (a) *Nyquist* plot of a $hRuO_2$ μ-SC device obtained from EIS measurements (±10 mV) at U = 0V in a two electrode configuration. Frequencies were ranged from 10^5 to 10^{-2} Hz. **(b)** Enlargement of (a) near the intercept with the real axis for determining the ESR and EDR values. **(c)** Specific capacitances as a function of the frequency, showing how τ_0 is determined at half of the maximum specific capacitance.

Furthermore, the ability of μ-SC devices to charge and discharge at high rates was confirmed through the characteristic response time (τ_0), which value represents the minimum time needed to discharge all the energy from the device with an efficiency of greater than 50% [16]. A value τ_0 of 250 ms was determined, making it one of the best devices reported regarding this parameter [4].

Table I. Calculated mean cell parameters from EIS measurements. The values presented are the average of three different cells and normalized for cell active surface area (A_{as}).

Cell C_s @ 50 mV·s⁻¹ ($\pm$ 0.7 mF·cm⁻²)	ESR ($\pm$ 0.03 Ω·cm²)	EDR ($\pm$ 0.01 Ω·cm²)	Cell P_s ($\pm$ 21 mW·cm⁻²)	τ_0 ($\pm$ 30 ms)
20.3	0.61	0.07	368	250

Finally, every μ-SC device tested sustained at least 2×10^4 CV on a 1 V potential window at a scan rate of 500 mV·s⁻¹, while maintaining more than 90 % of their initial capacitance. In some case, over 10^5 cycles were done with the same retention of capacitance.

CONCLUSIONS

Electrodeposition of $hRuO_2$ onto Ti interdigitated microarrays has been investigated for the first time. Such devices possess a a) high specific capacitance which is mostly conserved at high scan rates, b) an almost purely capacitive behavior at low frequencies and c) a very short characteristic response time. It was also demonstrated that the interdigitated configuration effectively lower the cell ESR, thus improving the resulting power. Furthermore, most of the initial capacitance is retained after several thousands of charge and discharge cycles. Further improvements of the specific energy and specific power of the μ-SC devices will be achieved through modulation of the interspace, i, between electrodes.

ACKNOWLEDGMENTS

This work was financially supported by the French National Research Agency (ANR) through the MIDISTOCK project, and the Canada Research Chair program. Collaborations between INRS and LAAS were supported by the Commission Permanente de Coopération Franco-Québécoise (CPCFQ).

REFERENCES

1. Z. L. Wang, *Adv. Mater.* **24**, 280-285(2012).
2. B. E. Conway, *Electrochemical Supercapacitors: Scientific Fundamentals and Technological Applications* (Kluwer Academic/Plenum Publishing, New-York, 1999).
3. R. Salot *et al., Appl. Surf. Sci.* **256**, 54-57 (2009).
4. P. Simon and Y. Gogotsi, *Nature Mat.* **7**, 845-854 (2008).
5. C.D. Lockhande, D.P. Dubal, O.-S. Joo, *Curr. Appl. Phys.* **11**, 255-270 (2011).
6. W. Olthuis, W. Streekstra and P. Bergveld, *Sens. Actuator B* **24-25**, 252-256 (1995).
7. J. A Rard, *Chem. Rev.* **85**, 1-39 (1985).
8. C.-C. Hu, W.-C. Chang and K.-H. Chan, *J. Electrochem. Soc.* **151**, A281-A290 (2004).
9. Y.S. Yoon *et al., J. Vac. Sci. Technol. B* **21**, 949 (2003)
10. D.S. Tsai *et al., Electrochim. Acta* **55**, 5768 (2010).
11. P. Soudan, J. Gaudet, D. Guay, D. Bélanger and R. Schulz, *Chem. Mater.* **14**, 1210-1215 (2002).
12. D. Pech *et al., J. Power Sources*, submitted.
13. D. Pech, M. Brunet, Taberna, P.Simon, N. Fabre, F. Mesnilgrente, V. Conédéra & H. Durou, *J. Power Sources* **195**, 1266-1269 (2010).
14. D. Pech, M. Brunet, H. Durou, P. Huang, V. Mochalin, Y.Gogotsi, P.-L. Taberna and P.Simon, *Nature Nanotech* **5**, 651-654 (2010).
15. Y. Shao-Horn, *Appl. Phys. Lett.* **88**, 083104 (2006).
16. P. L. Taberna, P. Simon and J. F. Fauvarque, *J. Electrochem. Soc.* **150**, A292-A300 (2003).

Mater. Res. Soc. Symp. Proc. Vol. 1494 © 2013 Materials Research Society
DOI: 10.1557/opl.2013.545

Seebeck Coefficient of Nanolayer Growth of Anatase TiO$_{2-x}$/Al-foil by Atomic Layer Deposition

Matthew Chamberlin[1], Renee E. Ahern[1], Costel Constantin[1]
[1]Department of Physics and Astronomy, James Madison University, Harrisonburg, VA 22807

ABSTRACT

Non-stoichiometric and impurity doped titanium dioxide materials are good candidates for use in high temperature thermoelectric devices. Nanolayers of non-stoichiometric (TiO$_{2-x}$) thin films were deposited on Al-foil by atomic layer deposition growth method. X-ray diffraction experiments showed anatase phase for these nanolayers. This crystal structure was maintained even after an annealing treatment of 600 °C for 60 minutes under an O$_2$ pressure of ~ 10 psi. This investigation presents for the first time how Al-foil can be functionalized by manipulating the Seebeck coefficient of these TiO$_{2-x}$ nanolayers.

INTRODUCTION

Due to ever increasing demand for energy consumption and storage for mobile applications, thermoelectric devices that efficiently convert heat into electricity are of major importance. Non-stoichiometric TiO$_{2-x}$ is a good candidate material for thermoelectric applications [1,2]. Titanium dioxide thin films have been widely used in photocatalysts, dyesensitised solar cells, gas sensors, and selfcleaning components because of its low absorption coefficient, and high dielectric constant. TiO$_2$ is an insulator in its intrinsic state; however, the electrical properties of TiO$_2$ can be controlled by oxygen vacancies. In the present study, atomic layer deposition growth method was employed to create potential thermoelectric devices by using non stoichiometric TiO$_{2-x}$ nanolayers that were deposited on Al-foil that we use in our kitchens which is ~ 13,000 cheaper than Silicon.

EXPERIMENTAL DETAILS

Thin films of Titanium Dioxide (TiO$_{2-x}$) were deposited on top of aluminium foil (Al-foil) (i.e. Reynolds heavy-duty purchased from Walmart) by atomic layer deposition (ALD). Our homemade ALD reactor is illustrated in Figure 1(a) and (b). A very nice comprehensive review on the advances of ALD has been recently published by Parsons *et al.*[3]. The design of our ALD reactor can be found elsewhere [4]. The ALD reactor used titanium isopropoxide [i.e. Ti(OCH(CH$_3$)$_2$)$_{4(l)}$] as a metal precursor, water as an oxygen precursor, and high purity nitrogen (i.e. 99.99%) gas as a carrier/purging gas. The x-ray diffraction (XRD) experiments were performed with a PANalytical MPD Powder X-ray diffractometer (XRD) that uses a monochromatic x-ray beam with a wavelength of $\lambda = 1.542$ Angstroms.

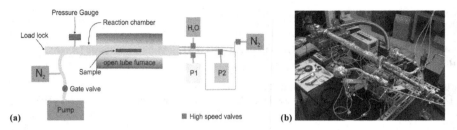

Figure 1. (a) Schematic diagram, and (b) photograph of the ALD reactor.

The Seebeck coefficient is defined as the negative ratio between the change in electromotive voltage and the change in temperature (i.e. $S = -\Delta V/\Delta T$). The S values were measured using a homemade instrument that is presented in Figure 2. The sample is set onto two blocks of Aluminum and the heating source is provided by a soldering iron through radiation [Figure 2]. The sample constitutes a layer of TiO_{2-x} thin film on top of Al-foil. To ensure that the temperature is transmitted from the Al blocks to the sample, we placed highly conductive Silicon based grease between the blocks and the sample. Two thermocouples K-type were imbedded into the Al blocks for measuring the difference of temperature between the hot and cold ends of the sample [Figure 2]. These thermocouples are imbedded into the Al blocks ~ 1.5 mm below the back of the sample. Our measurements show that there is no difference in temperature between sample surface and the embedded thermocouples. Two Nickel wires of gauge 30 were soldered on the surface of the substrates and the thermovoltage was measured by using a Keithley 2000 instrument [Figure 2].

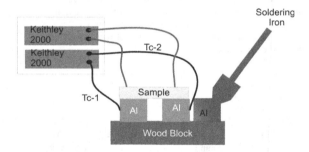

Figure 2. Homemade setup used to measure the Seebeck coefficients.

DISCUSSION

The chemical reaction that is taken place during the growth is presented in eq. 1.

$$Ti(OCH(CH_3)_2)_{4(l)} + 2H_2O \rightarrow TiO_{2(s)} + 4(CH_3)_2CHOH_{(g)} \quad (Eq. 1)$$

Essentially, in eq. 1, it can be seen that one part of Ti-isopropoxide (liquid) combined with two parts of water result in one part of TiO_2 (solid) and four parts of isopropyl alcohol (gas). Volume loss (%) as a function of Ti-isoporopoxide precursor temperature (°C) is shown in Figure 3. Before the deposition was performed with ALD, all the Al-foil substrates were first

272

cleaned with 10min (Acetone)/10 min (Isopropanol)/10 min DI water. The growth and the water precursor temperatures were kept constant at T_G = 250 °C and T_{H2O} = 40 °C. The precursor pulse times were t_{Ti-iso} = t_{H2O} = 2s, and t_{N2} = 5s. It is clear from Figure 3 that no growth was obtained for S4, on the other hand it has been observed an excessive growth behavior for sample S5. The volume consumption for S5 was V_L ~ 93%, which gave rise to a "flake-like" overgrowth of white-gray color. Although, we did not measure the thickness, it is believed that each individual flake had a thickness of more than few hundred nanometers. A volume consumption of V_L ~ 27% gave the best growth (i.e. sample S3) which had a thickness per number of cycle ratio of TCR ~ 0.12 nm/cycle for a total thickness of 84 nm. The TiO_{2-x} growth color of S3 was white non-transparent, which indicates a non-stoichiometric deposition with a ratio of Ti-atoms/O-atoms > 1 (i.e., Ti-rich behavior).

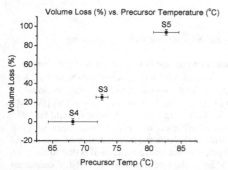

Figure 3. Total volume loss (%) vs. Titanium Isopropoxide Temperature.

In Figure 4, it is shown the XRD results obtained from S3 growth. The peaks at 2θ ~ 38.27° and 2θ ~ 44.5°, in Fig. 4, are from the Al-foil substrates and they represent Al(111) and Al(200) peaks, respectively. XRD measurements have been taken for sample S3 before and after annealing treatment—we exposed S3 to a temperature of 600 °C for 60 minutes under an O_2 pressure of ~ 10 psi. The peaks at 2θ ~ 25.15° and at 2θ ~ 47.85° observed for both annealed and not annealed samples, indicate the anatase phase for TiO_{2-x}. These peaks are in very good agreement with the standard spectrum (JCPDS no: 88-1175 and 84-1286) [5].

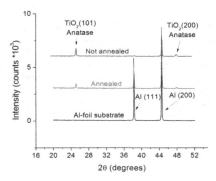

Figure 4. X-ray diffraction measurements for sample S3 with and without annealing.

In Figure 5 it is presented Seebeck coefficient measurements for both TiO_2/Al-foil (i.e. sample S3) and Al-foil. For a TiO_2 thickness of $\tau_{TO} \sim 84$ nm the Seebeck coefficient was $S_{TOA} = -23.30$ $\mu V/(°)$. The minus sign that is present in S_{TOA} indicates that the electro-thermal conduction is done through electrons. The Seebeck coefficient of Al-foil is a lot less [i.e. SA $= -3.51 \mu V/(°)$]. This result is very interesting in itself because of the following reasons. Firstly, it could be possible that what we measured was just the Seebeck coefficient of the TiO_2 layer, because it is well known that one can control the Seebeck coefficient through the manipulation of stoichiometry [6-8]. Secondly, it is that the Seebeck coefficient we measure it is due to the TiO_2 layer interaction with the top surface of Al-foil. Mainly, the electrons from the Al-foil encounter more scattering events, therefore, decreasing their drift velocity. According to Okinaka (*et al.*) [Ref. 6], the Seebeck coefficient of stoichiometric TiO_2 has been found usually in the range of 5.0 to 12.0 $\mu V/(°)$, and showed no particular dependence on temperature. On the other hand, non-stoichiometric $TiO_{1.1}$ gave large Seebeck coefficient values, with a general trend in which the value increased gradually from 400 $\mu V/(°)$ at 573 K to 1.0 mV/(°) at 1, 223K [6]. Our Seebeck coefficient is $\sim 2X$ the largest reported value mentioned above for the stoichiometric TiO_2, which implies that our sample S3 is non-stoichiometric. Although, we did not perform any electrical transport measurements, we believe that our sample S3 shows a metal behavior.

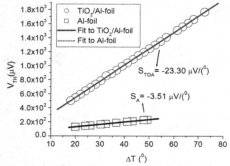

Figure 5. Seebeck Coefficients for TiO_2/Al-foil (open circles) and Al-foil (open squares).

CONCLUSIONS

The preliminary date presented in this study show that 84 nm of ALD grown TiO_{2-x} nanolayers on top of Al-foil produces a ~ 6.6 times increase in the Seebeck coefficient of just Al-foil. Although, the information presented in this paper represents just a snapshot of the research efforts undertaken by the authors. This study can be used as an initial effort to functionalize and use cheap materials such as Al-foil as great thermoelectric devices. Future experiments include careful stoichiometry measurements as a function of Seebeck coefficient of these TiO_{2-x} layers deposited on top of Al-foil. We will also measure the electrical transport properties of such non-stoichiometric nanolayers, so we can calculate the power factor of such materials.

ACKNOWLEDGMENTS

This collaborative work is supported by: Research Corporation Development Grant #7957, REU Grant DMR-0815367, and the Department of Physics and Astronomy at James Madison University. We also would like to thank Dr. Harry Hu for technical support.

REFERENCES

1. Grätzel, M. *Comments Inorg. Chem. 12*, 93–111 (**1991**).

2. Grätzel, M. *Nature 414*, 338–344 (**2001**). 3. G. N. Parsons, S. M. George, and M. Knez, *MRS Bulletin vol. 36, 865* (**2011**).

4. K. A. Vasquez, A. J. Vincent-Johnson, W. C. Hughes, B. H. Augustine, K. Lee, G. N. Parsons, G. Scarel, *J. Vac. Sci. Technol. A. 30, 01A105-1* (**2012**).

5. K. Thamaphat, P. Limsuwan, and B. Ngotawornchai, *J. (Nat. Sci.) 42, 357* (**2008**).

6. N. Okinaka, T. Akiyama, *ISIJ International 50, 1296* (**2010**).

7. T. Bak, and J. Nowotny, *J. Phys. Chem. 115, 9746* (**2011**).

8. M. Mikami, and K. Ozaki, *J. of Phys: Conf. Series 379, 012006* (**2012**).

Mater. Res. Soc. Symp. Proc. Vol. 1494 © 2013 Materials Research Society
DOI: 10.1557/opl.2013.810

Selective Etching of SnO₂:F Films with a Pulse Programmable Industrial Fiber Laser

M. Rekow[1], T. Panarello[1], W. S. Sampath[2]

[1]EOlite Systems, Inc. 275 Rue Kesmark, Dollard-des-Ormeaux, Québec H9B 3J1, Canada
[2]NSF I/UCRC for Next Generation Photovoltaics, Colorado State University, Fort Collins, CO, USA

ABSTRACT

In our work on laser scribing CdTe solar cells we have found what appears to be an unpublished laser material interaction that allows precise laser etching of SnO₂ films to an arbitrary thickness with high uniformity. This precise and efficient laser etching mechanism allows arbitrary reduction of the film thickness in a controlled manner on the scale of tens of nm. In addition to the fine depth selection, we find that there develops a pulse duration dependent microstructure on the surface. This micro microstructure results in a strong diffraction effect in the visible portion of the spectrum. In this work we propose a physical mechanism behind this novel depth selective laser interaction as well as the resultant micro-structure. Finally we demonstrate and propose some possible applications for this process.

INTRODUCTION

Fluorinated SnO₂ is one of the most common transparent conductive oxide (TCO) materials in use to today. Under the Pilkington brand name it is typically referred to as TEC series glass. Its primary use is in architecture glass as a heat reflector due to its low emissivity (Low-E) characteristics. Another large application is as an inexpensive substrate for CdTe solar cells and this material is under exploration for many energy production technologies. This SnO₂:F film is typically deposited on soda lime glass in a layer from a few hundred Å to 1 μm in thickness depending on the application. To make a device based on these TCO films typically the film must be patterned by some means. For example, in the case of scribing for CdTe solar cells the film is removed in periodic thin strips along the length of the substrate to create a series of electrically isolated pads. Likewise other applications generally required patterning of electrically isolated structures.

This work was originally aimed at demonstrating a novel way to perform the P1 scribe for CdTe solar cells. The intent was that the conductive layer could be removed while preserving the underlying insulating SiO₂ and intrinsic SnO₂ layers that exist as standard parts of the film deposition. These layers would then form a barrier to Na diffusion from the glass to the semiconductor. Our original work did show that the SiO₂ layer was preserved, but subsequent tests indicated that it did not significantly reduce the rates of Na diffusion [1]. The aim of this research then shifted to understanding this novel ablation mechanism and identifying other potential applications. Examples could include direct writing of holographic images, writing passive electronic components, and creating structures for active transparent electronic devices and the like [2]. In addition to precise thickness reduction we observe pulse duration and beam polarization dependent microstructure that results in a strong diffraction effect. This microstructure imparts apparent color to the film surface that can be tuned by controlling the pulse duration.

There are two mechanisms of laser ablation with pulse lasers that are commonly encountered in laser processing of materials [3]. The first mechanism generally consists of delivering enough energy into a volume of material that the atoms that make up the material become ionized. With the outer electrons stripped away the material lattice falls apart and repulsive electrostatic forces drive the remaining material away from the target surface. A second mechanism is inducing a phase change from solid to liquid and then from liquid to gas. The explosively expanding vapour then removes any remaining liquid material in an explosive manner. Both of these mechanisms have characteristic traits such as a ridge of recast material at the edge of an ablation pit as well as a pit topography that is reminiscent of the intensity profile of the beam (typically Gaussian). In our work on SnO_2:F we have observed an ablation pit morphology that is fundamentally different in character compared with these more well known processes of ablation. Namely, we find a pit profile with a flat bottom, no recast material, and a depth that appears to depend only on pulse duration and shows little dependence on the laser spot profile or even the energy contained within the pulse.

Our observations of this unusual ablation mechanism leads us to hypothesize that this laser ablation process is not laser ablation in the classical sense, but rather a laser induced chemical material transformation.

EXPERIMENTAL DETAILS

For this work we utilize the PyroFlex™-25 – 1064/532 nm pulsed fiber laser from EOlite Systems Inc. This laser provides up to 25 Watt of average power at 1064nm and up to 10 Watt at 532 nm. Unlike more traditional lasers, this fiber laser technology allows pulse durations to be varied from approximately 1 to several 100's of nanoseconds with 1 ns resolution. Furthermore the laser system allows independent control of the pulse amplitude ns by ns which allows the construction of arbitrary temporal pulse shapes. These pulse shapes are achieved independent of laser repetition rate, which can be varied up to 500 kHz. This temporal pulse shaping was key to developing this process.

The 1064 nm laser beam was expanded and collimated to a diameter of about 3 mm. It was then directed into a HurryScan II galvo scanning system with a 100 mm focal length f-θ telemetric scan lens giving a beam diameter on the work surface of about 50 μm. Scan speed and laser pulse repetition frequency (PRF) were varied depending on the type of experiment being performed, however typical values were 1 m/s scan speed and 20 kHz PRF. The substrate was oriented so that the laser beam was first incident on the TCO film and the distance between the focusing lens and the sample was adjusted so that the beam waist was coincident with the top surface of the sample.

The material used in this experiment was standard TEC 10 glass from Pilkington cut into 2.5"x3.0" squares. The film stack structure is shown in Figure 1 [4]. For the depth selective ablation described here the beam impinges from the film side of the substrate as shown.

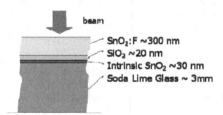

Figure 1: Multi-layer stack structure typical of TEC15 glass. The SnO_2:F layer is responsible for the high electrical conductivity. The TEC10 glass used in these experiments has a similar structure but with a thicker SnO_2:F layer.

This work utilized the following metrology equipment:

1. Optical microscope up to 1000x – Nikon OPTIPHOT-100
2. White light surface profilometer – Zygo NEWVIEW-600

RESULTS

Dependence on wavelength and substrate orientation

The laser pulse programmability was used to make an initial survey of the available laser parameter space. Table 1 below shows the initial experimental matrix. The data indicate that scribe depth control could not be achieved for both 532 and 1064 nm with the beam impinging on the glass side of the substrate. We find that under this scenario even very low pulse energies remove the film stack in its entirety. Shorter pulse durations tended to result in pitting of the glass substrate and longer pulses tend to result in the development of micro-cracks in the glass substrate. Figure 2 shows the result of single, 15 μJ, 2 ns pulse interactions with the film stack on TEC10 glass. In contrast with glass side processing, film side processing gave promising results for 532 nm and even more interesting results for 1064 nm.

Table 1: Preliminary experimental matrix.

Pulse Duration 2, 25, 50, 100, 200, 400 ns	Pulse Energy
532 nm Film Side	5 - 50μJ
532 nm Glass Side	5 - 50μJ
1064 nm Film Side	5 - 50μJ
1064 nm Glass Side	5 - 50μJ

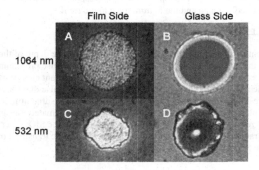

Figure 2: Single 15 μJ, 2ns pulse-film interaction morphology. A) 1064 nm film side interaction, pit depth stops well short of glass substrate. B) 1064 nm glass side interaction, pit depth extends to and stops at glass interface. C) 532 nm film side interaction, pit stops near glass interface, SnO_2:F layer appears to be explosively removed. D) 532 nm glass side interaction, SnO_2:F layer is explosively removed and a conical pit has formed in the glass substrate itself.

Pit morphology for 1064 nm film side ablation

At 1064 nm we observed two intriguing film side results. First, we find that at short pulse durations a single pulse appears to remove a very specific depth of the SnO_2:F layer. Secondly the bottom of the resulting ablation pit is flat in spite of the fact that the spot irradiance profile is nominally Gaussian. Figure 3 and Figure 4 illustrate these phenomena showing difference in ablation pit depth and appearance for three pulses of identical energy and different durations. Based on these observations, we speculated that removal of the SnO_2:F with 1064 nm is a time dependent phenomenon that may permit very accurate etching of the SnO_2:F layer by controlling the duration and the temporal shape of the laser pulse [5].

Figure 3: Appearance of pit bottom with increasing pulse duration. Left, 2 ns pulse shows granular morphology, 4 ns (middle) and 6 ns (right) show what appear to be "standing wave" patterns. The diameter of the left pit is about 25 μm.

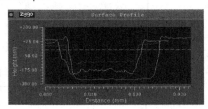

Figure 4: Profilometry indicating that the bottom of the pit is substantially flat in spite of the Gaussian irradiance profile of the beam. The top most trace corresponds to the left image of Figure 3, the middle trace to the middle image and the bottom trace to the right.

Pit depth as a function of pulse duration

The time dependent nature of this removal process implies that the depth of the pit can be controlled by controlling the pulse duration. To test this we utilize the pulse programmability of the PyroFlex™ laser to create a series of rectangular laser pulses of constant energy with duration increasing in 1 ns increments. Figure 5 shows pictorially how the laser pulse duration was changed for the experiment. The resulting pit "etch" depth for each pulse was measured utilizing the Zygo surface profilometer and plotted in Figure 6. The pit depth grows approximately linearly with pulse duration until the process reaches the interface between the SnO_2:F and the SiO_2 layer. At the interface there is a clear transition and reduction in etching rate. The slope transition matches the published thickness of the SnO_2:F layer which is typically about 370 nm.

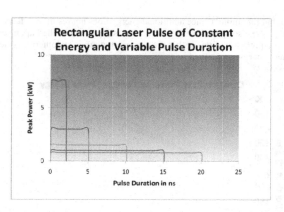

Figure 5: Cartoon depicting how the rectangular laser pulse duration was varied at constant total pulse energy for generating the data in Figure 6.

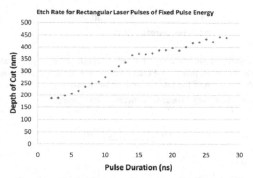

Figure 6: Pit depth as a function of pulse duration for rectangular shaped pulses. Note the change in slope at the SiO_2 / SnO_2:F interface.

Pit depth as a function of pulse energy

As the next step in this study we characterized the evolution of pit formation with time for both fixed peak power and fixed pulse energy. For a square temporal pulse shapes (as in Figure 5) of constant peak power, the total pulse energy increases in proportion to the pulse duration. Therefore the red trace (■) in Figure 7 represents a 5 fold increase in the pulse energy with pulse duration from 2 to 10 ns (15 to 75 uJ). The green trace (▲) on the other hand is fixed at 15 uJ for all pulse durations. This test gave a peculiar result. As Figure 7 clearly demonstrates, the pulse energy has little if any impact on the depth of the pit. The overall pulse duration appears to be the only important variable for determining the depth of the resulting pit.

Figure **8**, however, illustrates that the diameter of the pit is quite strongly dependent on total energy and only weakly dependent upon peak power, implying a threshold-based process in energy density (J/cm^2). The implication of this data is that the process is initiated by crossing an energy density threshold, but rate limited since additional energy does not result in a deeper pit or change in the pit morphology. We also speculate that the process is rapidly quenched upon cessation of the laser pulse. One interpretation of this is that the process at work is endothermic since it terminates as soon as the incident energy is removed.

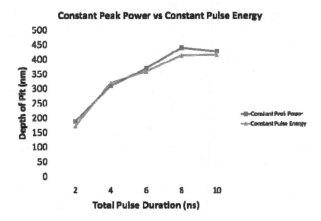

Figure 7: Impact of peak power and pulse energy on depth of single ablation pits. Data indicates that the process is independent of peak power and total energy over the range examined. The only variable that correlates with the depth of the pit is overall pulse duration.

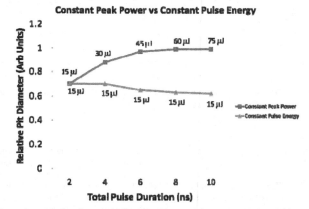

Figure 8: Effect of energy density on pit diameter. Process shows strong dependence on total energy density but only weak dependence on peak power.

Energy threshold for pit depth growth

Since we observed that once initiated this material removal process proceeds at a constant rate independent of pulse energy we postulated that reducing the energy content of the pulse after initiation could yield a valuable insight with regard to the physical mechanism at work. As a means to test this postulate, we constructed a set of "chair" shaped laser pulses with the goal to find the minimum peak laser power required to sustain the interaction. With a fixed amount of pulse energy, half of it always in the leading 2 ns of the pulse and half always in a fixed intensity tail of variable duration, we performed a series of experiments depicted in Figure 9 (bottom) over a range of total pulse energies. Figure 9 (top) illustrates the how the "chair" shape of the laser pulse was varied for each of the experiments. As before, we examined the resulting pits from the interaction of a single laser pulse with the substrate.

We confirmed that for the 2 ns pulse with no tail, the pit depth was independent of pulse energy (trace 1). The addition of a 4 ns tail (Pulse 2), shows an increased pit depth with only weak and slightly inverse energy dependence (trace 2). Laser pulse #3, with a 2 ns leading edge followed by an 8 ns tail, shows a puzzling inverse relationship to pulse energy. The depth of the ablation pit actually decreases substantially with pulse energy. At this point we do not have an explanation for the physical mechanism behind this behavior. For laser pulses 4 and 5 with the longest tail durations we see that all the material has been removed in spite of the fact that the peak energy in the tail has dropped to only 12% and 6% of the initial peak power respectively. Overall, this behavior points to a material removal process that, once initiated, requires very little energy input to keep it going.

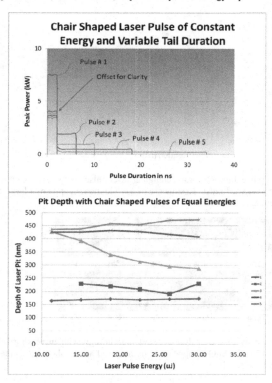

Figure 9: (Top) Cartoon depicting 5 different chair shaped pulses. (Bottom) Resulting pit depth from single laser pulse-film interactions for the chair shaped laser pulses depicted in the left figure.

To compare the chair shape pulse result with the rectangular laser pulse shape result, we modified the pulse duration vs pit depth test shown in Figure 6. In this experiment we used a "chair" pulse with a "seat" that was only 3% as high as the initial 2 ns leading edge and we varied the duration of the chair "seat" with a fixed chair "back" of 2 ns. Total energy was allowed to increase as the "seat" was lengthened, with a total increase in energy of 30% as the "seat" was increased from 2 to 22 ns. Figure 10 shows the comparison between the rectangular pulse shape result and the chair pulse shape result. Figure 11 shows how the morphology of the ablation pit evolves as the "seat" is lengthened. The striking feature is that despite significant differences in the distribution of energy between the "back" and the "seat", the pit depth increases in a systematic fashion. Of special interest is the fact that the SiO₂ and SnO₂ layers are reached at consistent overall pulse duration. Again this suggests that the material removal mechanism at play is rate limited and requires activation for about 15 ns to remove all SnO₂:F down to the SiO₂ layer. Furthermore, once activated, the removal mechanism requires only about 3% of the initial activation peak power to be sustained. It is our supposition that this is a thermally initiated chemical reaction that is either endothermic or perhaps slightly exothermic.

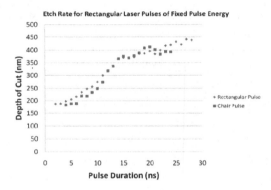

Figure 10: Etch depth vs. pulse duration for a rectangular laser pulse compared to etch depth vs. "seat" duration for a "chair" shaped pulse. Results are nearly identical in spite of a large difference in peak power and total energy delivered.

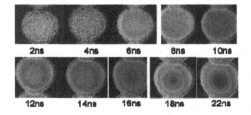

Figure 11: Optical microscopy of the evolution of pit morphology with increasing "chair" pulse "seat" duration. At about 12ns the SnO₂:F appears to be removed and at about 18 ns the remaining SiO₂ and SnO₂ layers are penetrated.

Origin of surface microstructure.

284

Another curious feature of this material removal mechanism is clearly seen in the pit morphology. Although the irradiance profile is spatially Gaussian, the bottom of pit is substantially flat with an appearance that changes from granular in texture to a texture that appears to have characteristics of a consistently aligned standing wave pattern. Other groups have reported similar standing wave patterns associated with ablation pits created with ps and fs laser sources [6], [7], [8] and the phenomenon appears to be universal across many different materials. The most common explanation appears to be that interaction of the beam with periodic or random features of the sample induce interference effects that ultimately modulate the laser intensity at the surface and hence create the microstructure. This does not seem a fully satisfactory explanation however since these materials are polycrystalline with no preferential lattice orientation and generally smooth on optical scales.

An insightful fact that we observe is that this pattern is dependent upon the polarization of the incoming beam. The ripple was found to orient itself perpendicular to the axis of polarization. This orientation dependence was observed both directly under a 1000x microscope as well as indirectly by observing the resultant angular dependency in diffractive effects (Figure 12). For left and right circular polarization we found ripple patterns that aligned at 45 degrees to the two orthogonal polarizations. This dependence on the electric field vector orientation has interesting implications. It may be possible that the periodic structure observed is due to resonance of the free electrons with the beam electric field. The electron oscillation may set up an out of phase electromagnetic field that forms a beat pattern with the incident beam and this modulates the intensity at the surface. Whatever the origin of this microstructure we observe that the periodicity results in a strong diffraction effect, so much so that the glass surface can be induced to have a colored appearance depending on viewing angle. Furthermore the apparent color, and by proxy, the mirco structure period is observed to be a function of pulse duration.

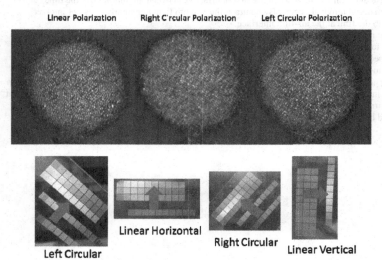

Figure 12: Orientation of surface microstructure depending upon polarization of the incoming beam (top). (Bottom) Arrays of ablation spots are written into 4 mm square boxes in order to create a bulk diffraction effect that imparts an apparent color to the film. The color effect is strongly dependent upon the orientation of the substrate relative to the viewer and the orientation of the optimal color effect is dependent upon the polarization beam used to create the pallet.

DISCUSSION

Implications of observations

At this point it is useful the recap the relevant features of this material removal mechanism. They are:

1. Pit depth is independent of pulse energy and peak power.
2. Bottom of pit is flat in spite of Gaussian irradiance profile.
3. Overall pit depth is a nearly linear function of pulse duration.
4. Pit diameter is a strong function of pulse energy but not peak irradiance.
5. Process appears to immediately halt up cessation of the pulse.
6. Once initiated, the process is sustained even when peak irradiance is reduced by 97%
7. Other than the appearance of surface microstructure, the bulk of the SnO_2:F seems to be unaffected by the process.

First of all we have observed that the material is highly transparent at 1064 nm, so much so that for a low power beam there is little noticeable absorption in the film. Furthermore, unlike the 532 nm result where the interaction appears to initiate at the interface between film and glass substrate, the 1064 nm interaction initiates at the outside surface and grows deeper into the SnO_2:F film with time. A clue to this behaviour can be found by comparing and contrasting the film side and the glass side results at 1064 nm as shown in Figure 2. We can see that in both cases the reaction appears to have initiated at first of the SnO_2:F surfaces the beam encounters. When the first surface is the interface between the glass and the SnO_2:F, the SnO_2:F film is completely removed at even the shortest pulse duration where as when the first interface is between air and SnO_2:F the time dependent material removal occurs. By contrast, the 532 nm light apparently does not interact strongly with the film at the air-film interface.

Next we must consider that the material is an N-type doped semiconductor [2] . The material is transparent to visible light because it has a very large band gap > 3.6 eV but at the same time is a conductor because of a populated conduction band. While we have no direct evidence, it seems reasonable to assume that high irradiance would tend to populate any inter-band defect states resulting in nonlinear absorption. Alternatively the photon energy at 1.17 eV of the laser at 1064 nm may excite intra-band transitions within the conduction band. Whatever the mechanism it seems that under high irradiance at 1064 nm there is significant absorption at the first SnO_2:F interface encountered by the beam.

SnO₂ physical and chemical properties

A calculation based on the enthalpy of formation for SnO_2 indicates that it should take about $4\mu J$ of energy to dissociate the volume of SnO_2 present in a single 25μm diameter pit from a single pulse interaction (Table 2). Recognizing that this calculation does not include the heat capacity, heat of fusion, or heat of vaporization, this is comparable with the actual energy (15μJ) delivered in the laser pulse. Furthermore, the reaction is endothermic which means as soon as the heat source is removed, the reaction will cease [9]. This also compares favorably with the fact that very little energy input is required to sustain the process and that the process terminates as soon as the energy input is discontinued. We believe that these observations suggest that a chemical dissociation of the SnO_2:F layer is occurring.

Table 2: SnO2 Thermodynamic Properties.

Enthalpy of Formation for SnO2 [9]	-577	kJ/mol
Typical Pit Volume (fully removed)	1.36E-10	cm^3
Density	6.95E+00	gm/cm^3
Mass	9.43E-10	gm
Molar Mass	150	gm/mol
Mole	6.29E-12	mol
Energy Released in Formation	4	uJ

The melting point of SnO_2 is reported as 1630 °C and the boiling point is reported to be between 1800 and 1900 °C with chemical dissociation reported at 1900 °C [13]. It is conceptually easy to view this as a thermally driven material dissociation. However, it is more difficult to understand why reaction proceeds uniformly in spite of the non uniform intensity profile of the irradiating beam. One clue comes through examining the material behavior in the far IR. The most common application for TEC glass is as far infrared reflectors for heat reflecting windows. This high reflectivity is known to arise due to interaction of far IR photons with electrons in the conduction band [10]. Furthermore higher electron mobility is associated higher electrical conductivity and with higher optical reflectivity at shorter wavelengths [11]. It reasonable to suppose that the intense oscillating electric field at the laser focus would have the effect of simultaneously promoting valence band electrons to the conduction band via defect states and promotion of conduction band electrons to more loosely bound conduction band states. However since we do not observe explosive ionization driven ablation we infer that ionization and ejection of electrons occurs only weakly or perhaps not at all.

Increased population of electrons in the conduction band and promotion of existing electrons to more loosely bound energy levels will increase the film conductance and mobility of the electrons. There are two clear effects that will occur as carrier concentration in the conduction band is increased and defect states are filled. First of all, optical absorption of the sample will increase [12]. However at the same time the reflectivity of the material should shift to shorter wavelengths as the higher mobility allows the electrons to respond resonantly to higher frequency electric fields. At this point we have no means to characterize these effects however we can analyze the observed material behavior and compare it to these two models.

As noted we observe that the laser interaction proceeds from the surface into the bulk of the material with no apparent damage to the underlying bulk material. If the immediate effect of the impinging laser energy was an increase in the absorption coefficient one would expect large energy deposition throughout the bulk of the film. As we have shown however, there is about 4 times more energy available in the laser pulse than would be required to completely dissociate the entire volume of the film yet what we observe is effectively a surface interaction that does not change its character even with very large changes in pulse energy. This observation is not consistent with the idea of bulk effects in the semiconductor.

Amalric-Popescu et al [12] showed that surface states and adsorbed species played a key role in determining the reflectivity and absorbtivity of SnO_2 films. Furthermore it makes sense that defect states would be more common on the surface of the material than in the bulk. Hence the surface of the film would respond more strongly to the incident radiation. The sudden increase in conduction band carrier concentration would then initiate a cascade that both increases absorption and reflectivity of the surface. This would reduce the energy density in the bulk thereby protecting it from damage and cause a rapid increase in temperature at the surface.

With these reference points a potential mechanism for this ablation mechanism becomes clear, namely that the surface interaction causes a chemical dissociation originating at the surface that progresses into the bulk as the duration of the interaction increases. Even with this mechanism it is difficult to understand how the pit depth can grow linearly independent of the intensity profile. If the transport mechanism is thermal conduction the reaction front would progress more quickly where the thermal gradient was highest but this is clearly not the case. The conclusion then is that this chemical process unfolds at a faster rate than the thermal conduction time.

Proposed material removal mechanism

We believe that the material removal mechanism can be encapsulated in the following model:

1. Defect states at the surface interact with the high intensity of the electric field to increase both the population and energy of the electrons in the conduction band.
2. Increased population of the conduction band increases the conductivity thereby increasing the reflectivity of the surface at shorter wavelengths. At the same time the increased temperature results in more surface absorption due to electron scattering.
3. These two effects reduce the intensity of the optical radiation in the underlying SnO_2 and increase the temperature at the surface.
4. The surface temperature rapidly reaches the thermal decomposition temperature for SnO_2 yielding Sn and O_2 gas.
5. Dissociation occurs at a faster rate than thermal conduction and the reaction front moves into the material faster than the rate of thermal conduction.
6. The reaction front velocity appears to be a constant that is independent of the total energy applied

CONCLUSIONS

This result clearly indicates that the depth of a laser pit in SnO_2:F films can be precisely controlled by simply controlling the pulse duration. The ability to precisely control the thickness of the TCO based on laser pulse duration alone offers interesting possibilities for TCO patterning. It offers a method to write passive electrical components into the film and may enable novel techniques for fabricating active devices. The interesting pulse duration dependent microstructure also has an obvious application in decorative glass marking. In order to test this possibility, arrays of spots were created with varying pulse duration. The resulting structured surface did indeed exhibit the expected diffractive effects with the apparent color of the surface showing a strong angular dependence. Figure 12 demonstrates the brilliant colors that were created on the surface of SnO_2:F coated glass by exploiting this micro structure.

We have not yet developed consistent physical quantitative model for the process at work in this laser-material interaction. However we believe that these observations suggest that a chemical dissociation of the SnO_2:F layer is occurring. It is therefore the reaction rate that regulates how rapidly the pit depth grows. While there is still more work to be done to understand this novel laser-material interaction, there are clearly interesting new possibilities to utilize this effect. Novel transparent electronic devices, energy production and storage, and display technologies are all areas where patterning of SnO_2 and other TCO films are routine. This novel laser material interaction promises to be a tool that can enable film structuring with a laser in a way not previously possible.

References

1. *Selective Removal of TCO Stack Layers for CdTe P1 Process with a Tailored Pulse Laser.* **M. Rekow, R. Murison, C. Dinkel, T. Panarello, S. Nikumb, W. S. Sampath.** Seattle : IEEE PVSC Proceedings, 2011.

2. **Wager, John F., Keszler, Douglas A., Presley, Rick E.** *Transparent Electronics.* s.l. : Springer, 2008. ISBN 978-0-387-72342-6.

3. **LIA.** *LIA Handbook of Laser Materials Processing.* Orlando : Magnolia Publishing Inc, 2001. 0-941463-02-8.

4. **K. VON ROTTKAY, M. RUBIN.** OPTICAL INDICES OF PYROLYTIC TIN-OXIDE GLASS. *Mater. Res. Soc. Symp.* 1996.

5. *Application of a Pulse Programmable Fiber Laser to a Broad Range of Micro-Processing Applications.* **Rekow, M, et al.** Anaheim : LIA, 2010.

6. *Modification of dielectric surfaces with ultra-short laser pulses.* **Costache, Florenta, Matthias Henyk, and Jürgen Reif.** 2002, Applied surface science, Vol. 186.1, pp. 352-357.

7. *Observations of higher-order laser-induced surface ripples on (111) germanium.* **Fauchet, P. M..Siegman, A. E.** 2, s.l. : Springer Berlin / Heidelberg, 11 1, 1983, Applied Physics A: Materials Science & Processing, Vol. 33. 0947-8396.

8. *Surface structures on crystalline silicon irradiated by 10 ps laser pulses at 694.3 nm.* **Lee, T. D., Lee, H. W.,Kim, J. K.,Park, C. O.** s.l. : Springer Berlin / Heidelberg, 1989, Applied Physics A: Materials Science & Processing. 10.1007/BF00619721.

9. **L. V. Gurvich, I. V. Veyts, C. B. Alcock.** *Thermodynamic Properties of Individual Substances: Part 1, Volume 2Elements C, Si, Ge, Sn, Pb and Their Compounds.* Moscow : Hemisphere Publishing Corporation, 1979. ISBN 0-89116-533-9.

10. **Mol, Antonius Maria Bernardus van.** Chemical Vapour Deposition. *Thesis.* s.l. : Technische Universiteit Eindhoven, 2003.

11. *Photoreflectance Probing of Below Gap States in Gan/Algan High Electron Mobility Transitor Structures.* **D. K. Gaskilla, O. J. Glembockia, B. Peresa, R. Henrya, D. Koleskea and A. Wickenden.** s.l. : MRS, 2002.

12. *Infrared studies on SnO2 and Pd/SnO2.* **D. Amalric-Popescu, F. Bozon-Verduraz.** 2001, Catalysis Today, Vol. 70, pp. 139–154.

13. *Thermal dissociation of compressed ZnO and SnO2 powders in a moving-front solar thermochemical reactor.* **Marc Chambon, Stéphane Abanades*, Gilles Flamant.** 8, s.l. : AIChE Journal, 2010, Vol. 57. DOI: 10.1002/aic.12432.

Mater. Res. Soc. Symp. Proc. Vol. 1494 © 2013 Materials Research Society
DOI: 10.1557/opl.2013.50

Growth of 130 μm Thick Epitaxial KNbO₃ Film by Hydrothermal Method

T. Shiraishi[1], H. Einishi[1], M. Ishikawa[1, 3], T. Hasegawa[2], M. Kurosawa[2] and H. Funakubo[1]

[1] *Department of Innovative and Engineered Materials, Tokyo Institute of Technology, J2-43, 4259*
  Nagatsuta-cho, Midori-ku, Yokohama, Kanagawa 226-8502, Japan
[2] *Department of Information Processing, Tokyo Institute of Technology, G2-32, 4259*
  Nagatsuta-cho, Midori-ku, Yokohama, Kanagawa 226-8502, Japan
[3] *Department of Clinical Engineering Faculty of Biomedical Engineering, Toin University of*
  Yokohama, 1614 Kurogane-cho, Aoba-ku, Yokohama, 225-8503, Japan

ABSTRACT

KNbO₃ thick films were deposited on $(100)_c$ SrRuO₃//(100)SrTiO₃ substrates at 240 °C for 3 h by hydrothermal method. Film thickness increased linearly with increasing the deposition number of times and 130 μm thickness was achieved by the 6 time deposition. XRD analysis showed the growth of epitaxial orthorhombic films with the mixture orientation of (100), (010) and (001). Cross-sectional SEM observation showed that the 130 μm-thick film was dense and no obvious voids inside the film. In addition, the crystal structure change along film thickness direction was not detected from the cross-sectional Raman spectral observation.

INTRODUCTION

Piezoelectric thick films are widely used, such as actuators of ink-jet printers and micro electro mechanical systems [MEMS] [1]. For variety applications, not only thin films, but also thick films on the order of submillimeter are required. In addition, taking into account of the strong crystal orientation dependency of the piezoelectric property, preparation of orientation-controlled films, especially epitaxial films, are highly required [2].

We concerned to the KNbO₃-based films as piezoelectric films because KNbO₃-based materials showed good piezoelectric property compared with other lead-free piezoelectric materials [3]. In fact, superior properties compatible to Pb(Zr,Ti)O₃ are reported for the KNbO₃-based ceramics [4]. Preparation of KNbO₃-based films has been reported for various methods including chemical vapor deposition, sputtering method and pulse lased deposition [5-7]. Although the epitaxial films are grown by these methods, thickness of the reported films is limited less than 10 μm. On the other hand, the aerosol deposition is suitable to prepare thick film, but the obtained films are polycrystalline one [8]. As a preparation method of the KNbO₃-based films, we pay attention to the hydrothermal method. This process can grow highly crystalline films including epitaxial films even at low temperature below 300 °C [9]. In fact, {100}-oriented epitaxial $KNbO_3$ thick films beyond 10 μm were grown at 240 °C on $(100)_c$ $SrRuO_3//(100)SrTiO_3$ substrates for 3 h by hydrothermal method [10]. These results suggest that the hydrothermal method is suitable preparation method for submillimeter-thick $KNbO_3$ film. In the present study, we report on the growth of {100}-oriented epitaxial $KNbO_3$ thick film over 100 μm by hydrothermal method.

EXPERIMENT

$KNbO_3$ films were deposited at 240 °C on $(100)_c$ $SrRuO_3//(100)SrTiO_3$ substrates by hydrothermal method. Deposition time was set to be 3 h because film was etched and the thickness decreased with the increase of the deposition time above 3 h [10]. The 50 nm - thick epitaxial $SrRuO_3$ layers were grown on $(100)SrTiO_3$ single crystal by sputtering method as the bottom electrode. 20 ml solution of 10 mol/dm^3 KOH (Kantoukagaku Co., Ltd.) and 9.4×10^{-4} mol of Nb_2O_5 powder (purity 99.9 %, Kantounkagaku Co., Ltd.) were put in an autoclave (PARR, 4748) as the source materials together with the $(100)_c$ $SrRuO_3//(100)SrTiO_3$ substrates.

Film thickness was measured by a scanning electron microscopy (SEM, HITACH S-4800) and a surface profile meter (Veeco DEKTAK 3ST). On the other hand, the deposition amount was obtained by the weight gain per unit area. Crystal structure and orientation of the films were investigated with X-ray diffractometry (XRD, Philips X'Pert MRD system). Raman spectroscopy was also used to evaluate the crystal structure from the cross section of the films. Backward scattering light was collected and dispersed by a subtractive single spectrometer using the 514.5 nm line from an Ar^+ laser. Pt top electrodes were deposited by electric beam evaporation to fabricate capacitor structures of Pt/ $KNbO_3$/$SrRuO_3$. The electric and ferroelectric properties were measured at room temperature using an impedance/gain-phase analyzer (Hewlett Packard, 4194A) and ferroelectric tester (Toyo Technica, FCE fast).

RESULTS AND DISCUSSION

Figure 1 shows the changes of the deposition amount and the film thickness with the deposition time. The deposition amount and film thickness linearly increased with increasing repeating the deposition time. The growth rate of $KNbO_3$ film was 10 μm per 1 time and the film thickness became about 130 μm after the 6 times deposition. It must be noted that all deposited films had the K/Nb ratio to be unity.

Figure 2 (a) shows the X-ray θ - 2θ patterns for the films measured at each time of the deposition. The series peaks assigned to be $\{100\}_c$ orientation was observed irrespective of the time of the deposition. $\{100\}_c$ denotes the (100) in pseudocubic notation. Figure 2 (b) shows the X-ray pole figure plots at a fixed 2θ angle corresponding to KNbO$_3$\{110\}$_c$ for 130 μm-thick KNbO$_3$ film. This film exhibited the four fold spots at an inclination angle of 45°, showing that 130 μm-thick KNbO$_3$ film showed in-plane alignment of the epitaxial film.

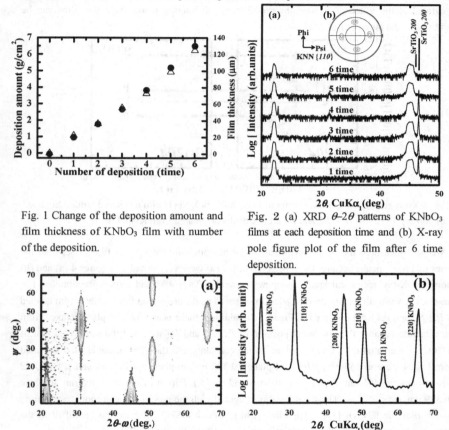

Fig. 1 Change of the deposition amount and film thickness of KNbO$_3$ film with number of the deposition.

Fig. 2 (a) XRD θ–2θ patterns of KNbO$_3$ films at each deposition time and (b) X-ray pole figure plot of the film after 6 time deposition.

Fig. 3 (a) X-ray 2θ-Ψ area mapping of 130 μm KNbO$_3$ thick film and (b) XRD θ-2θ patterns integrated along Ψ axis in Fig. 2 (a).

Figure 3 (a) shows the 2θ-Ψ area mapping of the 130 μm KNbO$_3$ thick film measured under ϕ-axis-rotation condition. This mapping leads us to know all diffraction peaks from the films. The details of the measurement methods are already described in Ref. 12. This mapping reveals *d*-spacing information along both the substrate surface normal and off-normal directions. The θ - 2θ pattern at ψ=0 correspond to the profiles shown in Fig. 2 (a). All peaks in Fig. 3 (a) can be identified as the $\{100\}_c$-oriented $KNbO_3$ film. Figure 3 (b) shows the θ - 2θ pattern integrated along Ψ axis. It was observed that all diffraction peaks from the films can be identified to the pure perovskite phase.

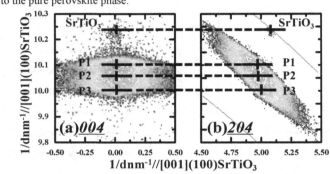

Fig. 4 X-ray reciprocal space mapping of 130 µm-thick $KNbO_3$ film measured around $SrTiO_3$ (a) *400* and (b) *240*. *P1*, *P2* and *P3* are the peaks originated from $KNbO_3$ film.

We next measured the X-ray reciprocal space mappings of the $KNbO_3$ film to separate the series of $\{100\}_c$ peaks because these peaks in Fig. 2 (a) were very broad . Figures 4 (a) and (b) show the X-ray reciprocal space mappings around $SrTiO_3$ *400* and *240* diffraction for 130 µm-thick $KNbO_3$ film, respectively. The three pared peaks originated from $KNbO_3$ film labeled as *P1*, *P2* and *P3* were observed in Fig. 4. In addition, these peaks were in-plane relaxed from the substrate because the horizontal position of *P1*, *P2* and *P3* were not the same with that of $SrTiO_3$. Calculated out-of and in-plane lattice constants and the internal angle of α obtained from Fig. 4 were as follows; *P1*(out-of-plane =0.398 nm, in-plane =0.402 nm and α=90°), *P2* (out-of-plane =0.401 nm, in-plane =0.401 nm and α=90°), *P3*(out-of-plane =0.405 nm, in-plane =0.399 nm and α=90°). These values almost agreed with the reported one at room temperature for rthorhombic $KNbO_3$ [11]. Therefore, 130 µm-thick $KNbO_3$ film can be identified as the orthorhombic symmetry with mixture orientations of (001), (010) and (100).

Figure 5 (a) shows the cross-sectional SEM image of 130 µm-thick $KNbO_3$ film. It was found that the dense film was obtained and there was no noticeable discontinuity irrespective of the repetition of the deposition. Figure 5 (b) schematically draws the measurement spots of the Raman spectroscopy along the cross section of the 130 µm-thick $KNbO_3$ film and the obtained spectrum are displayed in Fig. 5 (c). Spectrum numbers in Fig. 5 (c) correspond to those obtained at the marked points in Fig. 5 (b). Spectrum from the position (x) was different from the spectra from other positions because this position corresponded to the substrate. On the other hand, spectra from the position (i) to (ix) were basically the same and these were similar to the reported one for the orthorhombic KNbO₃ in good agreement with the results of XRD analysis [13]. Almost uniform spectra from the position (i) to (ix) shows that the relatively uniform films with similar crystallinity and the remained strain along the film thickness direction were ascertained to be obtained. These results also suggest that similar quality KNbO₃ was deposited irrespective of the deposition time.

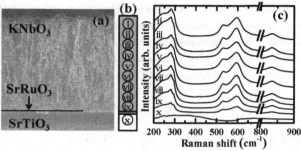

Fig. 5 (a) Cross-sectional SEM image of 130 μm-thick KNbO₃ film, (b) schematic drawing of SEM image and (c) Raman spectra at each position shown in (b).

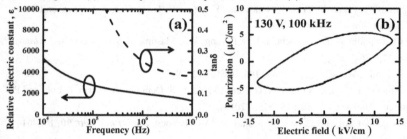

Fig. 6 (a) Frequency dependencies of the relative dielectric constant and tan δ and (b) P - E hysteresis loop of 130 μm-thick KNbO₃ film.

Figure 6 (a) shows the frequently dependencies of the relative dielectric constant and tan δ measured at room temperature for 130 μm-thick KNbO₃ film. The relative dielectric constant decreased from 5000 to 1000 with increasing frequency from 10^4 to 10^7 Hz. In addition, it showed relatively large dielectric loss, especially at low frequency. Figure 6 (b) shows the polarization- electric filed (P - E) hysteresis loop measured at room temperature and 100 kHz for 130 μm KNbO₃ thick film. The hysteresis loop was observed but included the contribution from the leakage. This is also suggested by the larger relative dielectric constant and tan δ at low frequency in Fig. 6 (a). Improvement of the leakage is under investigation by the post annealing process as already reported by other group [14].

CONCLUSIONS

The 130 μm-thick $KNbO_3$ film was deposited at 240 °C on $(100)_c$ $SrRuO_3$ //(100)$SrTiO_3$ substrates by the hydrothermal method. Thickness of $KNbO_3$ films was increased by repeating the deposition time. Epitaxial orthorhombic films with (100)/(010)/(001) orientations were obtained. Cross-sectional SEM observation showed that the 130 μm thick film was dense without obvious voids. In addition, the crystal structure change along film thickness direction was not detected from cross sectional Raman spectral observation.

REFERENCES

1. R.A. Dorey and R.W. Whatmore, *J. Electroceram.* **12**, 19 (2004).
2. C. S. Park, S. W. Kim, G. T. Park, J. J. Choi and H. E. Kim, *J. Mater. Res.* **20**, 243 (2005).
3. T. R. Shrout and S. J. Zhang, *J. Electroceram.* **19**, 111 (2007).
4. Y. Saito, H. Takao, T. Tani, T. Nonoyama, K. Takatori, T. Homma, T. Nagaya, and M. Akamura, *Nature* **432**, 84 (2004).
5. S. Chattopadhyay, B. M. Nichols, J. H. Hwang, T.O. Mason and B.W. Wessels, *J. Mater. Res.* **17**, 275 (2002).
6. J. Wu and J. Wang, *J. Appl. Phys.* **106**, 066101 (2009).
7. A. Rousseau, V. Laur, M. G. Viry, G. Tanné, F. Huret, S. Députier, A. Perrin, F. Lalu and P. Laurent, *Thin Solid Films* **515**, 2353 (2006).
8. J. Ryu, J. J. Choi, B. D. Hahn, D. S. Park and W. H. Yoon, *Appl. Phys. Lett.* **90**, 152901 (2007).
9. T. Shiraishi, H. Einishi, S. Yasui, M. Ishikawa, T. Hasegawa, M. Kurosawa, H. Uchida, Y. Sakashita, H. Funakubo, Jpn. *J. Appl. Phys.* **50**, 09ND11-1-4 (2011).
10. M. Ishikawa, K. Yazawa, T. Fujisawa, S. Yasui, T. Yamada, T. Hasegawa, T. Morita1, M. Kurosawa, H. Funakubo, Jpn. *J. Appl. Phys.* **48**, 09KA14-1-5 (2009).
11. A. W. Hewat, *J. Phys. C* **6**, 2559 (1973).
12. K. Saito, M. Mitsuya, N. Nukaga, I. Yamaji. T. Akai and H. Funakubo, *J. Appl. Phys.* **39**, 5489 (2000).
13. J.A. B. Saip, E. R. Moor and A.L. Cabrera, *Solid State Commun.* **135**, 367 (2005).
14. A. D. Handoko, G. K. L. Goh and R. X. Chew, *Cryst. Eng. Comm.* **14**, 421 (2011).

Devices and Applications

Mater. Res. Soc. Symp. Proc. Vol. 1494 © 2013 Materials Research Society
DOI: 10.1557/opl.2013.102

Low Temperature ZnO TFTs Fabrication with Al and AZO Contacts for Flexible Transparent Applications

Gerardo Gutierrez-Heredia[1,2], Israel Mejia[1], Norberto Hernandez-Como[1], Martha E. Rivas-Aguilar[1], Victor H. Martinez-Landeros[1,2], Francisco S. Aguirre-Tostado[2], Bruce E. Gnade[2] and Manuel Quevedo[1,C].
[1]Department of Materials Science and Engineering, University of Texas at Dallas, USA.
[2]Centro de Investigacion en Materiales Avanzados S. C., Campus Monterrey, Mexico.

ABSTRACT

Zinc Oxide (ZnO) Thin-Film Transistors (TFTs) using Aluminum (Al) and Aluminum-doped zinc Oxide (AZO) as Source-Drain (S-D) contacts are reported. The fabrication process was carried out using five photolithography steps with a maximum processing temperature of 100 °C, which makes the process compatible with flexible/transparent applications. The AZO and ZnO films were deposited using Pulsed Laser Deposition (PLD). Aluminum was deposited using e-beam. The devices showed mobilities >10 cm^2/V-s, threshold voltage in the range of 7 V and On/Off current ratios >10^5. The resistance analysis showed that AZO is a better contact with lower contact resistance as identified in the TFTs. The AZO and ZnO stacks characterized by UV-V shows an optical transmission >80 %.

INTRODUCTION

Compatibility with low temperature processing is one of the requirements for metal-oxide based devices used for transparent applications such as displays, health sensors, photodetectors and solar cells [1,2,3,4]. Low temperature process enables large area applications while maintaining lower fabrication costs compared to Silicon (Si) technologies.
Ideally, semiconductor materials with a wide bandgap (E_g > 3.0 eV) are required for transparent electronics [5,6,7]. Among those semiconductors, ZnO (E_g ~ 3.3 eV) has been proposed as a potential candidate given its compatibility with both requirements: low temperature processing and suitable with large area substrates, which are strict for flexible electronic applications. In addition, ZnO films can achieve Hall mobilities of ~200 cm^2/V-s [8]. Besides the semiconductor, Transparent Conductive Oxides (TCO) are required as contacts for transparent electronics. A promising TCO is AZO, which is a candidate to replace Indium Tin Oxide (ITO). Replacing ITO with AZO can potentially result in inexpensive devices while eliminating the need for a scarce element, such as indium [9].
In this paper, we demonstrate a low temperature photolithography process to fabricate ZnO-based TFTs. The entire fabrication process is carried out at a maximum temperature of 100 °C. In addition, detailed contact resistance analyses demonstrate that ZnO/AZO interface is more stable, resulting in lower contact resistance compared with the ZnO/Al system.

EXPERIMENTAL DETAILS

The ZnO TFTs were fabricated on 500 nm of thermally grown SiO_2. 100 nm of Chrome (Cr) was deposited by e-beam and patterned using wet etching processing to pattern the gate contact. Next, 90 nm of Hafnium Oxide (HfO_2) was deposited by Atomic Layer Deposition (ALD) at 100

°C. 45 nm of ZnO was then deposited by PLD followed by 500 nm of parylene-C (poly-p-xylylene) deposited by chemical vapor deposition (at room temperature) as hard-mask. The parylene hard-mask was patterned using Reactive Ion Etching (RIE) and the ZnO film was selectively removed using wet etching methods. Finally, RIE was used to open vias that define the S-D contacts. The S-D contacts were deposited using e-beam for Al (100 nm) and PLD for AZO (200 nm).

The ZnO and AZO PLD deposition was carried out at 100 °C. First, the deposition chamber was evacuated to 1×10^{-6} Torr and then backfilled with O_2 to reach the desired deposition pressure (30 mTorr for ZnO, 1 mTorr for AZO). A KrF excimer laser ($\lambda = 248$ nm) at a frequency of 10 Hz with an energy of ~1 J/cm^2 was used for the deposition with a target to substrate distance of 6.5 cm. The deposited films with this process have thicknesses of 50 nm for ZnO and 200 nm for AZO. Figure 1a shows the schematic cross-section for the resulting TFTs using the fabrication process described above. Figures 1b and 1c are optical images of the TFTs with Al and AZO contacts, respectively. TFTs with W= 40 µm and L= 20, 40 and 80 µm were evaluated.

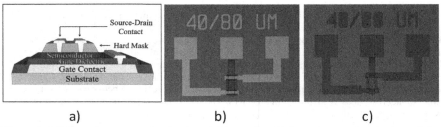

Figure 1.a) Schematic cross-section for the resulting TFTs. **(b)** Optical image for aluminum S-D contacts, and **(c)** Optical image for AZO S-D contacts.

DISCUSSION

Figure 2 shows the transmittance results for the HfO$_2$/ZnO stack and AZO in the 200-1000 nm wavelength range. The stacks showed ~82 % transparency at 555 nm, which corresponds to the wavelength where the human eye has the maximum sensitivity. The average transparency was 80 % in the visible range (390-750 nm). Bandgap extracted from these measurements were 3.2 and 3.5 eV for ZnO and AZO, respectively.

The TFTs were characterized with a Keithley 4200 semiconductor characterization system under dark and regular ambient conditions. Figure 3a shows the transfer curve for ZnO TFTs using Al and AZO as S-D contacts. TFTs dimensions are W = 40 µm and L = 20 µm. Both transistors show On/Off current ratios in the order of 10^5. Threshold Voltage (V$_{TH}$) was extracted from the plot of V$_{GS}$ vs. Sqrt (I$_{DS}$) and extrapolating a linear fit to the X axis. The slope of this fit was used to calculate the µ$_{FET}$ using the following equation 1.

$$I_{DS} = \left(\frac{C_i \mu_{FET} W}{2L} \right) (V_{GS} - V_{TH})^2 \qquad (1)$$

Where I$_{DS}$ is the measured current through S-D, W is the channel width, L is the channel length, C$_i$ is the dielectric capacitance density, and V$_{GS}$ is the voltage applied between the gate and source contacts. For V$_{DS}$=20 V (saturation), the ZnO TFTs with Al as S-D contact presented V$_{TH}$

~7 V and μ_{FET}=8.4 cm^2/V-s. For ZnO with AZO S-D, V_{TH} is ~7.5 V and μ_{FET}=10.1 cm^2/V-s. It is important to note that the model used for the mobility extraction does not consider the contact resistance (R_C) between the semiconductor and the S-D contacts. This effect is further discussed below in this paper. Output curves for both TFTs are shown in Figure 3b. The drain current for both TFTs are similar, but a slightly higher current for AZO S-D was observed. This was expected, given its slightly higher mobility.

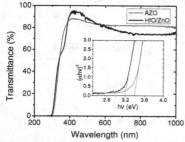

Figure 2. Transmittance of AZO and HfO$_2$/ZnO, Inset: band gap extracted from the transmittance spectra.

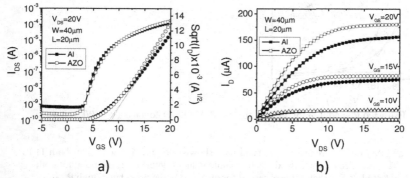

a)   b)

Figure 3a) Transfer and **b)** Output curves of ZnO TFTs using Al and AZO as S-D contact.

In order to further study the contact resistance in these devices, the total resistance (R_T) was extracted from the TFTs using V_{DS}=1 V at different V_{GS} (7.5, 10, 12.5, 15, 17.5 and 20 V). The total resistance in a TFT is due to contributions from R_C and channel resistance (R_{CH}): R_T=R_C+R_{CH}. Therefore, by plotting R_T vs. L, the R_C can be extracted by extrapolating to L=0. R_{CH} is a value known by subtracting R_T and R_C.

Figures 4a and 4b show R_T vs. L for Al and AZO S-D, respectively. TFTs with Al contacts show higher R_T than AZO TFTs at low V_{GS}. Furthermore, TFTs with Al show a non linear dependence R_T vs. L, which is likely due to an unstable Al/ZnO interface. Such instability likely results in Al oxidation (Al$_2$O$_3$), which increases R_C as reported by Kim et al [10]. Unlike Al, the

AZO film shows a much more stable behavior for the R_T vs. L plot, due to the material stability as previously reported [11].

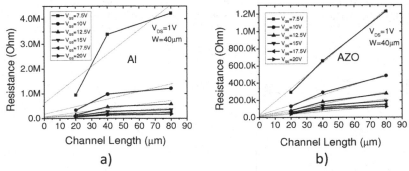

Figure 4. R_T extracted from ZnO TFTs with **a)** Aluminum and **b)** AZO as S-D contacts.

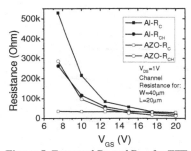

Figure 5. Extracted R_C and R_{CH} for TFTs with Al and AZO S-D contacts (W=40/L=20 μm).

The R_C and R_{CH} vs. V_{GS} trends for AZO and Al is shown in Figure 5. The R_{CH} for both TFTs are similar. This is expected since the dielectric-semiconductor interface is the same for both contacts. For AZO low R_C (that Al) is observed for all V_{GS}. On the other hand, high R_C is observed at low V_{GS} for Al contacts. However, at high V_{GS} (>12 V) the R_C is in the order of the R_{CH}. This indicates that Al_2O_3 is likely forming at the Al/ZnO interface.

CONCLUSIONS

Fabrication and electrical analysis of ZnO TFTs fabricated using Al and AZO for S-D contacts was shown. The entire fabrication uses temperatures <100 °C and photolithography techniques. Transmittance of the film stacks show transparency >80 % in the visible range.

The study of the ZnO TFTs with either Al or AZO indicates that AZO forms a better interface with ZnO, resulting in lower R_C. Al contacts with ZnO form a thin layer of Al_2O_3 which originates a high R_C for low V_{GS}. In summary, AZO represents an alternative indium-free TCO and is compatible with low temperature processing.

REFERENCES

1. S-H. K. Park, C-S. Hwang, M. Ryu, S. Yang, C. Byun, J. Shin, J-I. Lee, M. S. Oh and S. Im, Adv. Mater.Vol.21, 6, 678-682 (2009).
2. C. H. Park, S. Im, J. Yun, G. H. Lee, B. H. Lee and M. M. Sung, Appl. Phys. Lett. Vol. 95, 223506 (2009).
3. M. S. Oh, W. Choi, K. Lee, D. K. Hwang and S. Im, Appl. Phys. Lett. Vol. 93, 033510 (2008).
4. W.-C. Shin, K. Remashan, J.-H.Jang, M.-S. Oh ans S-J. Park, J. Korean Phys. Soc.Vol.55, 4, 1514-1518 (2009).
5. H. Q. Chiang, J. F. Wager, R. L. Hoffman, J. Jeong and D. A Keszler, Appl. Phys. Lett.Vol.86, 013503 (2005).
6. K. Nomura, H. Ohta, A. Takagi, T. Kamiya, M. Hirano and H. Hosono, Nature 432, 488-492 (2004).
7. E. Fortunato, P. Barquinha, A. Pimentel, A. Gonçalves, A. Marques, L. Pereira and R. Martins, Adv. Mater. Vol. 17, 5 (2005).
8. S. J. Pearton, C. R. Abernathy, M. E. Overberg, G. T. Thaler, D. P. Norton, N. Theodoropoulou, A. F. Hebard, Y. D. Park, F. Ren, J. Kim and L. A. Boatner, J. Appl. Phys. Vol. 93, 1 (2003).
9. Ü. Özgür, Ya. I. Alivov, C. Liu, A. Teke, M. A. Reshchikov, S. Doğan, C. Avrutin, S. J. Cho, and H. Morkoç, J. Appl. Phys. Vol. 98, 041301 (2005).
10. H.-K. Kim, K.-K. Kim, S.-J. Park, T.-Y. Seong and I. Adesida, J. Appl. Phys. Vol 94, 4225 (2003).
11. T. Miyata, Y. Ohtani, T. Koboi and T. Minami, Thin Film Solid 516, 1354-1358 (2008).

Mater. Res. Soc. Symp. Proc. Vol. 1494 © 2013 Materials Research Society
DOI: 10.1557/opl.2013.520

Fabrication of a p-NiO/n-Si Heterojunction Diode by UV Oxidation of Ni Deposited on n-Si

Dongyuan Zhang, Kazuo Uchida and Shinji Nozaki
Graduate School of Informatics and Engineering
The University of Electro-Communications
1-5-1 Chofugaoka, Chofu-shi, Tokyo 182-8585, Japan

ABSTRACT

The metallic nickel (Ni) deposited on an n-Si substrate with resistivity of $4 - 6$ Ω·cm was oxidized by the ultra-violet (UV) oxidation technique to form a p-NiO/n-Si heterojunction diode. The rectifying current-voltage (I-V) characteristic confirmed formation of a pn junction. The capacitance-voltage (C-V) characteristic further identified an abrupt p^+n junction between NiO and n-Si. The photocurrent increased with the increased wavelength of laser under illumination of the diode. The voltage-dependent photocurrent suggests that the carriers generated in the depletion region of Si was effectively collected but not outside the depletion region. A low diffusion length of holes was attributed to Ni diffusion in Si caused by the substrate heating during the UV oxidation.

INTRODUCTION

Nickel oxide NiO is a wide-band-gap semiconductor with a band gap of 3.8 eV. Its resistive switching phenomenon was extensively studied for a potential memory application [1]. We, however, paid a great attention to the p-type conductivity of NiO and have been motivated to form a heterojuntion pn junction of NiO with another n-type semiconductor. Although the method mostly commonly used to deposit metal oxide semiconductors such as ZnO and NiO is sputtering, the plasma damage often deteriorates the quality of metal oxide semiconductors. We have developed the UV oxidation technique to form NiO by oxidation of metallic nickel. In order to form semiconducting NiO, the UV oxidation parameters such as temperature and oxidation time were optimized using the Raman scattering measurement of the NiO films [2-3]. The UV oxidation of metal can be applied to form various oxide semiconductors including ZnO. As a preliminary study of heterojunction formation with oxide semiconductors, the p-NiO/n-Si heterojunction diode was made by UV oxidation of the metallic Ni on an n-Si substrate and characterized.

EXPERIMENT

Metallic nickel was deposited on n-Si substrate with resistivity of $4 - 6$ Ω·cm by electron-beam (e-beam) vacuum evaporation. Before the deposition, the n-Si substrate was ultrasonically cleaned in acetone, ethanol and deionized water each for 5 minutes subsequently. The pressure of the e-beam evaporation chamber was about $5 - 7 \times 10^{-6}$ torr during evaporation. The thickness of the deposited Ni film was controlled with a crystal monitor and confirmed with a dektak profilometer. The 50 nm-thick Ni film was formed by e-beam evaporation and then UV-oxidized in oxygen atmosphere at 350 °C under UV illumination using a metal halide lamp. The oxidation time of 1 hour was long enough to completely oxidize Ni, and the thickness of NiO measured using a dektak profilometer and ellipsometer was about 100 nm. Platinum was deposited to form a NiO ohmic contact with a diameter of 300 μm, and the back side of silicon was intentionally scratched to facilitate an ohmic contact to gold. Both front and back metal-semiconductor contacts were confirmed to be ohmic.

The I-V characteristics were measured in the dark and under illumination using laser diodes with various emission wavelengths. The C-V characteristics were also measured at different frequencies. In order to find out the influence of heating and UV irradiation upon n-Si, three samples of Ni/n-Si Schottky diodes were post-treated under different conditions: a) under UV irradiation in nitrogen atmosphere, b) under UV irradiation in nitrogen heated at 350 °C, and c) first treated in the same way as b) and then under UV irradiation in oxygen heated at 350 °C.

RESULTS AND DISCUSSION

The I-V characteristic shown in Fig. 1 shows rectification with a difference by two orders of magnitude between forward and reverse currents at 2 V. The reverse leakage current density was 2×10^{-5} A/cm^2 at 2 V. It is higher by one order than that reported earlier for the p-NiO/n-Si diode formed by thermal evaporation of NiO powder on n-Si [4] but much lower than that of the diode formed by sputter-depositing of NiO on n-Si [5]. The C-V characteristics were measured and the C^{-2} vs. V plots are made for various frequencies from 50 k to 1 MHz, as shown in Fig. 2.

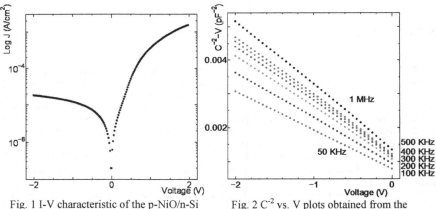

Fig. 1 I-V characteristic of the p-NiO/n-Si heterojunction.

Fig. 2 C^{-2} vs. V plots obtained from the C-V characteristics of p-NiO/n-Si heterojunction measured for various frequencies.

The C^{-2} vs. V plot shows good linearity. Such linearity was never reported for the p-NiO/n-Si diode either by thermal evaporation or by sputter deposition of NiO [4-6]. The impurity concentration calculated from the slope for 1 MHz with the assumption that the depletion region extends only in Si because of highly doped p⁺-NiO agrees with the donor concentration estimated from the resistivity of Si. The diode formed by UV oxidation of Ni seems to have an abrupt p⁺n junction. However, the C-V characteristic depends on the measurement frequency. Such a frequency-dependent capacitance is often observed in pn diodes and MOS capacitors and attributed to deep-level traps with the energy levels closer to the mid band gap. They may be physically located either at the NiO/Si interface or within the depletion region in Si. It is well known that nickel diffusion in Si causes a deep-level defect in Si [7, 8], as discussed next.

The I-V characteristics were also measured under illumination using laser diodes with various emission wavelengths. In Fig. 3, the reverse leakage currents in the dark, and under illumination with green and red lasers were shown. The photocurrent depends on the reverse bias, and its dependence is less for a shorter wavelength. Since most of photons with a shorter wavelength are absorbed near the surface and the metallurgical pn junction, the width of the depletion region, which increases with the increased reversed bias voltage, does not affect collection of the generated carriers. The photocurrent depends more on the bias voltage with the increased wavelength because of larger penetration depth of light. In contrast to normal Si pn diodes, the photocurrent seems to be determined more by the carriers generated in the depletion region rather than those within the diffusion length outside the depletion region. This is attributed to a short diffusion length of holes in n-Si.

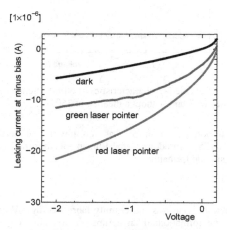

Fig. 3 Reverse I-V characteristics under illumination
with lasers of various wavelengths. The I-V in the
dark is also shown for comparison.

It was hypothesized that a small diffusion length of holes in n-Si might have been caused by Ni diffusion during UV oxidation since metal nickel is easily diffused into Si [7,8].

To confirm the hypothesis of Ni diffusion during UV oxidation, nickel was deposited on n-Si to make Ni/n-Si Schottky diodes. One sample was irradiated with UV light in N₂ atmosphere without heating and another with heating at 350 °C. The UV irradiation period was made the same as the UV oxidation to oxidize Ni on Si. As seen in Fig. 4, the latter sample shows larger leakage current. This suggests that heating the diode at 350 °C caused Ni diffusion. The latter sample was further annealed at 350 °C in oxygen atmosphere for another 1 hour under UV irradiation. As a result, the leakage current increased to 3 x 10^{-5} from 2 x 10^{-5} A at 2 V, as shown in Fig. 4. The increased leakage current may be attributed to either longer annealing time (a total of 2 hours) or oxygen atmosphere instead of nitrogen or both. However, it is clear that Ni diffusion was caused by heating, but not by UV irradiation.

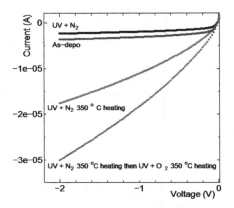

Fig. 4 Reverse I-V characteristics of Ni/n-Si Schottky diodes with and without annealing at 350 °C .

However, the concentration of Ni in Si was too low to be detected by the x-ray photoelectron spectroscopy (XPS). In addition, the Raman spectrum or XPS spectrum did not show an evidence of nickel silicide formation.

CONCLUSIONS

A p-NiO/n-Si heterojuction diode was successfully fabricated by UV oxidation of metallic nickel deposited on n-Si. The formation of an abrupt p^+n junction was confirmed by good linearity in the C^{-2}-V plot. The doping concentration calculated from the slope of the plot agrees with the doping concentration of Si. This suggests the highly-doped p-type NiO was formed by UV oxidation of Ni on Si. The reverse voltage-dependent photocurrent measured with laser of longer wavelength was attributed to Ni diffusion during UV oxidation. Ni diffusion was caused by heating of Ni during the UV oxidation. Nevertheless, UV oxidation of metal has proven to be a good technique to form metal oxide semiconductors.

ACKNOWLEDGMENTS

This research is partly supported by Ministry of Education, Culture, Sports, Science and Technology (MEXT) of Japan. Zhang also acknowledges the scholarship for his Ph.D. study.

REFERENCES

1. I. Hwang, M. -J. Lee, J. Bae, S. Hong, J. -S. Kim, J. Choi, X. L. Deng, S. -E. Ahn, S.-O. Kang and B. H. Park, IEEE Electron Device Lett. 33, 881 (2012)..
2. R. Srnanek, I. Hotovy, V. Malcher, A. Vincze, D. McPhail and S. Littlewood, ASDAM 2000. The Third International EuroConference on Advanced Semiconductor Devices and Microsystems. Smolenice Castle, Slovakia, 16-18 Octber 2000, pp. 303 – 306.
3. N. Mironova-Ulmane, A. Kuzmin, I. Steins, J. Grabis, I. Sildos and M. Pars. Jour. of Phys.: Conf. Seri. 93, 012039 (2007).
4. J-M. Choi and S. Im, Appl. Surface Science 244, 435 (2005).
5. Y. Nishi, T. Iwata and T. Kimoto, Jpn. J. Appl. Phys. 50, 015802 (2011).
6. Lenin Komsomol Voronezh State University. Translated from Izvestiya VUZ. Fizika, No. 12, pp. 151-152, December, 1973. Original article submitted November 5, 1971; revision submitted November 9, 1972.
7. H. Kitagawa, S. Tanaka, H. Nakashima and M. Yoshida, Journal of Electronic Materials 20, 441 (1991).
8. H. Indusekhar and Vikram Kumar, J. Appl. Phys. 61, 1449 (1987).

Mater. Res. Soc. Symp. Proc. Vol. 1494 © 2013 Materials Research Society
DOI: 10.1557/opl.2013.239

The Effects of Structure on the Formation of Schottky Barriers at Nanoparticle-Oxide Interfaces

Ramsey Kraya[1] and Laura Y. Kraya[1]
[1]Department of Materials Science and Engineering, University of Pennsylvania,
Philadelphia, PA, 19104 U.S.A.

ABSTRACT

The surface structure of oxide materials may be the limiting factor in controlling switching properties at interfaces. Here we investigate and correlate the surface structure and electronic properties of $BaTiO_3$ substrates. By using low energy electron diffraction and scanning tunneling microscopy we are able to identify surface reconstructions based on annealing treatments. We then investigate the effect of contact size on the transport properties on oxide surfaces utilizing atomic force microscopy. Our results show the critical importance of controlling surface structure to optimize electronic properties at oxide interfaces.

INTRODUCTION

Understanding atomic mechanisms of surface reactivity in ferroelectric oxides has garnered much interest in recent years because of the potential impact to nanolithography, device memory, and catalysis.[1] The physical properties (elastic, dielectric, piezoelectric, optical, etc) of many ferroelectric oxides are already well documented. However, the structure and electronic properties of defects is not understood. Due to the trend towards miniaturization of devices, an understanding of these relationships is crucial because ferroelectric oxides display a high sensitivity to structural variations and defects. As the surface science of metal oxides has developed over the past two decades, contributions have been made to the understanding of nanoparticle-oxide interfaces. The number of studies connecting atomic resolution SPM with interface studies on oxides is still quite limited and on ferroelectric oxides is nearly non-existent.

Low Energy Electron Diffraction (LEED) and scanning tunneling microscopy (STM) are used to investigate the surface and electronic structures of $BaTiO_3$ (001) of various reconstructed surfaces. The coexistence of the $(\sqrt{5}x\sqrt{5})R26.6°$ and (3x1) reconstructions have been observed using LEED, and corresponding STM images of the surface reveal the presence of these reconstructions in addition to defects, including Ba adatoms, oxygen vacancies, under-coordinated atoms, and other impurities. The presence of defects influences both the surface and electronic structure of the reconstructions, and it is important to understand the effect of these defects on the electronic structure of highly oriented surfaces.

EXPERIMENT

Polished BaTiO3 (001) (Princeton Scientific Co.) with sample dimensions 2.5 mm x 2.5 mm x 0.6 mm are used, and experiments were carried out in an ultrahigh vacuum STM Omicron chamber. The sample was annealed at 1000°C for 45 minutes and was characterized by LEED to determine the structure at the surface. After cooling, the sample was transferred to the STM stage. The STM measurements were obtained with etched tungsten tips at a bias voltage of 2.0 V and 0.2 nA.

The ($\sqrt{5}$x$\sqrt{5}$)R26.6° and (3x1) surface structures were found to coexist using both LEED and STM (Figures 1 and 2b). Because the (3x1) and ($\sqrt{5}$x$\sqrt{5}$)R26.6° can form together on the same surface, they are related to one another in that they have similar structure and energies and the formation of the two reconstructions can be understood as follows. The (3x1) reconstruction occurs at higher temperature anneals and has higher Ti concentrations than the ($\sqrt{5}$x$\sqrt{5}$) R26.6° reconstruction.[2] Its Ti concentration is greater than stoichiometry although oxygen is deficient with respect to stoichiometry. It is likely that the (3x1) has formed by an increase of Ti at the surface. The higher temperature allows this to happen as it increases the mobility of Ti. The LEED pattern indicates that the surface first forms a ($\sqrt{5}$x$\sqrt{5}$) reconstruction, and then as the temperature increases, transforms to a (3x1), suggesting that the (3x1) forms out of the ($\sqrt{5}$x$\sqrt{5}$). The ($\sqrt{5}$x$\sqrt{5}$) and (3x1) reconstructions were found to coexist (figure 2(b)). The periodicity of the (3x1) surface is 1.2 nm and that of the ($\sqrt{5}$x$\sqrt{5}$) is 0.83 nm, consistent with previous results of the Ti adatom structure.[2] The bright spots on the STM image of the (3x1) reconstruction are attributed to Ti^{2+}, whereas the Ti-O deficiencies on the ($\sqrt{5}$x$\sqrt{5}$) reconstruction are composed of Ti^{3+}. The Ti adatoms in each of these reconstructions influence the electronic structure of the surface and contribute to the in-gap state in the local density of states at different positions below the Fermi level.

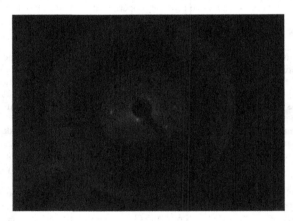

Figure 1. LEED pattern of a combined ($\sqrt{5}$x$\sqrt{5}$)R26.6° and (3x1) reconstruction obtained with incident beam energy of 127 eV

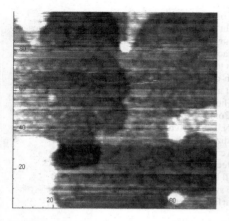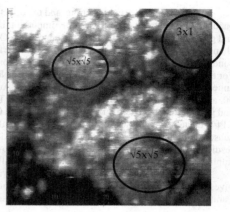

Figure 2(a). STM image of 100 nm BaTiO₃ (001) surface after annealing at 1000°C for 45 minutes. Figure 2(b). STM image of 40 nm x 40 nm BaTiO₃ (001) reduced surface after annealing at 1000°C for 45 minutes.

In addition to the reconstructions observed in the LEED pattern and STM image, several types of defects including adatoms, low-coordinated atoms, and impurities are present on the surface (Figure 2). The STM image of 100 nm x 100 nm region of the combined (3x1) and ($\sqrt{5}$x$\sqrt{5}$)R26.6° is shown in Figure 2(a). Though the surface is atomically flat, it possesses several impurities, as evidenced by the bright points. A higher resolution image (Figure 2(b)) reveals atomic resolution images of both surface reconstructions, along with evidence of step edges, adatoms, oxygen vacancies, and other impurities.

DISCUSSION

The electronic structure of materials is largely controlled by defects and impurities, and experimental studies are complicated by the presence of surface defects. Typical defects in oxides are point defects (missing atoms, e.g. oxygen vacancies), intersitital atoms (additional atoms incorporated on sites other than substitutional sites), and step edges. Each defect type is accompanied with a perturbation of the electronic structure, producing states in the band gap and exposing under-coordinated and chemically unsaturated atoms.[3] In addition, under-coordinated surface sites possess a smaller Madelung potential with respect to the bulk lattice position due to the missing quantity of lattice neighbors. As a result, the local gap size at the surface decreases, and the defects, which are the preferred trapping and recombination centers for electron-hole pairs, offer an attractive binding environment for adsorbates. It is very important to understand the electronic structure of each of the surface reconstructions and the influence of defects, adatoms, and grain boundaries in order to open the door for understanding the influence of nanoparticles and interfaces on these surfaces.

Understanding reconstructions and the effect of the reconstruction has on the bandbending at the interface is not only important, but vital to controlling ferroelectric properties and general transport conditions at the nanoscale. Our recent results have shown that small changes in interface diameter at reduced oxide interfaces can fundamentally affect the transport conditions with transitions from thermionic emission to tunneling to ohmic contact behavior. For these samples single-side epitaxially polished Nb-doped (0.02 at. % Nb) $SrTiO_3$ (100) were annealed in flowing hydrogen for 1 h at 900 °C. Au metal nanoparticles from solutions of 20, 50, and 100 nm sizes were boiled onto the surface. The samples were then annealed in oxygen and ground to gold coated pucks using silver paint. Contacts to the nanoparticles were made with a Pt/Ir coated AFM tip with a nominal force of 60 nN. Current-voltage measurements were subsequently taken. [4] Other measurements on oxidized surfaces did not show this type of result, and maintained similar transport properties over a large range of sizes.[5-7]The implications are that surface structure can readily affect the size dependent behavior.

In the future these types of measurements will be performed on $BaTiO_3$ substrates to investigate the properties that contact size has on the switching properties of these interfaces. Paramount in these studies will be a control of the surface structure.

CONCLUSIONS

The reconstructions on $BaTiO_3$ surfaces were analyzed using LEED and STM measurements and the effect of defects on the surface were discussed. We investigated and correlated the surface structure and electronic properties of $BaTiO_3$ substrates. Size dependent measurements on $SrTiO_3$ showed a strong correlation between size and structure and are an important avenue to investigating switching behavior of $BaTiO_3$.

REFERENCES

[1]   Y. Xu, *Ferroelectric Materials and Their Applications*: North Holland, 1991.

[2]   L. Y. Kraya and R. Kraya, "Determination of the electronic structure of ferroelectric surfaces by scanning tunneling microscopy," *Journal of Applied Physics*, vol. 111, 2012.

[3]   N. Nilius, "Properties of oxide thin films and their adsorption behavior studied by scanning tunneling microscopy and conductance spectroscopy," vol. 64, pp. 595–659, 1 December 2009 2009.

[4]   R. Kraya and L. Kraya, "The role of contact size on the formation of Schottky barriers and ohmic contacts at nanoscale metal-semiconductor interfaces," *Journal of Applied Physics*, vol. 111, p. 4, 2012.

[5]   R. A. Kraya and D. A. Bonnell, "Determining the Electronic Properties of Individual Nanointerfaces by Combining Intermittent AFM Imaging and Contact Spectroscopy," *Ieee Transactions on Nanotechnology*, vol. 9, pp. 741-744, Nov 2010.

[6]   R. Kraya, L. Y. Kraya, and D. A. Bonnell, "Orientation Controlled Schottky Barrier Formation at Au Nanoparticle-SrTiO3 Interfaces," *Nano Letters*, vol. 10, pp. 1224-1228, Apr 2010.

[7]   R. A. Kraya and L. Y. Kraya, "Controlling the Interface Dynamics at Au Nanoparticle - Oxide Interfaces," *Nanotechnology, IEEE Transactions on*, vol. 11, p. 3, 2012.

Mater. Res. Soc. Symp. Proc. Vol. 1494 © 2012 Materials Research Society
DOI: 10.1557/opl.2012.1710

Synthesis and Photocatalytic Properties of High-Surface-Area Mesoporous TiO$_2$ Nanoparticle Assemblies

Ioannis Tamiolakis[1], Ioannis N. Lykakis[2] and Gerasimos S. Armatas*[1]
[1]Department of Materials Science and Technology, University of Crete, Heraklion 71003, Greece
[2]Department of Chemistry, Aristotle University of Thessaloniki, Thessaloniki 54124, Greece.

ABSTRACT

Mesoporous TiO$_2$ nanoparticle assemblies have been synthesized *via* a surfactant-assisted aggregating process. The products feature a three-dimensional network of interconnected anatase-TiO$_2$ NPs with large internal BET surface area (ca. 142–152 m^{2}g^{-1}) and uniform pores (ca. 7–8 nm). Preliminary catalytic experiments indicated that these mesophases exhibit excellent catalytic activity in UV-visible light oxidation of 1-phenylethanol with molecular oxygen.

INTRODUCTION

The self-assembly of nanoparticles (NPs) into a well-defined porous nanostructure is one of the greatest challenges in materials chemistry. Although individual metal oxide nanoparticles feature unique catalytic, magnetic and electronic properties, their low surface area and tendency for clustering limit the applicability of these nanomaterials, especially in adsorption and catalysis. Three-dimensional assemblies of nanoparticles with large internal surface area and open-pore structure promise a different kind of porous materials with advantageous characteristics. These materials are expected to achieve excellent catalytic and optoelectronic properties and yet possess additional characteristics such as size-selective adsorption and molecular recognition of substrate.

Among the simple metal oxides, titanium dioxide (TiO$_2$) has been predominantly used for photocatalytic, photovoltaic, and electrochromic applications because of its inherent superior photocatalytic and opto-electronic properties [1]. Recent efforts to improve photoactivity of TiO$_2$–based materials have been focused on synthesis of nanometer-sized TiO$_2$ particles, which possess high crystallinity and large surface area [2]. The surfactant-assisted templating of inorganic species provides a simple and effective method for producing mesoporous ensembles of nanoparticles. In this context, several organic amphiphilic molecules, like charged small molecules and block copolymers, have been employed as templates to direct the assembly of Ti molecular compounds (TiCl$_4$, Ti(OR)$_4$, etc.) or pre-synthesized TiO$_2$ nanocrystals into various nanoscale architectures [3]. Interestingly, modulation of the interactions between inorganic species and surfactant molecules provides a way of controlling the morphology and pore geometry of resulting structures. These materials show enhanced photocatalytic activity and Li-storage capability granted by the nanoscopic framework components [4]. Although significant progress in this area has been achieved, developing high-surface-area mesoporous assemblies from well crystalline TiO$_2$ NPs is a worthwhile goal.

Herein, we report the synthesis of high surface area, high crystalline, nanostructured anatase-titania *via* a surfactant-assisted self-assembly of TiO_2 NPs. The synthesis was accomplished in one-step chemical process in which the crystallization and assembly of TiO_2 NPs into three-dimensional assembled mesostructures takes together. The obtained materials consist of interconnected spherical NPs and exhibit large and accessible pore surface, as evidenced by XRD, SAXS, TEM and N_2 physisorption measurements. Interestingly, control of the experimental conditions, such as template and time of polymerization, greatly affects the textural properties (i.e. pore width, surface area, and nanoparticle diameter) of the resulting structures. We also show that these TiO_2 mesophases can be used as photoactive materials for the UV-visible-light-driven oxidation of 1-phenyethanol with molecular oxygen, exhibiting high reactivity.

EXPERIMENT

The synthesis of mesoporous TiO_2 nanoparticle assemblies (MTA) was accomplished by a surfactant-directed evaporation induced self-assembly method [5]. In a typical experiment, 1 g of diblock copolymer surfactant was dissolved in anhydrous ethanol at room temperature. Consequently, $TiCl_4$ and titanium(IV) propoxide (TPP) precursors were slowly added to the surfactant solution under vigorous stirring and the resulting mixture was kept at room temperature for 2 h. The molar composition of the sol was 0.9 surfactant : 171 ethanol : 1.6 $TiCl_4$: 5.6 TPP. The transparent yellow colored solution was then transferred in an oven at 40 °C and left undisturbed during aging for 7 days. To efficiently remove the template and promote crystallization and adequate interparticle connection, the resulting monolithic gel was dried at 100 °C for 12 h under vacuum and gently calcined in air for 6 h at 260 °C and, then for 4 h to 350 °C (heating rate of 0.5 °C min^{-1}). A series of MTA materials with different textural parameters was prepared on using various structure-directing agents with different molecular weights and block fractions (see Table I). For the preparation of MTA-1a and MTA-1b samples, a similar to MTA-1 procedure was used, except that the sol solution was aged for 5 and 14 days, respectively.

Table I. Different surfactant used for the synthesis of MTA materials.

Sample	ICI trademark	Name	Chemical formula	Average M_n
MTA-1	Brij 58	Polyethylene glycol hexadecyl ether	$C_{16}H_{33}(OCH_2CH_2)_nOH$ n~20	1124
MTA-2	Brij S10	Polyethylene glycol octadecyl ether	$C_{18}H_{37}(OCH_2CH_2)_nOH$ n~10	711
MTA-3	Brij S20	Polyethylene glycol octadecyl ether	$C_{18}H_{37}(OCH_2CH_2)_nOH$ n~20	1152
MTA-4	Brij C10	Polyethylene glycol hexadecyl ether	$C_{16}H_{33}(OCH_2CH_2)_nOH$ n~10	683

DISCUSSION

The mesoporous structure of MTA was investigated by small-angle X-ray scattering (SAXS), X-ray diffraction (XRD) and transmission electron microscopy (TEM). The SAXS pattern revealed an intense diffraction peak at scattering wave vector q ($=4\pi \sin\theta/\lambda$, where 2θ is the scattering angle) range 0.45–0.47 nm^{-1}, indicating mesoscopic order (Figure 1A). The mean interparticle distance ($d = 2\pi/q$) of mesoporous structures, calculated from the angular position of the intense peak, was found to be ~13.4–14 nm. The diameter of nanoparticles was determined using the Guinier approximation in the low q-region of the scattering curves [6]. Analysis of the scattering data in this term revealed spherical nanoparticles with an average diameter of ~6.8–7.7 nm. These results indicate that the particle size of TiO$_2$ nanocrystals become larger as the hydrophilic character of surfactants increases (MTA-1 and MTA-3 materials). More likely the hydrophilic domains (-OCH$_2$CH$_2$-) of the surfactant micelles interact strongly with the inorganic species (TiO$_x$(OH)$_y$)$^{n-}$), resulting in the formation of large nanoparticles [5].

The wide-angle XRD pattern, in Figure 1B, confirmed the nanocrystalline nature of the MTA materials, showing a series of broad diffraction peaks that correspond to the anatase structure (PDF #21-1272) of TiO$_2$. By means of this technique (Scherrer analysis), the average grain size of the crystallites was estimated to be 9.1–9.8 nm that is reasonably comparable to the nanoparticle diameters.

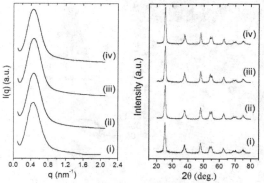

Figure 1. (a) SAXS pattern and (b) wide-angle XRD of MTA materials: (i) MTA-1, (ii) MTA-2, (iii) MTA-3 and (iv) MTA-4.

Table II. Textural parameters of MTA materials.

Sample	Surface area, S$_p$ (m^{2}g^{-1})	Pore volume[†] (cm^{3}g^{-1})	Pore diameter (D$_p$) (nm)	Intraparticle distance (d$_{SAXS}$) (nm)	Mean NP diameter[§] (D$_{SAXS}$) (nm)	Crystalline size[‡] (nm)
MTA-1	149	0.27	7.1	13.7	7.6	9.1
MTA-2	142	0.26	7.5	14.0	6.8	9.8
MTA-3	152	0.30	7.9	13.6	7.7	9.4
MTA-4	146	0.29	7.8	13.4	7.1	9.7

[†]Total pore volume at P/P$_o$=0.98. [§]Mean diameter of NPs according to the Guinier law: $I(q)=I(0)\exp(-q^2R_g^2/3)$, where R$_g$ is radius of gyration (for spherical particle: D$_{SAXS}$=2R$_g$/√(3/5)). [‡]The average crystalline size of TiO$_2$ calculated by the Scherrer formula d_p=0.9λ/Bcosθ, where λ is source wavelength (λ_{cu}=0.154 nm) and B is the full-width at half maximum of the Bragg diffraction peak centered at 2θ degrees.

Typical TEM images obtained from mesoporous MTA-1 sample are shown in Figure 2. TEM images show a porous, foam-like, network of interconnected nanoparticles with remarkable monodispersity (7.8 ±0.3 nm) (Figure 2A, B). Meanwhile, high resolution TEM (HRTEM) image reveals a single crystalline morphology throughout the nanoparticles, showing well-resolved lattice fringes that associated with a d-spacing of 0.35 nm (Figure 2C). This interplanar distance coincides with the d-spacing of (101) lattice plane in anatase TiO_2. The crystallinity of TiO_2 nanoparticles was also investigated with selected area electron diffraction (SAED). The MTA-1 feature a well-defined multi-crystalline diffraction pattern, revealing a series of broad Debye–Scherrer diffraction rings, which can be indexed to the (101), (004), (200), (211), (204), (215) and (224) reflections of anatase TiO_2 phase, Figure 2D. These results are consistent with XRD data. Note that highly crystalline anatase nanostructure is desirable in photocatalysis and solar cells because of the high mobility of photogenerated electrons and limited recombination of electron-hole pairs [7].

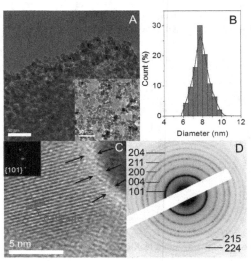

Figure 2. (A) TEM image, (B) particle size distribution (based on 200 counts), (C) high-resolution TEM (HRTEM) of TiO_2 nanocrystal, and (D) SAED pattern of the MTA-1.

Figure 3 shows the nitrogen adsorption–desorption isotherms and the corresponding density functional theory (DFT) pore size distribution plot of MTA materials. All these materials displayed typical type-IV adsorption curves with a H_2 hysteresis loop (according to IUPAC classification), characteristic of three-dimensional interconnected mesoporous solids. Analysis of the adsorption data by the Brunauer-Emmet-Teller (BET) method gave surface areas in the range of 142–152 m^2g^{-1} and total pore volumes in the 0.26–0.30 cm^3g^{-1} range. The MTA mesostructures feature a quite narrow DFT pore size distribution with an average interparticle edge-to-edge distance of ~7.1–7.9 nm. On the basis of SAXS interparticle distance and DFT pore size, we derived an average pore wall thickness of about 6–7 nm. These values are very close to the nanoparticle diameter, suggesting that the pore walls of MTA are made up of a single layer of TiO_2 NPs. Table II summarizes all the textural parameters of mesoporous MTA samples.

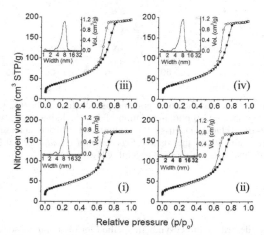

Figure 3. N_2 adsorption–desorption isotherms (77K) and the corresponding DFT pore size distribution (inset) of mesoporous (i) MTA-1, (ii) MTA-2, (iii) MTA-3 and (iv) MTA-4 samples.

Interestingly, the pore size of MTA can be controllably varied by changing the aging time of the TiO_2/surfactant sol. Specifically, the MTA-1 materials obtained after being aged for 5 (MTA-1a) and 14 (MTA-1b) days displayed BET surface areas of 135 and 155 m^2g^{-1} and pore sizes of 6.3 nm and 8.3 nm, respectively. Furthermore, direct observation with TEM revealed mesoscopic porous structures of interconnected TiO_2 NPs with an average diameter of 7.5 ±0.5 nm for MTA-1a and 8.1 ±0.3 nm for MTA-1b materials.

The photocatalytic activity of MTA has been evaluated using UV-visible-light-driven oxidation of 1-phenylethanol as a probe reaction. For comparison purposes, we also evaluated the catalytic efficiency of three commercial TiO_2 photocatalysts, i.e. Degussa P25 (S_p~ 53 m^2g^{-1}, D~ 20-30 nm), Hombikat UV-100 (S_p~ 250 m^2g^{-1}, D~ 10 nm) and Aldrich anatase (S_p~ 45-55 m^2g^{-1}, D < 25 nm). As shown in Figure 4A, the MTA-1 exhibited superior catalytic performance compared to those of commercial catalysts, reflected to the higher conversion of 1-phenylethanol to acetophenone (~78 % for MTA-1 *vs.* ~69% for Degussa P25, ~68% for Hombikat UV-100 and ~21% for Aldrich anatase, in 1h). Indeed, the photo-oxidation of 1-phenylethanol proceeded faster over MTA-1 than the other examined catalysts, as indicated from the relevant pseudo first-order photo-oxidation rate constants (k) (0.0238 min^{-1} for MTA, 0.0206 min^{-1} for Hombikat UV-100, 0.0032 min^{-1} for Aldrich anatase, and 0.0188 min^{-1} for Degussa P25), see Figure 4B. The superior catalytic performance of MTA-1 is attributed to the small grain size of anatase crystallites and three-dimensional open porous structure. It is suggested that mesoporous structures of TiO_2 with large and accessible pore surface can favor fast mass-transfer kinetics as well as efficient trapping of light *via* multiple backscattering effects [8].

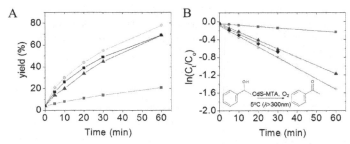

Figure 4. (A) Time courses and (B) the corresponding pseudo first-order (C_o, C_t are the initial and final concentrations, respectively) photo-oxidation of 1-phenylethanol catalyzed by MTA-1 (green), Hombikat UV-100 (black), Degussa P25 (blue) and Aldrich anatase (yellow). The corresponding red lines are fit to the data. (*Experimental conditions*: 0.1 mmol of 1-phenylethanol, 45 mg of catalyst, 25 mL min^{-1} O_2, 3 mL CH_3CN, 5 °C).

CONCLUSIONS

In conclusion, we describe a direct synthetic method for obtaining mesoporous titania *via* a surfactant-assisted aggregating assembly of TiO_2 NPs. SAXS, TEM and N_2 physisorption showed assembled mesostructures consisting of monodisperse TiO_2 NPs with large surface area and open-pore structure. Meanwhile, the pore size and particle diameter can be altered by varying the reaction conditions (aging time, surfactant). These mesophases demonstrated great application perspectives in photocatalytic oxidation of 1-phenylethanol with molecular oxygen.

ACKNOWLEDGMENTS

Financial support by the European Union (MC-IRG, No. 230868) and the Greek Ministry of Education (ERC-09, MESOPOROUS-NPs) is kindly acknowledged.

REFERENCES

1. A. L. Linsebigler, G. Q. Lu and J. T. Yates, *Chem. Rev.* **95**, 735 (1995).
2. X. Chen and S. S. Mao, *Chem. Rev.* **107**, 2891 (2007).
3. J. Polleux, M. Rasp, I. Louban, N. Plath, A. Feldhoff and J. P. Spatz, *ACS Nano* **5**, 6355 (2011).
4. W. Yue, C. Randorn, P. S. Attidekou, Z. Su, J. T. S. Irvine and W. Zhou, *Adv. Funct. Mater.* **19**, 2826 (2009).
5. I. Tamiolakis, I. N. Lykakis, A. P. Katsoulidis and G. S. Armatas, *Chem. Commun.* **48**, 6687 (2012).
6. O. Glatter and O. Kratky in *Small-Angle X-Ray Scattering*, (Academic Press, New York, 1982).
7. a) L. Kavan, M. Gratzel, S. E. Gilbert, C. Klemenz and H. J. Scheel, *J. Am. Chem. Soc.* **118**, 6716 (1996). b) B. Ohtani, Y. Ogawa and S. J. Nishimoto, *Phys. Chem. B* **101**, 3746 (1997).
8. a) A. T. Bell, *Science*, **299**, 1688 (2003). b) J. C. Yu, X. C. Wang and X. Z. Fu, *Chem. Mater.* **16**, 1523 (2004).

Mater. Res. Soc. Symp. Proc. Vol. 1494 © 2013 Materials Research Society
DOI: 10.1557/opl.2013.118

Improved Switching Response of VO_2 Devices Deposited on Silicon Nitride Membranes

Yan Wang[1, 2] and John F. Muth[1]
[1]Department of Electrical and Computer Engineering, North Carolina State University,
Raleigh, North Carolina 27695, USA
[2]Department of Physics, North Carolina State University, Raleigh, North Carolina 27695, USA

ABSTRACT

VO_2 films were deposited on sapphire, ITO glass and 200 nm thick silicon nitride membranes by Pulsed Laser Deposition (PLD). The electrical and optical properties have been investigated. Joule heating devices fabricated on silicon nitride membranes switches from semiconductor phase to metal phase by applying a constant voltage across two metal contacts. Compared to the devices fabricated on the normal substrates, such as sapphire, silicon or glasses, the switching speed of the devices on membrane is an order of magnitude faster. Decreasing the area and thickness of VO_2 on top of thinner membranes allows kHz bandwidth to be achieved.

INTRODUCTION

VO_2 is well known for its semiconductor to metal transition at 68 °C, which is associated with an abrupt change of resistance and a decrease in infrared wavelength transmittance [1-4]. These properties make VO_2 an interesting material for novel electronic and optical device applications, such as optical switches, reconfigurable antennas and smart window [5].

In this work, highly oriented crystalline VO_2 films were deposited on different substrates by Pulsed Laser Deposition (PLD). The results show the transition temperature, the width of the hysteresis loop, and the amplitude of the transition depend strongly on the substrates, which affect the strain on VO_2 films. The VO_2 thin films on sapphire show large amplitude transition (4 orders) and narrow hysteresis, about 8 °C. The transition temperature of heating and cooling are 70 °C and 62 °C respectively. VO_2 thin films were also deposited on top of SiN_x membrane to construct VO_2 devices. On the 200 nm thick SiN_x membrane, the VO_2 films show 3 orders resistance change and a 15 °C wide hysteresis loop during the semiconductor to metal transition. The device switches from semiconductor phase to metal phase by applying a constant voltage across two metal contacts. The transition is caused by joule heating from the current flowing through the VO_2 thin films. A 300 μm × 300 μm area/100 nm thick VO_2 device takes about 1000 μs to reach the fully "on" state. The transition time of the devices can be controlled by changing the area and the thickness of VO_2 thin films.

EXPERIMENT

Pulsed Laser Deposition (PLD) was used to deposit the VO_2 films on different substrates. A KrF excimer laser at a wavelength of 248 nm was focused on a homemade V_2O_5 target. The repetition rate and energy level of laser pulses were kept at 10 Hz and 200 mJ during deposition. The base pressure of the deposition chamber was pumped to 10^{-7} Torr. The depositions were performed in an ambient of 20 mTorr oxygen and the substrate temperature was maintained at 800 °C. After the deposition, the samples were cooled down at the deposition oxygen pressure to room temperature. We deposited VO$_2$ films on different substrates including sapphire, ITO glass and SiN$_x$ membranes. The crystalline structure of the as-deposited VO$_2$ thin films was determined by x-ray diffraction (XRD) measurements. The result of θ–2θ x-ray scan of VO$_2$ thin films on a (0001) sapphire substrate is shown at Figure 1. The film shows four peaks corresponding to (020), (220), (130) and (040) of monoclinic VO$_2$ [6, 7].

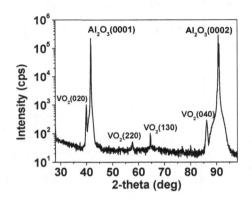

Figure 1. θ–2θ x-ray scan for a VO$_2$ thin film on sapphire substrate.

Different devices with different size were fabricated for comparison. As shown in Figure 2. The first device is VO$_2$ (100 nm) on sapphire with gold contact. The area of VO$_2$ is 100 μm X 400 μm. The second device is VO$_2$ (100 nm) on 200 nm SiN$_x$ membrane with gold contact. The area of VO$_2$ on top of SiN$_x$ is 300 μm X 300 μm and 500 μm X 500 μm. We switch the device from semiconductor phase, which is more transparent to light, to metal phase, which is more reflective to light, by applying a constant voltage. The transition happens due to joule heating mechanism. The optical properties of these devices were measured at the temperatures between insulating and metallic phase. A tungsten light source with 430-900 nm wavelength range was used to provide a collimated light beam and normally incident onto the VO$_2$ film. The transmission light was collected by an Ocean Optics USB 4000 spectrometer and a BWTEK BTC261 infrared spectrometer. The switching time has been investigated by connecting our transmission setup to a photo detector and an oscilloscope. We send out the square pulse voltage and collect the transmission response from a photo detector which converts the light signal into voltage. The photo detector is DET10A Si detector which is from Thorlabs with 1ns raise time.

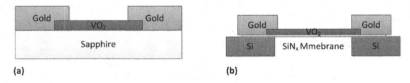

Figure 2. Device structure fabricated on (a) sapphire, (b) SiN$_x$ membrane.

DISCUSSION

VO$_2$ thin films were deposited by PLD under the condition mentioned previous with thickness of 100 nm on sapphire, ITO glass and SiN$_x$. The electrical and optical properties have been investigated. The results show the transition temperature, the width of the hysteresis loop, and the amplitude of the transition depend strongly on the substrates. The VO$_2$ thin films on sapphire showed large amplitude transition and narrow hysteresis which is attributed to the highly oriented columnar structure and highly single crystal composition of the VO$_2$ grains. As shown in Figure. 3a, the resistivity has changed up to four orders through the insulating to metallic transition. The thermal hysteresis is about 8 °C. The derivative of the resistivity was plotted in Figure 3b. The transition temperature of heating and cooling are 70 °C and 62 °C respectively. But for VO$_2$ thin films on SiN$_x$ and ITO glass, the hysteresis is broader and the transition has lower amplitude. In the case of VO$_2$ on SiN$_x$, the transition temperature during heating up and cooling down are 76 °C and 61 °C respectively, resulting a 15 °C wide hysteresis. Higher resistivity in the metallic phase and lower resistivity in the insulating phase cause the lower amplitude. The different transition properties were induced by the effect of strain on VO$_2$. Because of the lattice mismatch of the substrates with the VO$_2$ films, substrates can affect the strain on VO$_2$, thereby changing the transition properties. The lattice mismatch between C-cut plane (0001) sapphire and VO$_2$ films are in order of 5%, but the other two substrates have quite larger mismatch with VO$_2$ films. The larger strain effect causes the modification of the transition properties [8] [9].

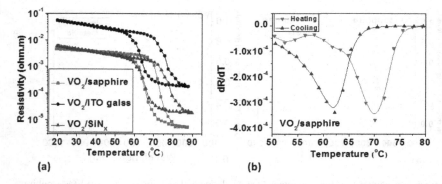

Figure 3. (a) Resistivity of VO$_2$ thin films on different substrates as temperature change; (b): Derivative of resistivity over temperature as temperature change for VO$_2$/sapphire.

The change of the optical property has strong potential to be used in future electronic devices as VO$_2$ go through the insulating to metallic transition. We have studied the optical transmission in the wavelength range from 400 nm to 1350 nm. VO$_2$ on sapphire and ITO glass have been studied. For both samples, the transmittance has decreased as much as 30% in the infrared range as the temperature change from 20 °C to 90 °C. In the case of VO$_2$/ITO glass, the transmittance has almost no changed in the visible range 550 nm-650 nm, but there was nearly 10% decrease for VO$_2$/sapphire as temperature increase from 20 °C to 90 °C. (Figure 4)

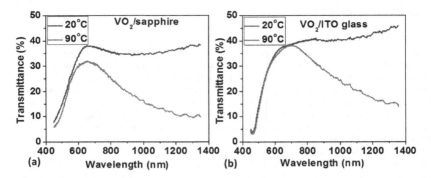

Figure 4. Transmittance at 20 °C and 90 °C for (a) VO₂/sapphire (b) VO₂/ITO glass from 450 nm to 1350 nm.

Joule heating devices were fabricated on sapphire and 200 nm thick SiN_x membranes with 100 nm thick VO_2. We measured the transmission of these devices with voltage on and off. The results in Figure 5 show the transmittance decrease in the visible range from 500 nm to 800 nm as VO_2 change from insulating state to metallic state.

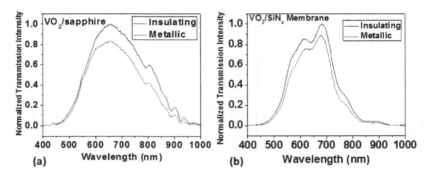

Figure 5. Normalized transmission intensity at insulating and metallic state for (a) VO₂/sapphire (b) VO₂/SiN$_x$ from 450 nm to 900 nm.

For the device fabricated on sapphire substrates, as we increase the voltage, the transition happens at 8 V and takes about 30 ms to fully reach the metallic state, as shown in Figure 6a. The VO_2 film was heated up due to the applied voltage. The higher and lower voltage of the photo detector corresponds to the insulating and metallic state respectively.

To increase the performance of the joule heating devices, we fabricated them on 200 nm thick SiN_x membranes. For the devices with area 500 μm X 500 μm, it takes 5 ms to switch from insulating state to metallic state. As shown in Figure 6(b) (c). The results also show that for a 300 µm X 300 µm device, the transition time is about 3 ms. Comparing to the devices fabricated on sapphire, the switching speed is an order of magnitude faster. These devices were fabricated based on the joule heating mechanism. With 500 µm thick sapphire substrate underneath the VO$_2$ films, more heat is needed to heat up the films to its transition temperature, hence longer switching time. However, same thickness of VO$_2$ films on 200 nm thick SiN$_x$ membrane requires much less heat to reach the critical temperature, which increase the transition speed of the devices.

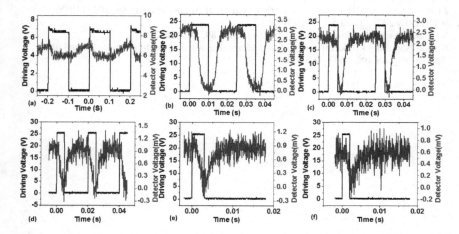

Figure 6. Transmission changes for different devices at different voltage pulse period,(a) devices on sapphire at 5Hz, (b) devices on SiN$_x$(500 µm X500 µm), 10 ms voltage pulse, (c) devices on SiN$_x$(500 µm X500 µm), 5 ms voltage pulse, (d) devices on SiN$_x$(300 µm X300 µm), 5 ms voltage pulse, (e) devices on SiN$_x$(300 µm X300 µm), 3 ms voltage pulse, (f) devices on SiN$_x$(300 µm X300 µm), 2 ms voltage pulse.

CONCLUSIONS

This study shows variation of the transition temperature, the width of the hysteresis loop, and the amplitude of the transition of VO$_2$ thin films on different substrates. Those properties depend strongly on the film morphology and stoichiometry. Joule heating devices were fabricated on SiN$_x$ membranes. The switching time from insulating to metallic state of the devices can be controlled by changing the area and the thickness of VO$_2$ thin films. The switching speed of the devices on membrane has been improved an order of magnitude faster comparing to the devices fabricated on the normal substrates.

REFERENCES

1. F. Morin, Physical Review Letters **3**, 34–36 (1959).
2. D.H. Kim and H.S. Kwok, Applied Physics Letters **65**, 3188 (1994).
3. N.R. Mlyuka, G. a. Niklasson, and C.G. Granqvist, Physica Status Solidi (a) **206**, 2155 (2009).
4. H.W. Verleur, Journal of the Optical Society of America **58**, 1356 (1968).
5. C. Granqvist, Thin Solid Films **193**, 730–741 (1990).
6. J.F. De Natale, P.J. Hood, and a. B. Harker, Journal of Applied Physics **66**, 5844 (1989).
7. ICDD (PDF2009). Card No. 00-019-1398
8. Y. Muraokaa and Z. Hiroi, Applied Physics Letters **80**, 583 (2002).
9. Kazuki Nagashima, Takeshi Yanagida, Hidekazu Tanaka, and Tomoji Kawai, Journal of Applied Physics **100**, 063714(2006).

Mater. Res. Soc. Symp. Proc. Vol. 1494 © 2013 Materials Research Society
DOI: 10.1557/opl.2013.179

Effect of Dispersal of Pd Nanocatalysts on H_2 Sensing Response of SnO_2 Thin Film Based Gas Sensor

Manish Kumar Verma[1], Neha Batra[1], Monika Tomar[2], and Vinay Gupta[1]
[1]Department of Physics and Astrophysics, University of Delhi, Delhi, 110007, INDIA.
[2]Physics Department, Miranda House, University of Delhi, Delhi, 110007, INDIA.

ABSTRACT

SnO_2 based sensor structures prepared by rf magnetron sputtering technique have been studied for detecting H_2 gas. Pd catalyst was integrated onto the SnO_2 thin film in the form of clusters and nano-particles to obtain enhanced sensing response characteristics. The prepared sensor structures have been studied over a temperature range of 50-250°C for sensing response towards 500 ppm H_2 gas. The sensor with Pd catalyst dispersed in the form of nanoparticles was found to exhibit an enhanced sensing response of 1.9×10^3 at a relatively low operating temperature of 150°C with a fast response time of 2 s and recovery time of 65 s towards 500 ppm H_2 gas. The origin of enhanced sensing response is identified in the light of the enhanced spill over of H_2 gas molecules on the uncovered surface of SnO_2 thin film.

INTRODUCTION

Recently there has been a lot of interest in hydrogen sensors, due to it being the prospective fuel of future. Hydrogen is a colourless, explosive, and extremely flammable gas having low minimum ignition energy (0.017 mJ), high heat of combustion (142 kJ/g H_2) and wide flammable range (4−75%), as well as a high burning velocity, detonation sensitivity and an ignition temperature of 560°C. Hydrogen gas is important in the synthesis of ammonia and methanol, the hydration of hydrocarbons, the desulphurization of petroleum products and the production of rocket fuels, in metallurgical processes etc [1-3]. Being an explosive, colourless and odourless gas, hydrogen cannot be detected by human senses. Therefore, detection of hydrogen at low concentration is very essential to prevent disaster.

Semiconducting tin oxide (SnO_2) thin films are the most popular amongst the various semiconductor materials used for H_2 gas sensing because of their capability of adsorption and desorption of oxygen from its surface. However, SnO_2 based gas sensors lack the selectivity and give poor response which is being improved by use of metal and metal oxide catalysts. Different metal and metal oxides have been integrated with sensing SnO_2 thin film for improvement in response and selectivity. Pd is the most widely used catalyst to improve selectivity towards H_2 due to its strong affinity towards H_2. Film morphology and integration of suitable catalyst along with the dispersal mechanism plays major role in the enhancement of sensing response characteristics.

In the present work, Pd catalyst has been integrated with SnO_2 thin film in the form of clusters and nanoparticles dispersed over the surface. The fabricated sensor has been studied for sensing response towards 500 ppm H_2 gas. The effect of way of dispersal has been studied on the sensor response towards 500 ppm H_2 gas.

EXPERIMENTAL

SnO$_2$ thin films of 90 nm thickness were deposited by rf sputtering technique using Sn metal target (99.999%) under a reactive ambient of 20% O$_2$ and 80% Ar and rf power of 300 W, onto corning glass substrates having interdigital electrodes of platinum patterned on it. The sputtering pressure was kept high at 50 mTorr to obtain the rough and porous morphology of the films. SnO$_2$ thin films were annealed in air at 300°C for two hour to improve crystallinity and achieve stable resistance. Pd clusters of 6 nm thickness were deposited using rf sputtering through a shadow mask having uniformly distributed pores of 200 μm diameter. Pd nanoparticles were prepared by conventional polyol synthesis method and were dispersed over SnO$_2$ thin film using spin coating technique [3, 4]. SnO$_2$ thin films after integration of Pd catalysts were annealed in air at 300°C to stabilize the sensor resistance. The pure SnO$_2$ sensor is labelled as S0, Pd clusters (6 nm)/SnO$_2$ sensor was labelled SI and Pd NPs (4 nm)/SnO$_2$ sensor was labelled as SII.

Crystallographic structure of the SnO$_2$ thin films was studied using X-ray Diffraction studies (Bruker D8-Discover). Sizes of the synthesized nanoparticles were measured using high resolution transmission electron microscope (FEI TECNAI G^{2}30-U-TWIN) using carbon coated copper grids. Sensing response characteristics of the prepared sensor structures were studied in a test gas chamber using Digital multimeter (Keithley 2002) interfaced with a PC for data acquisition. The sensing response is defined as $S = R_a/R_g$, where R_a is defined as the sensor resistance measured in the presence of dry air, and R_g is the sensor resistance measured in the presence of reducing (H$_2$) gas (Purity 99.9%). At a specific temperature, the sensor was first stabilised to attaining a stable resistance (R_a). Thereafter, the test chamber was filled with desired concentration of target H$_2$ gas using calibrated leaks through needle valves. At the time of recovery study, the target gas was flushed out of the test chamber and clean dry air was introduced.

RESULTS AND DISCUSSION

As deposited SnO$_2$ thin films were found to be strongly adherent to the substrates and optically transparent in the visible region. The films were amorphous and became polycrystalline after post deposition annealing treatment. Broad and well defined XRD peaks were observed at $2\theta = 26.5°$, 33.9°, 38.0°, and 51.8° corresponding to (110), (101), (200), and (211) planes respectively of the rutile structure of SnO$_2$ (Figure 1), confirming the formation single phase polycrystalline SnO$_2$ thin film after post deposition annealing treatment at 300°C [3,5].

SnO$_2$ thin films were found to be transparent in the visible region with a sharp absorption edge around 380 nm (Inset of Figure 2 a). The bandgap of annealed SnO$_2$ thin films calculated from the UV-visible data was found to be 4.01 eV which is close to the one reported by other workers [5-7]. A broad peak in the absorption spectra of Pd nanoparticles, shows the formation of reduced Pd particles (Figure 2 a). A small hump around 325 nm indicates presence of residual Pd^{2+} [8]. The synthesised Pd nanoparticles were found to be monodisperse, spherical in shape, having a size of 4 nm (Figure 2b).

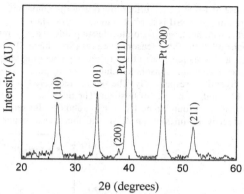

Figure 1. XRD spectra of annealed SnO_2 thin film

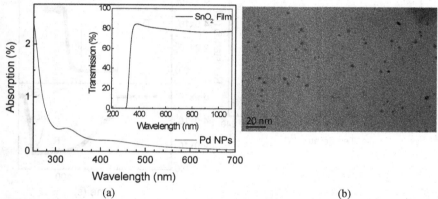

(a)   (b)

Figure 2. (a) Absorption spectra of Pd nanoparticles (Inset shows the transmission spectra of SnO_2 thin film), (b) TEM image of Pd nanoparticles.

Sensing response characteristics

Sensing response of all the sensors (S0, SI, and SII) were studied over a wide temperature range of 50°C to 220°C towards 500 ppm H_2 gas. Sensing response of all the sensors was found to increase with increasing temperature, attaining a maximum response at their operating temperature and thereafter show a decrease in sensing response with further increase in temperature (Figure 3 a). Bare SnO_2 thin film sensor (S0) showed a maximum response of 4.6 towards 500 ppm H_2 gas at an operating temperature of 170°C. A significant improvement in response characteristics of the sensor with Pd cluster/SnO_2 sensor (SI) was observed and found to be 1.0×10^3, at a slightly lower operating temperature of 160°C. Pd NPs/SnO_2 sensor (SII) showed a maximum response of 1.9×10^3 at a further low operating temperature of 150°C. The observed enhanced response of sensor SII is attributed to the high surface area of the catalyst Pd nanoparticles, which interact with the target gas molecules to produce a high response. The variation in sensor resistance of all the three sensor structures (S0, SI and SII) both with the interaction of 500 ppm H_2 gas and its removal at their respective operating temperature are shown in Figure 3 (b). Time taken by the sensors to achieve 90% response were found to be 50 s, 1 s and 2 s for sensors S0, SI and SII respectively. The sensors recovered to their original resistance after removal of H_2 and flushing in of fresh air. The time taken by the sensors to attain 90% of their original resistance was found to be 150 s, 130 s and 65 s for sensors S0, SI, and SII respectively. Response and recovery times of sensor SII was found to be better than the other sensor structures.

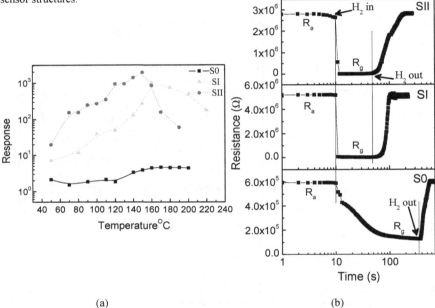

(a)   (b)

Figure 3. (a) Sensing response towards 500 ppm H_2 gas for all the prepared sensors (S0, SI and SII), (b) Variation of resistances of all prepared sensors (S0, SI and SII) in the presence of 500 ppm H_2 and after its removal from the sensing chamber.

Figure 4 (a) shows the variation of resistance of all the sensors in air (R_a) as a function of temperature. It can be observed that the resistance of all the sensors decrease initially with increasing temperature in consistence with the semiconducting behaviour of the films. It may be noted from Figure 4 (a) that the sensor SI exhibits a higher value of R_a in comparison to the corresponding value obtained for bare SnO_2 thin film sensor (S0) over the entire measured temperature range. Since the work function of Pd (5.12 eV) is higher in comparison to that of SnO_2 (4.7 eV), the transfer of large number of electrons from the conduction band of underlying

SnO$_2$ thin film towards catalytic clusters at the interface occurs [3]. The formation of large space charge region near the SnO$_2$-Pd catalyst interface results in the observed increase in the value of R$_a$ for sensor SI. However, the value of R$_a$ for SII was found to be lower in comparison to that obtained for bare SnO$_2$ thin film (S0) at lower temperatures (<120 °C) despite the presence of Pd NPs catalyst (Figure 4 a). This can be attributed to the presence of PVP along with Pd NPs on the surface of sensing SnO$_2$ thin film, which results in transfer of electrons from C = O group present in the PVP towards Pd nanoparticles and hence resulting in smaller value of R$_a$ at low temperatures (Figure 4 a) [3]. For all the sensors an increase in resistance is observed at higher temperatures with increasing temperature. The increase is more for sensors SI and SII, and is due to increased oxygen adsorption activity of the Pd catalyst. Pd catalyst adsorb and spill over oxygen over the uncovered surface of sensing SnO$_2$ thin film, resulting in removal of conduction electrons by the adsorbed oxygen and hence increase resistance in air.

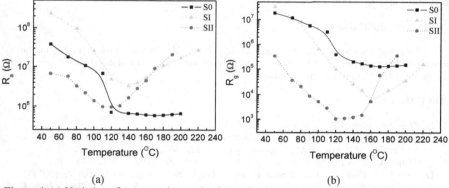

(a)   (b)

Figure 4. (a) Variation of sensor resistance in air as a function of temperature, (b) Variation of sensor resistance in presence of 500 ppm H$_2$ as a function of H$_2$.

Figure 4 (b) shows the variation of resistance of all the sensors in the presence of 500 ppm H$_2$ (R$_g$) as a function of temperature. A decrease in sensor resistance is observed for all the sensors after interaction with target H$_2$ gas. The decrease in resistance was small for pure SnO$_2$ sensor (S0). It can be observed that the value of R$_g$ decreased initially with increasing temperature. However, at higher temperature the R$_g$ value increased with increasing temperature due to competing desorption phenomenon becoming pronounced at higher temperatures. It is to be noted that the fall in resistance was greater for sensors SI and SII as compared to the sensor S0. This is expected as the presence of Pd catalysts on the surface of SnO$_2$ leads to enhanced H$_2$ adsorption and spilled over after dissociation on the uncovered SnO$_2$ thin film and its reaction with the adsorbed oxygen. Response and recovery characteristics of SnO$_2$ thin films with Pd catalysts in the form of nanoparticles is improved because Pd nanoparticles easily adsorbed oxygen in large amount from the atmosphere after removal of reducing H$_2$ gas and introduction of clean air. The observed high response (1.9×10^3) for sensor SII is because of the favourable dispersal of Pd catalyst in the form of nanoparticles providing sufficient uncovered surface area of sensing SnO$_2$ thin film facilitating the accelerated spillover of dissociated H$_2$ molecules by the catalyst onto the uncovered surface of SnO_2 thin film for fast reaction with adsorbed oxygen species in large amount.

CONCLUSIONS

The sensors comprising of catalyst Pd particles in the form of clusters and nanoparticles are studied for the response towards 500 ppm H_2. The sensor structure with Pd catalysts dispersed in the form of nanoparticles over sensing SnO_2 thin film was found to give enhanced response of 1.9×10^3 with a fast response time of 2 s and recovery time of 65 s at a relatively low operating temperature of 150°C. Enhancement in sensing response was found to dependent on the way the catalysts are incorporated into the sensing film. Nanoparticulate catalysts provided larger surface area for interaction with the target gas and hence enhanced spill over of target gas molecules, leading to high sensing response.

ACKNOWLEDGMENTS

The authors acknowledge the financial support provided by the National Programme for Micro and Smart Systems (NPMASS) and Department of Electronics and Information Technology (DIT), Government of India.

REFERENCES

1. N. Yamazoe, *Sens. Actuators B*, **108**, 2, (2005).
2. T. Hubert, L. B. Brett, G. Black and U. Banach, *Sens. Actuators B*, **157**, 329, (2011).
3. M. K. Verma and V. Gupta, *IEEE Sensors Journal*, **12**, 2993, (2012).
4. L. J. Chen, C. C. Wan, and Y. Y. Wang, *J. Colloid Interf. Sci.*, **297**, 143, (2006).
5. D. Haridas, A. Chowdhuri, K. Sreenivas, and V. Gupta, *Sens. Actuators B*, **153**, 89, (2011).
6. A. Bouaine, N. Brihi, G. Schmerber, C. Ulhaq-Bouillet, S. Colis, A. Dinia, J. Phys. Chem. C **111**, 2924, (2007).
7. E. R. Leite, M. I. B. Bernardi, E. Longo, J. A. Varela, and C. A. Thin Solid Films, **449**, 67, (2004).
8. L. D'Souza and S. Sampath, *Langmuir*, **16**, 8510, (2000).

Mater. Res. Soc. Symp. Proc. Vol. 1494 © 2013 Materials Research Society
DOI: 10.1557/opl.2013.259

Copper oxide nanoparticles for thin film photovoltaics

Maurice Nuys[1], Jan Flohre[1], Christine Leidinger[1], Florian Köhler[1] and Reinhard Carius[1]
[1]Institute of Energy and Climate Research 5 -Photovoltaics-, Forschungszentrum Jülich GmbH, D-52425 Jülich Germany

ABSTRACT

Commercially available tenorite (CuO) nanoparticles (NPs) were investigated in particular with respect to their suitability for photovoltaic applications. NPs with a diameter of about 30 nm were step wise annealed up to 1000°C in nitrogen atmosphere. The influence of the annealing treatment on the structural and electronic properties was investigated by Raman, photo-luminescence (PL) and photothermal deflection spectroscopy (PDS) as well as X-ray diffraction measurements. Size, shape, and phase of the untreated NPs are analyzed by TEM measurements. The PL and PDS results show a strong increase of the tenorite band edge emission at about 1.3 eV accompanied by a decreasing sub gap absorption with increasing annealing temperature up to 700°C. According to literature, a phase transition from tenorite to cuprite (Cu_2O) was expected and observed after annealing at 800°C. Strong cuprite band edge emission at about 2 eV accompanied by very weak defect and possibly tenorite band edge emission was found for samples annealed at 800°C and 1000°C.

INTRODUCTION

The long term objective for the world wide installation of photovoltaic systems is far beyond 10TW. A sustainable development asks for photovoltaics based on almost unlimitedly available elements. Up to now, silicon based materials already fulfill this requirement but further improvements in efficiency and cost reduction can only be achieved by multi-band gap thin film solar cells. We have investigated copper oxide based nanoparticles (NPs) for their feasibility as a new absorber material in thin film solar cells. Tenorite and cuprite show optical band edges at about 1.35 eV to 1.5 eV and 2.1 eV, respectively, which is in the range for a reasonable photo-voltaic application [1-4]. Especially cuprite had been intensively investigated regarding its po-tential as an absorber material in solar cells due to its suitable optical band gap, cheapness, and abundance [5, 6]. However, poor material properties due to high defect concentration and the formation of metallic copper at the surface were crucial parameters leading to a fading interest in cuprite for photovoltaic application. Recently, copper oxide regained attention as absorber mate-rial in thin film solar cells [7]. Using NPs instead of thin films is a promising approach to over-come problems of the past. For instance, phase pure and almost defect free NPs are prepared more easily compared to thin films. Moreover, NPs may favor a self-purification process, where the impurities and material defects tend to diffuse to the surface upon thermal annealing. In the present study, tenorite NPs were step wise annealed in nitrogen atmosphere. The impact of this treatment on the electro-optical and structural properties was investigated by Raman and photo-luminescence (PL) spectroscopy as well as photothermal deflection spectroscopy (PDS) and X-ray diffraction (XRD) measurements.

EXPERIMENT

Commercially available tenorite (CuO) NPs with a nominal diameter of 40 nm to 80 nm were purchased as powder from IoLiTec GmbH (Germany). These NPs are dispersed in deionized water by treatment in an ultrasonic bath. For XRD, Raman, PL, and PDS measurements a small volume of this dispersion was applied to quartz substrates (8 mm x 15 mm) and allowed to dry in air. The NPs form larger agglomerates on the substrate leading to an inhomogeneous coverage. For TEM measurements a TEM grid was coated with NPs by dipping it into the dispersion and subsequent drying in air. The NPs on quartz substrate were step wise annealed in nitrogen atmosphere for 30 min up to an annealing temperature of 1000°C. After each annealing step multiple Raman, PL, and PDS measurements were performed. PDS was used, since it is a sensitive method to measure the absorbance of thin films. Moreover, it is weakly influenced by scattering, e.g. of agglomerates and by inhomogeneity caused e.g. by a partial coverage of the substrate [8]. In contrast to the PDS measurement which averages over an area of about 0.5 mm^2, a microscope is used to collect the Raman and PL signals. Here, a local spot with a diameter of about 1 µm is analyzed which allows investigating local differences. Furthermore, the same setup is used for Raman and PL measurements, so that the corresponding signals arise from the same nanoparticle ensemble. The micro-Raman and PL setup offers the opportunity to switch between an InGaAs and Si detector so that the PL signal can be measured in the range of about 0.8 eV to 2.3 eV. The PL signals are corrected for the spectral response of the experimental setup. XRD was measured after selected annealing steps in order to confirm the expected phase transition and to monitor changes in the microstructure. TEM measurements were performed to analyze the initial size, shape and phase of the NPs.

DISCUSSION

Structural properties (TEM, Raman, XRD)

Figures 1a and b show TEM bright field images of the untreated tenorite NPs. The NPs are not uniform and exhibit a large variation in size and shape. According to TEM, mainly NPs with a diameter about 30 nm and few larger NPs with diameter above 100 nm are found. Figure 1c illustrates a representative diffraction pattern corresponding to the bright field in figure 1b. The diffractogram (figure 1d) was calculated from the angular integration of the pattern in figure 1c and shows a multitude of reflexes resulting from different crystallographic orientations of the NPs, similar to a powder pattern. Reflexes referring to characteristic tenorite lattice distances are observed [9]. There is no indication for the presence of cuprite (Cu_2O). Different NP agglomerates were investigated and apart from slight differences in the relative peak intensities, similar diffractograms are obtained. Within the limits of the TEM measurements, the NPs seem to be highly crystalline and purely present in the tenorite phase.

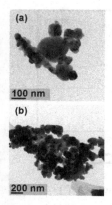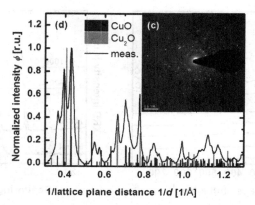

Figure 1. (a) and (b) TEM bright field images of tenorite NPs at different sample positions; (c) diffraction pattern of tenorite NPs corresponding to the bright field image as shown in figure 1(b); (d) angular integrated diffractogram of the pattern shown in figure 1(c). Reference data for tenorite (CuO) and cuprite (Cu₂O) is indicated by black and grey columns, respectively.

Figure 2 depicts representative Raman spectra for the step wise annealed tenorite NPs. The as-prepared sample shows three typical tenorite Raman modes at about 298 cm^{-1}, 345 cm^{-1}, and 632 cm^{-1} [10]. The weak feature at about 320 cm^{-1} as well as the signal around 500 cm^{-1} can be assigned to the quartz substrate. Besides, no cuprite Raman signal is detected. With increasing annealing temperature up to 700°C the full width at half maximum (FWHM) decreases and the peak intensity increases. Analyzing several positions of the samples, the FWHM of the most intense peak at 298 cm^{-1} is found to be reduced from about 15 cm^{-1} to 9 cm^{-1} accompanied by a doubling of the maximum peak intensity. The reduction of the FWHM and the increasing Raman intensity indicate a reduction of the density of structural defects and a release of stress.

At annealing temperatures of 800°C and 1000°C the tenorite NPs are transformed into cuprite which is confirmed by the detection of cuprite Raman modes, i.e. at about 144 cm^{-1}, 200 cm^{-1}, and 220 cm^{-1} [11]. No Raman modes of the tenorite phase are visible after annealing at 800°C and 1000°C. While the mode at 464 cm^{-1} is explained by the crystallization of the quartz substrate after the high temperature annealing steps, the origin of the broad Raman signal at 300 cm^{-1} is unknown.

Figure 3 shows the diffractograms of the NPs in the as-prepared state and after annealing at 700°C, 800°C, and 1000°C. In the as-prepared sample the observed reflexes can be assigned to characteristic tenorite lattice planes [9]. After instrumental correction, the signal obtained from this sample shows a crystallite size of about 29 nm +/-3 nm and a microstrain value $\Delta d/d0 = 0.22$ +/- 0.03, where d0 is the undistorted lattice spacing, as derived from the Double-Voigt approach (e. g. [12]) using the software TOPAS [13]. For this evaluation, Lorentzian broadening was assigned to the crystallite size whereas Gaussian broadening was attributed to microstrain. After annealing at 700°C, the signal broadening is significantly limited by the experimental setup leading to values with large error bars of 220 nm +/- 80nm for the crystallite

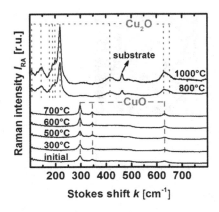

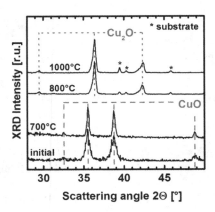

Figure 2. Representative Raman spectra of step wise annealed copper oxide NPs.

Figure 3. XRD diffractograms of step wise annealed copper oxide NPs.

size and 4% +/- 2% for the microstrain. These structural changes might be caused by partial sintering of the particles as well as the annealing of structural defects like e. g. stacking faults, in line with the conclusion derived from the Raman data. After annealing at 800°C there is no significant indication for a remaining tenorite phase. Characteristic cuprite reflexes are observed while further peaks could be assigned to the crystallized quartz substrate. Further annealing at 1000°C does not show further changes.

Photoluminescence and optical absorption at room temperature

PL spectra taken from the same spot wherefrom the Raman spectra are presented inFigure 4a-b. Figure 4a illustrates representative PL signals up to an annealing temperature of 700°C. Representative PDS signals are given in figure 4b. In as-prepared tenorite NPs the PL spectrum at room temperature exhibits a peak at about 1.28 eV and a FWHM of about 0.35 eV with an unsymmetrical shape, i.e. a slightly stronger decay towards higher energies. The PDS signal (figure 4b) slightly increases between 0.5 eV and 1.4 eV, and then increases further by a factor of three in the region between 1.4 eV and 1.6 eV before it saturates for higher energies. The value just below the saturation corresponds well to the optical band gap reported in literature for tenorite bulk material [1, 2]. The onset of the PL is at about 1.6 eV in good agreement to the band edge determined by PDS and is consequently interpreted as a band edge emission. PDS shows significant sub gap absorption extending to low energies indicating a high defect concentration in the initial state of the NPs. Stepwise annealing reduces the sub gap absorption continuously and concomitantly the PL intensity increases by about one order of magnitude. The decreasing sub gap absorption confirms the reduction of the defect concentration which leads to the conclusion that the increasing PL signal is due to the reduction of non-radiative recombination. Band edge emission from tenorite was not reported so far [7]. Therefore, our results show that tenorite NPs exhibit superior electronic properties in comparison with bulk material. Figure 5a-b shows PL and PDS signals for samples after annealing at 800°C and 1000°C. In both cases the decrease of the absorbance between 1.9 eV and 2.2 eV by almost two orders of magnitude is in good agreement with the optical band gap of 2.1 eV for cuprite bulk material [3, 4]. Furthermore, an absorption tail extending to low energies is observed and attributed to defect absorption from copper and oxygen vacancies [14]. After annealing at 800°C there still seems to be some amount of tenorite phase present leading to a slight increase of the PDS signal at 1.4 eV. No contribution of a tenorite phase is observed in the PDS signal after annealing at 1000°C. The PL spectra exhibit a pronounced emission signal at about 2 eV which is in good agreement with the optical band edge deduced from PDS and is therefore attributed to cuprite band edge emission. The fine structure of this peak which is also visible in the fine structure of the PDS signal is due to excitonic transitions [15]. Both PL signals show a weakly pronounced PL signal in the range between 1.0 eV and 1.6 eV. The origin of these features is ambiguous. Remaining tenorite band edge emission and well known defect emission coming from copper and oxygen vacancies could be responsible for the emission. Three emission signals at 1.72 eV, 1.53 eV, and 1.35 eV have been attributed to doubly charged oxygen, singly charged oxygen, and copper vacancies, respectively [16, 17]. The exact peak positions, and intensities strongly depend on the preparation procedure.

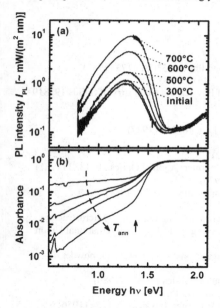 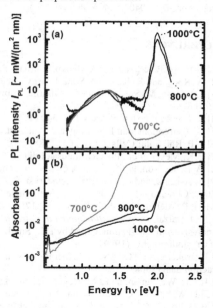

Figure 4. PL spectra (a) and PDS signals (b) of tenorite NPs step wise annealed up to an annealing temperature of 700°C.

Figure 5. PL spectra (a) and PDS signals (b) of tenorite and cuprite NPs step wise annealed up to an annealing temperature of 1000°C.

CONCLUSIONS

Band edge emission of phase pure tenorite NPs is observed at room temperature. The structural and electronic properties of tenorite NPs are improved by oven annealing in nitrogen atmosphere up to an annealing temperature of 700°C. A phase transition from tenorite to cuprite phase occurs after annealing at 800°C. Strong band edge emission accompanied by weak defect emission and possibly tenorite band edge emission is observed. After annealing at 1000°C very likely phase pure cuprite is prepared. The high PL intensity demonstrates the high quality of the tenorite and cuprite NPs that can be achieved by proper annealing and indicates the suitability of this material for photovoltaic applications. In comparison to results reported for tenorite bulk materialNPs include fewer defects and therefore show superior opto-electronic properties. The cuprite NPs exhibit excellent properties at least similar to high quality bulk material.

ACKNOWLEDGMENTS

The authors would like to thank O. Thimm and M. Hülsbeck for technical assistance. This work was partially funded by the German Federal Ministery of Education and Research (BMBF) project nr. 03SF0402A „NADNuM".

REFERENCES

[1] F. P. Koffyberg, and F. A. Benko, Journal of Applied Physics 53(2), 1173 (1982)
[2] A. Roos, and B. Karlsson, Solar Energy Materials 7, 467 (1983)
[3] J.B. Grun, M. Sieskind, and S. Nikitine, Journal of Physics and Chemistry of Solids 19, 819 (1961)
[4] R. Elliott, Physical Review 124, 340 (1961)
[5] L.O. Grondahl, and P.H. Geiger, J. Am. Inst. Elec. Eng. 46, 215 (1927)
[6] B.P. Rai, Solar Cells 25(3), 265 (1988)
[7] B.K. Meyer *et al.*, Physica status solidi (b), 1 (2012)
[8] W.B. Jackson, N.M. Amer, A.C. Boccara, and D. Fournier, Applied Optics 20, 1333 (1981)
[9] S. Asbrink, and L.J. Norrby, Acta Cryst. B26, 8 (1970)
[10] J. Chrzanowski, and J.C. Irwin, Solid State Communications 70, 11 (1989)
[11] J. Reydellet, Phys. Stat. Sol. (b) 52, 175 (1972)
[12] D. Balzar, *Voigt-function model in diffraction line-broadening analysis.* – Microstructure Analysis from Diffraction, edited by R.L. Snyder, H.J. Bunge, and J. Fiala, International Union of Crystallography, (1999)
[13] Bruker AXS TOPAS V4: General profile and structure analysis software for powder diffraction data (2008)
[14] F.L. Weichman, and J.M. Reyes, Canadian Journal of Physics 58, 325 (1980)
[15] D.W. Snoke, A.J. Shields, and M. Cardona, Phys. Rev. B 45, 693 (1992)
[16] J. Bloem, A.J. Van der Houven van Oordt, and F.A. Kröger, Physica 22, 1254 (1956)
[17] T. Ito, and T. Masumi, J. Phys. Soc. Jpn. 66, 2185 (1997)

Mater. Res. Soc. Symp. Proc. Vol.1494 © 2013 Materials Research Society
DOI: 10.1557/opl.2013.236

Simulating Constant Current STM Images of the Rutile TiO$_2$ (110) Surface for Applications in Solar Water Splitting

F. F. Sanches[1], G. Mallia[1], and N. M. Harrison[1,2]

[1] Thomas Young Centre, Department of Chemistry,
Imperial College London, South Kensington London SW7 2AZ, UK

[2] STFC Daresbury Laboratory, Daresbury, Warrington, WA4 4AD, UK

ABSTRACT

Solar water splitting has shown promise as a source of environmentally friendly hydrogen fuel. Understanding the interactions between semiconductor surfaces and water is essential to improve conversion efficiencies of water splitting systems. TiO$_2$ has been widely adopted as a reference material and rutile surfaces have been studied experimentally and theoretically. Scanning Tunneling Microscopy (STM) is commonly used to study surfaces, as it probes the atomic and electronic structure of the surface layer. A systematic and transferable method to simulate constant current STM images using local atomic basis set methods is reported. This consists of adding more diffuse p and d functions to the basis sets of surface O and Ti atoms, in order to describe the long range tails of the conduction and valence bands (and, thus, the vacuum above the surface). The rutile TiO$_2$ (110) surface is considered as a case study.

INTRODUCTION

Solar water splitting has shown a lot of promise as an environmentally friendly source of hydrogen fuel. However, solar-to-fuel efficiencies have to be improved significantly for this to become a viable alternative to fossil fuels. A fundamental understanding of photoanode surface and water interaction could be essential in improving these efficiencies. TiO$_2$ is a commonly used semi-conductor for solar water splitting [1, 2]. Whilst the large bandgap of TiO$_2$ is somewhat prohibitive for widespread use, it has been widely adopted as a model material for experimental and theoretical study. In practice, nanostructured, predominantely anatase TiO$_2$ is most commonly used [3–5]. These systems are difficult to study at an atomic scale experimentally as well as computationally. However, pristeen clean surfaces of TiO$_2$ can be investigated with a number of experimental and theoretical techniques. These are useful as model systems, where the surface structure can be analysed, as well as the interaction of surfaces with adsorbates[6, 7]. Scanning Tunneling Microscopy (STM) is commonly used to study surfaces, as it probes the atomic and electronic structure of the surface layer.

The rutile (110) TiO$_2$ surface has been studied using STM previously [8, 9]. Experimentally, the observed bright spots are attributed to the surface undercoordinated Ti atoms. Theoretical studies played an important role in predicting/confirming this observation. The LDA approximation to DFT and the plane wave (pseudopotential) approach was used to reproduce these experimental results computationally by relaxing the surface structure and then simulating STM images [9].

It has been previously shown that the use of hybrid exchange functionals (where a proportion of Fock exchange is included in the exchange functional) gives an accurate description of the structural energetics and of the electronic structure (i.e. band gap and band offset) for periodic systems [10–21], particularly for transition metal oxides. The implementation of hybrid-exchange functionals using local atomic basis sets, as in the CRYSTAL code, is computationally efficient also for large periodic systems [22, 23]. Furthermore, local basis sets allow for a chemical description of molecular charge densities (eg. adsorbates on the surface). Di Valentin has demonstrated that using an atom-centered Gaussian basis set optimised for the ground state energy does not describe the long range tails of the valence and conduction bands in the vacuum above the surface sufficiently accurately [24]. In this study, the contrast in the STM image of the (110) surface is reversed with respect to previous calculations and experimental observations [9] (bright spots above the O atoms). In an attempt to remedy this and exploit the advantages of hybrid exchange functionals, additional s functions (called 'ghost functions') were added 2 Å above the x and y coordinates of the surface Ti and O (with the z axis perpendicular to the surface). The addition of these functions allows for more accurate description of the vacuum above the surface. This work has established the need to improve the description of the conduction band states for simulations of STM images. The approach of adding non-atom centred functions is unsatidfactory, however, as its application to complex surface geometries requires ad hoc decisions about positioning of the additional

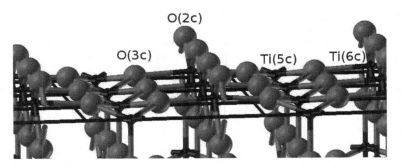

FIG. 1: 3D view of the rutile TiO$_2$ (110) surface. Small (large) spheres are titanium (oxygen) atoms. The surface cell lattice cell is also drawn. O(3c) and O(2c) are the three-fold (planar) and two-fold (bridging) coordinated oxygen ions. Ti(6c) and Ti(5c) are the six-fold and five-fold coordinated titanium ions.

basis functions.

Here, we propose an alternative method to improve the description of the long range tails of the valence and conduction bands and thus the description of the charge density in the vacuum. In principle, adding more diffuse functions to the atomic basis sets for the surface undercoordinated atoms should sufficiently improve the description of long range tails and achieve a similar effect described in Ref. [24]. This method could be more systematic and, perhaps, more transferrable to other surfaces/materials.

COMPUTATIONAL DETAILS

All calculations were performed using the CRYSTAL09 code[22], based on the periodic *ab initio* linear combination of atomic orbitals (LCAO). The hybrid exchange and correlation functional B3LYP [25] was adopted. A previous *ab initio* study of bulk rutile and anatase TiO$_2$ using various hybrid exchange correlation functionals showed that the B3LYP functional gave accurate descriptions of structural and electronic properties [26]. The atoms were described using local basis set (BS) consisting of atom centred Gaussian orbitals. Both the Ti and O atoms are described by a triple valence all-electron BS: an 86-411G** contraction (one s, four sp, and two d shells) and an 8-411G* contraction (one s, three sp, and one d shells), respectively; the most diffuse $sp(d)$ exponents are α^{Ti}=0.3297(0.26) and α^{O}= 0.1843(0.6) Bohr^{-2}. Integration was carried out over reciprocal space using a shrinking factor of 8 to form a Pack-Mockhorst mesh of k points. This grid converges the integrated charge density to an accuracy of about 10^{-6} electrons per unit cell. The Coulomb and exchange series are summed directly and truncated using overlap criteria with thresholds of 10^{-7}, 10^{-7}, 10^{-7}, 10^{-7} and 10^{-14} as described previously [22, 27]. The self-consistent field (SCF) algorithm was set to converge at the point at which the change in energy, ΔE, was less then 10^{-7} Hartree.

DISCUSSION

Rutile (110) Surface

The rutile structure belongs to the $P4_2/mnm(D_{4h}^{14})$ tetragonal space group and the unit cell is defined by the lattice vectors $\mathbf{a_B}$ and $\mathbf{c_B}$ (the subscript B denoting the bulk phase) and contains two TiO$_2$ units with Ti ions at (0,0,0) and $(\frac{1}{2},\frac{1}{2},\frac{1}{2})$ and O ions at $\pm(u,u,0)$ and $\pm(u+\frac{1}{2},\frac{1}{2}-u,\frac{1}{2})$ [28, 29]. The predicted structural parameters, with the deviation from those observed [30] in parenthesis, are: $\mathbf{a_B}$ = 4.639Å(1.20%), $\mathbf{c_B}$ = 2.979Å(0.88%), u = 0.306(0.00%)

and $V_{\mathbf{B}} = 64.120\text{Å}^3 (3.32\%)$. This structure is consistent with that predicted in previous calculations [28].
Each Ti is octahedrally coordinated to six O ions. The TiO_2 octahedron is distorted, with the length of the apical
Ti-O_{ap} bonds slightly longer than equatorial, Ti-O_{eq}, bonds. The calculated (observed) lengths (in Å) being 2.009
(1.983) for Ti-O_{ap} and 1.959 (1.946) for Ti-O_{eq} [30].

The (110) surface displayed in Fig. 1 exposes three-fold (planar) and two-fold (bridging) coordinated oxygen ions,
labelled as O(3c) and O(2c), and six-fold and five-fold coordinated titanium ions, labelled Ti(6c) and Ti(5c). A stoichiometric and non-polar termination is obtained by terminating such slabs on a bridging oxygen layer and only considering slabs consisting of multiples of 3 atomic layers according to the following sequence of planes: $O - Ti_2O_2 - O$.
The effect of slab thickness on the surface structure and energy of formation has been carefully analysed and a full
relaxation of slabs containing up to 36 atomic layers has been carried out.

Electronic Structure

The density of states (DOS) of the bulk rutile is displayed in Fig. 2(bottom), in which the contributions from two
bands are evident: the top of the valence band and the conduction band. The valence band has predominantly O-2p
character but is hybridised with Ti-3d orbitals. The conduction band is mainly derived from Ti-3d atomic orbitals
(with some hybridisation with O-2p orbitals). Fig. 2(top) shows the DOS of the relaxed (110) surface. The calculated
fundamental bandgap is very similar in both cases (3.41eV and 3.55eV for the bulk and surface, respectively). This
is an indication that no surface states form in the band gap. The bulk band gap measured using optical techniqes is
~3eV [31, 32]. The optical bandgap includes a contribution from excitonic binding which is not accounted for in the
fundamental bandgap computed here. Therefore, the calculated values are in good agreement with this. The B3LYP
functional gives a better estimate of fundamental band gaps in semi conductors than the commonly used LDA and
GGA approaches [16, 33, 34].

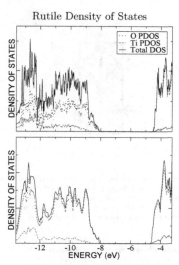

FIG. 2: (Colour on-line) Projected Density of States for bulk rutile (bottom) and the relaxed (110) surface (top). In both
cases the continuous (black) line represents the total DOS, whilst the dotted (red) and dashed (blue) lines represent the DOS
projected on all O and Ti atoms, respectively. The top of the VB of the bulk rutile was shifted to align with that of the (110)
surface.

Simulated STM images

A systematic approach to optimising the basis sets for describing the long range tails of the conduction and valence bands (and, thus, the vacuum above the surface) has been adopted. This was achieved by adding more diffuse p and d functions to the basis sets of surface O and Ti atoms, respectively. Here we present a case study of the rutile TiO_2 (110) surface. We have also reproduced the calculations performed by Di Valentin for comparison [24].

The STM images were produced based on the Tersoff-Hamann approximation [35]. The interaction of tip and surface is ignored. The current is approximated as a charge density corresponding to the states in the lower part of the conduction band; an energy window, or bias potential, of 1V above the bottom of the conduction band is considered.

(110) STM Images

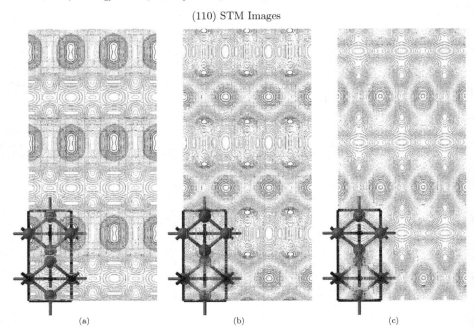

(a)   (b)   (c)

FIG. 3: (Colour on-line) Simulated constant current STM images for various basis set enhancements and the original BS. These are contour maps of the charge density isosurface at 5×10^{-6} electrons/bohr^3 and 1V sample bias. (a) Original Basis set (described in the computational details) (b) Original Basis set plus added s functions above the surface (c) Original Basis set with additional diffuse p and d functions in the surface O and Ti basis sets, respectively. In these images the largest values of height (red contours), comparable to bright spots on STM images, are located above the O atoms in (a) and in the region above the Ti atoms in (b) and (c).

In constant current STM images bright spots correspond to areas where surface topography and/or charge density has pushed the tip away from the sample (i.e. the height of the tip is increased). The contour plots shown in Fig. 3 are constant current contour plots. Red contours represent larger values of height (or bright spots on an STM image), whilst blue represents the lowest heights (dark areas). Examination of the position of the 'bright' areas on the three plots in Fig. 3 reveals that in a) the bright spots would be directly above the bridging oxygen atoms O(2c). This is in disagreement with the experimentally observed contrast, as well as previous (plane wave) calculations[8, 9]. This discrepancy is purely due to the inability of the basis set to describe long range conduction and valence band tails with sufficient accuracy. With the additional functions and enhanced atomic basis sets applied, (b) and (c) display

(110) Charge Density Isosurfaces

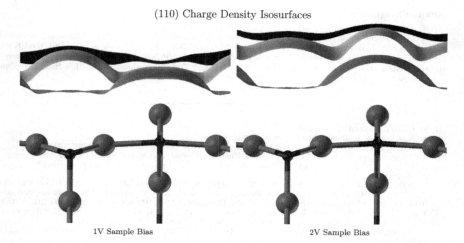

1V Sample Bias   2V Sample Bias

FIG. 4: (Colour on-line) Small (black) spheres represent Ti atoms and large (red) spheres represent O atoms. Traces above the atoms are charge density isosurfaces at 5×10^{-6} electrons/bohr^3. Above the titanium atom the traces represent from bottom to top: the isosurface obtained with the original BS (light blue); the isosurface obtained obtained after adding additional s functions above surface atoms (gray); the isosurface obtained after addition of diffuse p and d orbitals on the surface O and Ti BS respectively (black), respectively.

the expected contrast (bright spots above the Ti and dark(er) areas above the oxygen atoms.) Therefore, the method developed here produces results comparable to previous calculations [8, 24].

Analysis of the charge density isosurfaces at 2V applied bias in Fig. 4 reveals that in both the case of the enhanced atomic basis sets as well as the 'ghost atom' approach, the highest point remains above the O(2c). We propose that sampling 2V above the bottom of the conduction band still includes regions where the description of the long range tails of the CB and VB are not described adequately. However, experimentally, a sample bias of 1V is commonly used to obtain high resolution STM images of TiO_2 surfaces [8, 9, 36]. Therefore, the use of 1V applied bias in the calculations is reasonable.

The isosurfaces at 1V applied bias reveal that using the basis set enhancements produces a broad peak centered around the Ti(5c) and a trough directly above the O(2c). In contrast, the 'ghost atom' approach produces two peaks above the Ti(5c) and the O(2c), where the latter peak is considerably smaller than the former. This would indicate bright spots above the titanium atoms in both cases. These observations are in line with the constant current contour plots in Fig. 3. Qualitative comparison of these isosurfaces with similar images in Ref. [8] appear to show better agreement between our method and the plane wave calculation than with the 'ghost atom' approach. The images produced by Diebold et. al. also show a peak above the titanium atom and a trough above the oxygen atom at comparable currents.

CONCLUSIONS

We have proposed an alternative approach to simulate constant current STM images using local atomic basis set methods. By enhancing the atom centred basis sets of the undercoordinated Ti and O atoms we avoid the need for additional functions above the surface. The approach was tested on the relaxed (110) surface and was found to produce similar contrast as in Ref [24]. The advantage of the approach we have presented here is that it provides a systematic and transferable method for the simulation of constant current STM images.

Acknowledgments

This work made use of the high performance computing facilities of Imperial College London and - via membership of the UK's HPC Materials Chemistry Consortium funded by EPSRC (EP/F067496) - of HECToR, the UK's national high-performance computing service, which is provided by UoE HPCx Ltd at the University of Edinburgh, Cray Inc and NAG Ltd, and funded by the Office of Science and Technology through EPSRC's High End Computing Programme.

---

[1] M. Gratzel, Nature **414**, 338 (2001).

[2] A. Fujishima and K. Honda, Nature **238**, 37 (1972).

[3] J. Tang, J. R. Durrant, and D. R. Klug, Journal of the American Chemical Society **130**, 13885 (2008).

[4] A. Cowan, J. Tang, W. Leng, J. Durrant, and D. Klug, The Journal of Physical Chemistry C **114**, 4208 (2010).

[5] A. M. Peir, C. Colombo, G. Doyle, J. Nelson, A. Mills, and J. R. Durrant, The Journal of Physical Chemistry B **110**, 23255 (2006).

[6] M. Patel, G. Mallia, and N. M. Harrison, in *NSTI NANOTECH 2011, TECHNICAL PROCEEDINGS - MICROSYSTEMS, PHOTONICS, SENSORS, FLUIDICS, MODELING, AND SIMULATION* , edited by Laudon, M and Romanowicz, B, (CRC PRESS-TAYLOR & FRANCIS GROUP, 6000 BROKEN SOUND PARKWAY NW, STE 300, BOCA RATON, FL 33487-2742 USA, 2011), ISBN , Nanotechnology Conference and Expo (Nanotech 2011), Boston, MA, JUN 13-16, 2011.

[7] M. Patel, G. Mallia, L. Liborio, and N. M. Harrison, Phys. Rev. B **86**, 045302 (2012).

[8] U. Diebold, J. Lehman, T. Mahmoud, M. Kuhn, G. Leonardelli, W. Hebenstreit, M. Schmid, and P. Varga, Surface science **411**, 137 (1998).

[9] U. Diebold, J. F. Anderson, K.-O. Ng, and D. Vanderbilt, Phys. Rev. Lett. **77**, 1322 (1996).

[10] D. Muñoz, N. M. Harrison, and F. Illas, Phys. Rev. B **69**, 085115 (2004).

[11] J. Muscat, A. Wander, and N. Harrison, Chemical Physics Letters **342**, 397 (2001).

[12] G. Mallia and N. M. Harrison, Phys. Rev. B **75**, 165201 (2007).

[13] N. C. Wilson, S. P. Russo, J. Muscat, and N. M. Harrison, Phys. Rev. B **72**, 024110 (2005).

[14] G. Mallia, R. Orlando, M. Llunell, and R. Dovesi, in *Computational Materials Science*, edited by C. Catlow and E. Kotomin (IOS Press, Amsterdam, 2003), vol. 187 of *NATO SCIENCE SERIES, III: Computer and Systems Sciences*, pp. 102–121.

[15] C. Di Valentin, G. Pacchioni, and A. Selloni, Phys. Rev. Lett. **97**, 166803 (2006).

[16] F. Corà, M. Alfredsson, G. Mallia, D. Middlemiss, W. Mackrodt, R. Dovesi, and R. Orlando, in *Principles and Applications of Density Functional Theory in Inorganic Chemistry II*, edited by N. Kaltsoyannis and J. McGrady (Springer Berlin / Heidelberg, 2004), vol. 113, pp. 171–232.

[17] G. C. De Fusco, L. Pisani, B. Montanari, and N. M. Harrison, Phys. Rev. B **79**, 085201 (2009).

[18] L. Liborio, G. Mallia, and N. Harrison, Phys. Rev. B **79**, 245133 (2009).

[19] C. L. Bailey, L. Liborio, G. Mallia, S. Tomić, and N. M. Harrison, Phys. Rev. B **81**, 205214 (2010).

[20] L. M. Liborio, C. L. Bailey, G. Mallia, S. Tomic, and N. M. Harrison, Journal of Applied Physics **109**, 023519 (pages 9) (2011).

[21] E. A. Ahmad, L. Liborio, D. Kramer, G. Mallia, A. R. Kucernak, and N. M. Harrison, Phys. Rev. B **84**, 085137 (2011).

[22] R. Dovesi, V. Saunders, C. Roetti, R. Orlando, C. Zicovich-Wilson, F. Pascale, B. Civalleri, K. Doll, N. Harrison, I. Bush, et al., Università di Torino (Torino, 2006).

[23] I. J. Bush, S. Tomi, B. G. Searle, G. Mallia, C. L. Bailey, B. Montanari, L. Bernasconi, J. M. Carr, and N. M. Harrison, Proceedings of the Royal Society A: Mathematical, Physical and Engineering Science **467**, 2112 (2011), http://rspa.royalsocietypublishing.org/content/early/2011/04/06/rspa.2010.0563.full.pdf+html.

[24] C. Di Valentin, The Journal of chemical physics **127**, 154705 (2007).

[25] A. Becke, Chem. Phys **98**, 5648 (1993).

[26] F. Labat, P. Baranek, C. Domain, C. Minot, and C. Adamo, The Journal of chemical physics **126**, 154703 (2007).

[27] C. Pisani, R. Dovesi, and C. Roetti, *Hartree-Fock ab initio Treatment of Crystalline Systems*, vol. 48 of *Lecture Notes in Chemistry* (Springer Verlag, Heidelberg, 1988).

[28] J. Muscat, V. Swamy, and N. Harrison, Physical Review B **65**, 224112 (2002).

[29] G. Cangiani, A. Baldereschi, M. Posternak, and H. Krakauer, Physical Review B **69**, 121101 (2004).

[30] U. Diebold, Applied Physics A: Materials Science & Processing **76**, 681 (2003).

[31] D. W. Fischer, Phys. Rev. B **5**, 4219 (1972).

[32] K. Hellwege and O. Madelung, Electron Paramagnetic Resonance, Springer-Verlag, Berlin (1984).

[33] R. Martin and F. Illas, Physical review letters **79**, 1539 (1997).

[34] I. de PR Moreira, F. Illas, and R. Martin, Physical Review B **65**, 155102 (2002).

[35] J. Tersoff and D. Hamann, Physical Review B **31**, 805 (1985).

[36] W. Hebenstreit, N. Ruzycki, G. Herman, Y. Gao, and U. Diebold, Physical Review B **62**, 16334 (2000).

Mater. Res. Soc. Symp. Proc. Vol. 1494 © 2012 Materials Research Society
DOI: 10.1557/opl.2012.1646

Al-Doped ZnO Film as a Transparent Conductive Substrate in Indoline-Sensitized Nanoporous ZnO Solar Cell.

B. Onwona-Agyeman[1], M. Nakao[2], G.R.A. Kumara[3] and T. Kitaoka[1]
[1]Department of Agro-environmental Sciences, Graduate School of Bioresource and Bioenvironmental Sciences, Kyushu University, 6-10-1 Hakozaki, Higashi-ku, Fukuoka 812-8581, Japan.
[2]Department of Basic Science, Faculty of Engineering, Kyushu Institute of Technology, 1-1 Sensui, Tobata-ku, Kitakyushu 804-8550, Japan.
3Department of Chemistry, Peradeniya University, Peradeniya 20400, Sri Lanka.

ABSTRACT

We have deposited porous ZnO films on aluminum-doped ZnO (ZnO/AZO) and fluorine-doped tin oxide (ZnO/FTO) transparent substrates, and annealed both in air at 500°C. X-ray diffraction measurements of the nanoporous ZnO films after heat treatment showed that, ZnO/AZO film exhibited a dominant (002) diffraction while the ZnO/FTO showed mixed diffraction peaks with the (100) and (101) being dominant. Dye-sensitized solar cells (DSC) based on the sensitization of the porous ZnO films on AZO and FTO substrates with an indoline dye were constructed. The photoaction spectrum, which is a measure of the degree of sunlight harvesting, was broad and higher in the ZnO/AZO DSC than that of the ZnO/FTO DSC. Conversion efficiency of 7.3 % was obtained for the ZnO/AZO DSC while 4.5 % was recorded for the ZnO/FTO. The superior photovoltaic performance of the ZnO/AZO DSC is attributed to better ZnO film orientation after thermal treatment and the higher sunlight harvesting.

INTRODUCTION

Light absorption is an important parameter in solar energy devices because the absorption process starts the conversion of solar energy to electricity. In dye-sensitized solar cells (DSC), the light absorption is achieved by surface modification of wide bandgap porous semiconductors such as TiO_2, ZnO and SnO_2 with visible-light absorbing dye molecules [1-4]. Under illumination of a DSC, an electron is injected from the excited state of the dye into the conduction band of a porous semiconductor followed by a hole transfer to the electrolyte [5]. The injected electrons then travel through the porous semiconductor to the transparent electrode, where they are collected and directed to the external circuit. These processes lead to the direct conversion of sunlight to electricity in the DSC. The efficiency of a DSC in solar energy conversion therefore depends on the performance of the individual components in the solar cell. In this work, we have investigated the role of two important components in a DSC, that is, the porous semiconductor and the transparent conductor. The porous semiconductor used was ZnO and was deposited on two transparent conductors, AZO and FTO. An indoline dye (D358), a metal-free dye was used to sensitize the two phoelectrodes, and the resultant DSCs were constructed and their solar cell parameters were measured under AM 1.5 (1000 W m^{-2}) irradiation using a calibrated solar cell evaluation system.

EXPERIMENT

The AZO transparent films on glass substrates were prepared by radio frequency (rf) magnetron sputtering technique using ceramic target consisting of ZnO and 2.5 wt. % aluminum. Pure argon was used as the sputtering gas, the substrate temperature and rf power were kept at 100°C and 3 Pa respectively, during the sputtering process. The AZO films obtained under these sputtering conditions yielded films with thickness of about 1.3 μm, sheet resistance of 9 Ω/Sq and optical transmittance of about 83 % within the wavelength 400-800 nm. The FTO-coated glass substrates were obtained from Nihon Sheet Glass Company with sheet resistance of 10 Ω/Sq and optical transmittance of 80 % within the wavelength 400-800 nm. Porous ZnO films were deposited on the AZO and FTO by spray pyrolysis technique. ZnO powder (particle size less than 30 nm in diameter, Wako Chemicals, Japan), few drops of acetic acid and ethanol were mixed and ultrasonically dispersed for about 10 min. The solution was then sprayed onto heated (150°C) AZO and FTO transparent substrates, and the resultant films were then heated in air at 500°C for 30 min. X-ray diffractometry (XRD) was used to study the structural properties of the porous ZnO films on the two substrates after the thermal treatment. The two photoelectrodes were then coated with the D358 dye (Mitsubishi Paper Mills, Japan) by dipping the films in 3×10^{-4} mol dm^{-3} dye solutions in *tert*-butanol: acetonitrile (1:1 by vol) for 12 h. The dye-coated photoelectrodes (active area: 25 cm^2) were sandwiched with a platinum-sputtered glass as counter electrode and the intervening space filled with an electrolyte [1].

DISCUSSION

Figure 1 illustrates the chemical structure of the indoline dye D358 used in the sensitization of the two porous ZnO films on AZO and FTO. Figure 2 shows the XRD patterns of

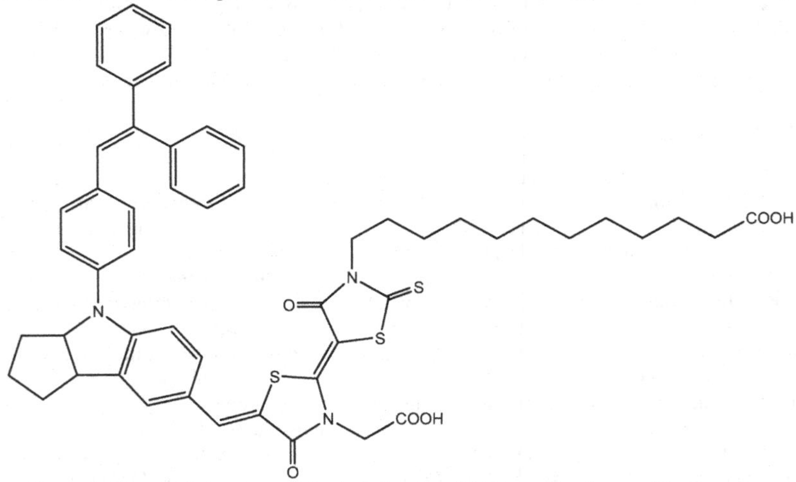

Figure 1. Chemical structure of the indoline dye (D358) used in the sensitization of porous ZnO

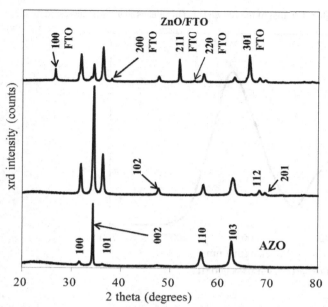

Figure 2. XRD patterns of annealed AZO, ZnO/AZO and ZnO/FTO films.

the AZO, ZnO/AZO and ZnO/FTO films. From the XRD patterns of the AZO and ZnO/AZO films, the (002) was the dominant diffraction peak while (100) and (101) were the main dominant peaks in the case of ZnO/FTO. When the porous ZnO was deposited on AZO and heated, the ZnO film still maintained the same crystallographic alignment with the AZO substrates. This is very important in the successful operation of a DSC as photo-injected electrons moving through the porous ZnO can be easily collected by the AZO transparent electrode. On the other hand, when the porous ZnO was deposited on the FTO and heated, the crystallographic alignment of the porous ZnO was dominated by the (100) and (101) orientations. In general, films with many different crystallographic orientations contain many grain boundaries and electron movements within such films are difficult because of the defect sites along the grain boundaries act as electron traps [6, 7].

Figure 3 shows the absorption spectrum of the D358 dye in dimethylformamide (DMF). Figure 4 is the incident photon-to-current conversion efficiency (IPCE) of the ZnO/AZO and ZnO/FTO electrodes sensitized with the D358. The absorption band maximum of the D358 dye in the DMF was about 530 nm and this significantly shifted to higher wavelengths when the dye was coated on the ZnO/AZO and ZnO/FTO, respectively, as shown in Fig. 4. Indoline dyes are known to form dye aggregates when they are coated on oxide semiconductors and as a results shifting the absorption band towards higher wavelength [8, 9]. The IPCE spectrum of the D358 sensitized ZnO/AZO was broader than the ZnO/FTO electrode with the maximum absorption peak almost

80 % and was also higher than that of the ZnO/FTO. This means that, light harvested by the

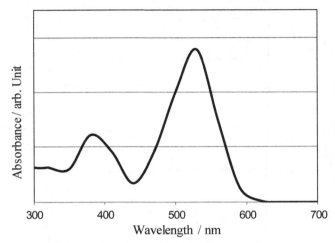

Figure 3. Absorption spectrum of D358 in DMF.

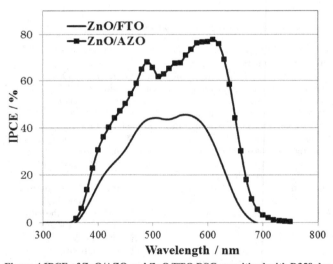

Figure 4 IPCE of ZnO/AZO and ZnO/FTO DSCs sensitized with D358 dye.

ZnO/AZO sensitized electrode for the crucial light-to-current conversion process is higher than the ZnO/FTO electrode. Also, the shifting of the absorption band of the ZnO/AZO towards higher wavelength than the ZnO/FTO may be due to D358 dye aggregates on the better crystallographic ZnO/AZO after the heat treatments. This behavior of the D358 dye on the two electrodes needs further research. Table 1 summarizes the I-V parameters of the ZnO/AZO and ZnO/FTO

Table 1 I-V parameters of ZnO/AZO and ZnO/FTO electrodes sensitized with D358 dye.

Electrode	Jsc (mA/cm^2)	Voc (V)	FF	Efficiency (%)
ZnO/AZO	21.2	0.64	0.54	7.3
ZnO/FTO	17.5	0.53	0.49	4.5

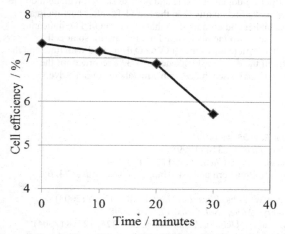

Figure 5. Conversion efficiency of ZnO/AZO DSC measured at different times.

electrodes sensitized with the D358 dye. From Table 1, the superior photovoltaic properties of the ZnO/AZO DSC are attributed to the better crystalline structure and higher sunlight harvesting of the porous ZnO film on the AZO transparent substrate. When the porous ZnO was deposited on FTO and heated at 500°C, XRD measurement revealed that, the film showed diffraction peaks oriented differently. Such films were made up of different crystallographic grains with many grain boundaries. The grain boundaries are made up of defect sites and recombination traps that impede the flow of electrons. In a DSC, the photo-electrode (porous ZnO in this work) accepts photo-injected electrons from the excited dye and the electrons then travel through the porous semiconductor where they are collected. The porous ZnO on the AZO substrate showed mainly the (002) diffraction peak, indicating that the film is composed of few grain boundaries and the effect of the grain boundaries on the flow of electrons will be less. The lower photocurrent observed for the ZnO/FTO DSC can be attributed to low sunlight harvested compared with ZnO/AZO DSC. Also, the lower Voc and FF values recorded for the ZnO/FTO DSC are due to the recombination of the photo-injected electrons as the electrons have to overcome many grain boundaries found in the porous ZnO film. The efficiency recorded for the ZnO/AZO DSC was 7.3 %, and we measured the efficiency again at an interval of 10 min. to check the stability. Figure 5 shows the rate of decrease in the efficiency of the ZnO/AZO DSC with time. Organic dyes are known not to be very stable when used to sensitized wide bandgap semiconductors and recently, addition of chenodeoxylicholic acid during the preparation of the photo-electrode seems to maintain the efficiency [10]. We are planning to investigate the effect of this reagent on the efficiency of porous ZnO when sensitized with D358.

CONCLUSIONS

We have prepared dye-sensitized solar cells based on the deposition of porous ZnO films on aluminum-doped zinc oxide and fluorine-doped tin oxide and both sensitized with an indoline dyes. The ZnO/AZO exhibited better crystalline quality than the ZnO/FTO after heating in air. The IPCE of the ZnO/AZO DSC was higher and broader than that of the ZnO/FTO, indicating better sunlight harvesting capability. These two factors resulted in the superior solar cell performance of the ZnO/AZO DSC. The efficiency of the ZnO/AZO decreased with time and the decreased was attributed to instability of the dye. The efficiency of 7.3 % recorded for the ZnO/AZO DSC is very high among ZnO DSCs sensitized with a metal-free organic dye.

REFERENCES

1. M. Grätzel, *Nature* **414**, 338 (2001).
2. A. Hagfeldt and M. Grätzel, *Chem. Rev.* **95**, 49 (1995).
3. M. Grätzel, *Curr. Opin. Colloid Interface Sci.* **314** (1999).
4. K. Kalyanasundaram and M. Grätzel, *Coord. Chem. Rev.* **177**, 347 (1998).
5. S.G. Yan, L.A. Lyon, B.I. Lemon, J.S. Preiskorn and J.T. Hupp, *J. Chem. Educ.* **74**, 657 (1997).
6. M. Birkholz, B. Selle, F. Fenske and W. Fuhs, *Phys. Rev. Lett. B* **68**, 205414 (2003).
7. S. Major and K.L. Chopra, *Sol. Energy Mater. Sol. Cells*, **17**, 319 (1998).
8. T. Horiuchi, H. Miura, K. Sumioka and S. Uchida, *J. Am. Chem. Soc.* **126**, 12218 (2004).
9. T. Horiuchi, H. Miura and S. Uchida, *Chem. Commun.* 3036 (2003).
10. S. Ito, H. Miura, S. Uchida, M. Takata, K. Sumioka, P. Liska, P. Comte, P. Pechy and M. Grätzel, *Chem. Commun.* 5194 (2008).

Mater. Res. Soc. Symp. Proc. Vol. 1494 © 2013 Materials Research Society
DOI: 10.1557/opl.2013.238

Development of High Temperature Optical Interference Filters

Thomas C. Parker[1] and John D. Demaree[1]
[1]U.S. Army Research Lab, 4600 Deer Creek Loop, Aberdeen Proving Grounds, MD 21005, U.S.A.

ABSTRACT

Oblique angle deposition (OAD) is a self-organizing physical vapor deposition (PVD) technique that has been used to grow sculpted 3D nanostructures including helices, slanted rods, and zigzag structures, and other shapes. OAD structures can be fabricated from virtually any material that can be deposited using PVD including: polymers, metals, semiconductors, oxides, and nitrides. The control over the nano-scale structural anisotropy of these materials allows one to tailor their electrical, magnetic, mechanical, crystalline, and optical properties. Through the careful design of the OAD structure and material selection this technique can be used to create photonic materials (1D, 2D, and 3D) with unique properties. We will discuss ongoing work using OAD to develop oxide thin film interference filters that can withstand extreme temperatures (800-1000° C) at mTorr vacuum levels, which are being developed for thermal photovoltaic applications.

INTRODUCTION

The development of efficient, portable thermal photovoltaic (TPV) power systems depends on the ability to maximize the power transfer from a hot radiating emitter to a photovoltaic element. Thin film optical filters may be used to achieve this goal by rejecting long wavelength light which will not be efficiently captured by the particular photovoltaic material, effectively "recycling" these photons to increase the efficiency of the system. In terrestrial-based light weight TPV systems (<50 W) the emitter and optical filter materials must be able to withstand temperatures as high as 900 to 1400° C [1] as well as resist oxidation due to the imperfect vacuum present between the emitter and photovoltaic. Optical filters that can withstand these conditions are not currently commercially available due to the coefficient of thermal expansion (CTE) differences between the thin film layers and the substrate.

In this paper we investigated the use of the oblique angle deposition (OAD) technique to grow oxide-based porous optical filters. The porous columnar structure of the OAD films help to reduce the stress from CTE mismatch. This property allows the OAD filters to withstand temperature cycles from room temperature to 1000° C in air at atmospheric pressure, and may form a critical component of future portable TPV power systems.

The interference filter was fabricated from an OAD deposited SiO_2 and then a normal deposition of Y_2O_3. Both materials were chosen for their high temperature stability in an oxidizing environment. A large factor in the overall performance of an interference filter in the index contrast between adjacent layers. The OAD SiO_2 was chosen as the low index material, due to its extremely low index of 1.05[2].

EXPERIMENTAL

A cyrogenically pumped four pocket electron beam (MDC Frame 3) deposition system was used to deposit the filters in this study . Two of the ebeam pockets were loaded with SiO_2 (K. J. Lesker, purity 99.99%) and Y_2O_3 (K. J. Lesker, purity 99.99%) pieces, respectively. The deposition rate and total thickness attained was monitored using a Sycon STM-100 monitor and a Sycon VSO-100 quartz crystal microbalance (QCM) positioned next to the sample. The base pressure of the system was 1×10^{-7} Torr and the pressure during deposition was approximately 7×10^{-7} Torr; no effort was made to control the partial pressure of oxygen during deposition by backfilling the chamber during deposition. The distance between the ebeam source and the substrate was fixed at 40 cm, and the angle between the sample normal with respect to the incoming atomic flux was 82° for the porous SiO_2 layer and 0° for the subsequent Y_2O_3 capping layer. Polished silicon and sapphire substrates were placed in the chamber adjacent to each other for simultaneous deposition. After deposition each sample was cleaved into smaller pieces, one of which was annealed at 1000° C in a quartz tube furnace for 90 minutes in a dry air flow of 300 Sccm.

The filters were examined before and after annealing using Rutherford backscattering spectrometry (RBS) and reflectance spectrophotometry. The RBS was performed using a 2 MeV He^+ ion beam from a National Electrostatics 5SDH-2 pelletron accelerator with a backscattering angle of 170°, and the results interpreted using the simulation program SIMNRA. Reflectance spectrophotometry measurements were performed using a Perkin-Elmer Lambda 950. Some samples were also examined by Variable Angle Spectroscopic Ellipsometry (VASE) using a J.A. Woollham VASE® instrument.

RESULTS AND DISCUSSION

In Figure 1 the RBS spectrum of a deposited filter on a silicon substrate is shown (black circles), with a best fit simulated spectrum obtained using SIMNRA (red line). The RBS analysis showed that both the SiO_2 and Y_2O_3 layers were stoichiometric. The thickness of the SiO_2 layer was found to be ~530 nm (after correcting for the porosity and the angle of growth, as in Ref. [3]) and the thickness of the Y_2O_3 was ~95 nm. There were no significant differences in the spectra between films grown on sapphire or silicon substrate after deposition. After annealing, however, there was evidence of significant oxidation underneath the deposited film on a silicon substrate. A bare silicon substrate placed in the tube furnace also oxidized to a similar extent, and both RBS and VASE examination of these samples confirmed an oxide thickness of approximately 42 nm underneath the deposited layer after annealing.

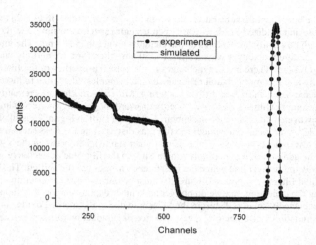

Figure 1. RBS spectrum of the as-deposited film stack, Y_2O_3(100nm)/SiO_2(550 nm, porous) on a Si substrate.

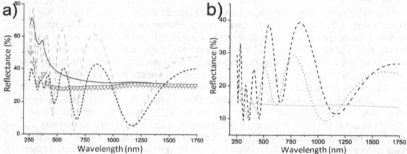

Figure 2. a) measured reflectance spectroscopy of Si substrate samples, the green curve (long dash) sample has the film stack Y_2O_3(100nm)/SiO_2(550 nm, porous) on Si as deposited, blue curve (short dash) is the same sample annealed as 1000° C in air for 90 minutes, red curve (downward open triangles) is a Si wafer annealed at 1000° C in air for 90 minutes, and the black curve (solid) is the as received Si wafer. b) the measured reflectance spectroscopy of Al_2O_3 substrate samples, the black curve (long dash) has the film stack on Al_2O_3 as deposited, red curve (dots) is the film stack on Al_2O_3 annealed at 1000° C in air for 90 minutes, and the green curve (solid) is the as received Al_2O_3 substrate.

Figure 2 shows the reflectance spectra of the films before (long dash line) and after annealing (short dash line) for (a) films deposited on silicon and (b) films deposited on sapphire with the as deposited (long dash line) and post anneal (short dash line), along with reference spectra from bare substrates. The bare sapphire substrate is shown in Fig. 2b (solid line). In Fig. 2a the Si (solid line) substrate and oxidized Si (downward open triangle) spectra are similar to those found in literature [3], with no significant features at wavelengths longer than 500 nm. The interference fringes evident in the figure at longer wavelengths (>350 nm), therefore, are entirely due to the deposited SiO_2/Y_2O_3 filter. There are slight changes in the optical properties of the filters after annealing, primarily a reduction in overall reflectance on all samples, likely due to increased surface roughness and hence increased diffuse scattering. Most spectral features remain unchanged in wavelength, with very small shifts either toward longer wavelengths (samples on silicon) or shorter wavelengths (samples on sapphire). The red shift in Figure 2a is likely due to the growth of the 42 nm thermal oxide under the filter, as discussed in the RBS section above. In effect the original low index porous SiO_2 layer is no longer sandwiched between the Y_2O_3 and the Si substrate. The addition of the thermally grown SiO_2 to the film stack effectively increases the height of the low index layer and hence the interference fringes are red shifted. The blue shift noted in the annealed filters on sapphire (Figure 2b) must be due to some compaction of the porous SiO_2 during the annealing, but we do not yet have other data to support this supposition. Similarly, the unexpectedly small red shift in the Si sample due to the thermal oxide, as will be discussed in the simulation section below, gives evidence of compaction of the porous SiO_2 layer.

Optical Simulation

The reflectance of the thin films stacks were simulated using the *Essential Macleod* software. In modeled we used perfectly smooth films and interfaces were assumed. The optical properties of the Si substrate were taken from literature [3] and the optical properties of Al_2O_3 were determined from reflectance and transmission spectrophotometry. In Figure 3 the black curve (long dash) shows the reflectance with respect to wavelength for a Si wafer with a native oxide of 2.7 nm followed by a 550 nm thick / 65% porous SiO_2 layer with a 100 nm thick Y_2O_3 capping layer. The SiO_2 porosity and film thickness were calculated using values from literature [4]. Utilizing the effective medium approximation [5] yields an optical index of 1.16 for the porous SiO_2. The index of the porous SiO_2 layer is the combination of the volume fraction of air (65%, n=1) and SiO_2 (35%, n=1.46), where the contribution from air is 0.65 and from the SiO_2 is 0.51 for a total index of 1.16. The index of refraction for the Y_2O_3 used in the simulation was ~1.9 over the spectral range (250-1750 nm), as measured by ellipsometry (data not shown). The model for the blue curve (solid line) uses the same film stack with the addition of a 42 nm thick thermal oxide to represent the annealed sample A, i.e. Si substrate sample. The red curve (short dash) shows the same film stack as the black curve but without the native oxide (SiO_2) on an Al_2O_3 (sapphire) substrate. Similar to that shown in Figure 2a for filters on silicon (pre and post anneal), the simulated blue and black curves show a red shift but much more pronounced than was observed experimentally. We believe that, similar to filters on sapphire, filters on silicon must also be undergoing compaction of the porous SiO_2, the blue shift from compaction is being masked by the larger red shift due to substrate oxide growth.

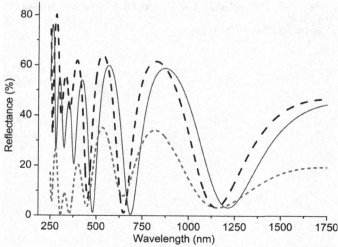

Figure 3. Simulations of the near normal reflectance spectroscopy of the Y_2O_3(100nm)/SiO_2(550 nm, porous)/(Si or Al_2O_3) filters. The black curve (long dash) is for the film stack on a Si wafer with a 2.7 nm native oxide, the blue curve (solid line) is for the film stack on a Si wafer with a 42 nm thick oxide, and the red curve (short dash) is the film stack on an Al_2O_3 wafer.

CONCLUSIONS

Oblique angle deposition has been used to deposit oxide-based optical filters for possible use in thermal photovoltaic systems. These OAD fabricated filters have been shown to withstand high temperatures (1000° C) in air for 90 minutes without delamination or significant degradation of their optical properties, even when the substrate underneath the filter is undergoing significant oxidation. The basic spectral characteristics of the reflectance data were preserved after the samples were annealed, though we have hypothesized that the small shifts seen in the reflectance spectra may be attributed to compaction of the porous OAD layer.

ACKNOWLEDGMENTS

This project was supported in part by an appointment to the Internship/Research Participation Program for the U.S. Army Research Laboratory administered by the Oak Ridge Institute for Science and Education through an agreement between the U.S. Department of Energy and the USARL.

REFERENCES

1. K. Qiu and A.C.S. Hayden, Solar Energy Materials & Solar Cells **91**, 588 (2007).
2. J.-Q. Xi, J. K. Kim, E. F. Schubert, D. Ye, T.-M. Lu, and S.-Y Lin, Optics Letters **31**, 5, 601(2006).
3. T. Minemoto, H. Takakura, Y. Hamakawa, Solar Energy Materials and Solar Cells **90**, 20, 3576 (2006).

4. D. J. Poxson, F. W. Mont, M. F. Schubert, J. K. Kim, and E. F. Schubert, Applied Physics Letters **93**, 101914 (2008).
5. D. E. Aspnes, Thin Solid Films **89**, 249 (1982).

AUTHOR INDEX

SUBJECT INDEX

Printed in the United States
by Baker & Taylor Publisher Services